INFECTIOUS DISEASE

INFECTIOUS DISEASE

A SCIENTIFIC AMERICAN READER

THE UNIVERSITY OF CHICAGO PRESS • CHICAGO AND LONDON

The University of Chicago Press, Chicago 60637

The University of Chicago Press, Ltd., London

© 2008 by Scientific American, Inc.

All rights reserved. Published 2008

Printed in the United States of America

17 16 15 14 13 12 11 10 09 08   1 2 3 4 5

ISBN-13: 978-0-226-74263-2   (cloth)

ISBN-13: 978-0-226-74264-9   (paper)

ISBN-10: 0-226-76463-6   (cloth)

ISBN-10: 0-226-74264-4   (paper)

Library of Congress Cataloging-in-Publication Data

Infectious disease : a Scientific American reader.

   p. ; cm.

  Includes bibliographical references.

ISBN-13: 978-0-226-74263-2   (cloth : alk. paper)

ISBN-10: 0-226-74263-6   (cloth : alk. paper)

ISBN-13: 978-0-226-74264-9   (pbk.: alk. paper)

ISBN-10: 0-226-74264-4   (pbk.: alk. paper)

  I. University of Chicago Press.   II. Scientific American.

  [DNLM: 1. Communicable Diseases—Collected Works.   2. Bacterial Infections—Collected Works.   3. Immune System—microbiology—Collected Works.   4. Virus Diseases—Collected Works.   WC 5 I427 2008]

  RA643.I652   2008

  614.5—dc22

2007051724

♾ The paper used in this publication meets the minimum requirements of the American National Standard for Information Sciences—Permanence of Paper for Printed Library Materials, ANSI Z39.48-1992.

CONTENTS

THE IMMUNE SYSTEM

GLOBAL MANAGEMENT AND TREATMENT ISSUES

Evolution and the Origins of Disease

RANDOLPH M. NESSE AND GEORGE C. WILLIAMS

ORIGINALLY PUBLISHED IN NOVEMBER 1998

Thoughtful contemplation of the human body elicits awe—in equal measure with perplexity. The eye, for instance, has long been an object of wonder, with the clear, living tissue of the cornea curving just the right amount, the iris adjusting to brightness and the lens to distance, so that the optimal quantity of light focuses exactly on the surface of the retina. Admiration of such apparent perfection soon gives way, however, to consternation. Contrary to any sensible design, blood vessels and nerves traverse the inside of the retina, creating a blind spot at their point of exit.

The body is a bundle of such jarring contradictions. For each exquisite heart valve, we have a wisdom tooth. Strands of DNA direct the development of the 10 trillion cells that make up a human adult but then permit his or her steady deterioration and eventual death. Our immune system can identify and destroy a million kinds of foreign matter, yet many bacteria can still kill us. These contradictions make it appear as if the body was designed by a team of superb engineers with occasional interventions by Rube Goldberg.

In fact, such seeming incongruities make sense but only when we investigate the origins of the body's vulnerabilities while keeping in mind the wise words of distinguished geneticist Theodosius Dobzhansky: "Nothing in biology makes sense except in the light of evolution." Evolutionary biology is, of course, the scientific foundation for all biology, and biology is the foundation for all medicine. To a surprising degree, however, evolutionary biology is just now being recognized as a basic medical science. The enterprise of studying medical problems in an evolutionary context has been termed Darwinian medicine. Most medical research tries to explain the causes of an individual's disease and seeks therapies to cure or relieve deleterious conditions. These efforts are traditionally based on consideration of proximate issues, the straightforward study of the body's anatomic and physiological mechanisms as they currently

1

exist. In contrast, Darwinian medicine asks why the body is designed in a way that makes us all vulnerable to problems like cancer, atherosclerosis, depression and choking, thus offering a broader context in which to conduct research.

The evolutionary explanations for the body's flaws fall into surprisingly few categories. First, some discomforting conditions, such as pain, fever, cough, vomiting and anxiety, are actually neither diseases nor design defects but rather are evolved defenses. Second, conflicts with other organisms—*Escherichia coli* or crocodiles, for instance—are a fact of life. Third, some circumstances, such as the ready availability of dietary fats, are so recent that natural selection has not yet had a chance to deal with them. Fourth, the body may fall victim to tradeoffs between a trait's benefits and its costs; a textbook example is the sickle cell gene, which also protects against malaria. Finally, the process of natural selection is constrained in ways that leave us with suboptimal design features, as in the case of the mammalian eye.

EVOLVED DEFENSES

Perhaps the most obviously useful defense mechanism is coughing; people who cannot clear foreign matter from their lungs are likely to die from pneumonia. The capacity for pain is also certainly beneficial. The rare individuals who cannot feel pain fail even to experience discomfort from staying in the same position for long periods. Their unnatural stillness impairs the blood supply to their joints, which then deteriorate. Such pain-free people usually die by early adulthood from tissue damage and infections. Cough or pain is usually interpreted as disease or trauma but is actually part of the solution rather than the problem. These defensive capabilities, shaped by natural selection, are kept in reserve until needed.

Less widely recognized as defenses are fever, nausea, vomiting, diarrhea, anxiety, fatigue, sneezing and inflammation. Even some physicians remain unaware of fever's utility. No mere increase in metabolic rate, fever is a carefully regulated rise in the set point of the body's thermostat. The higher body temperature facilitates the destruction of pathogens. Work by Matthew J. Kluger of the Lovelace Institute in Albuquerque, N.M., has shown that even cold-blooded lizards, when infected, move to warmer places until their bodies are several degrees above their usual temperature. If prevented from moving to the warm part of their cage, they are at increased risk of death from the infection. In a similar study by Evelyn Satinoff of the University of Delaware, elderly rats, who can no longer achieve the high fevers of their younger lab companions, also instinctively sought hotter environments when challenged by infection.

A reduced level of iron in the blood is another misunderstood defense mechanism. People suffering from chronic infection often have decreased levels of blood iron. Although such low iron is sometimes blamed for the illness, it actually is a protective response: during infection, iron is sequestered in the liver, which prevents invading bacteria from getting adequate supplies of this vital element.

Morning sickness has long been considered an unfortunate side effect of pregnancy. The nausea, however, coincides with the period of rapid tissue differentiation of the fetus, when development is most vulnerable to interference by toxins. And nauseated women tend to restrict their intake of strong-tasting, potentially harmful substances. These observations led independent researcher Margie Profet to hypothesize that the nausea of pregnancy is an adaptation whereby the mother protects the fetus from exposure to toxins. Profet tested this idea by examining pregnancy outcomes. Sure enough, women with more nausea were less likely to suffer miscarriages. (This evidence supports the hypothesis but is hardly conclusive. If Profet is correct, further research should discover that pregnant females of many species show changes in food preferences. Her theory also predicts an increase in birth defects among offspring of women who have little or no morning sickness and thus eat a wider variety of foods during pregnancy.)

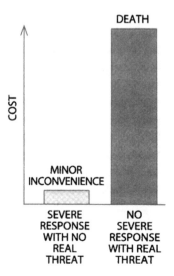

Much suffering is unnecessary but inevitable because of the smoke-detector nature of our defenses. The cost of a false alarm—a strong reaction such as vomiting in the absence of a true threat of life—is temporary unpleasantness. But the cost of no alarm, in the presence of a true threat, such as a food toxin, could mean death. A lack of defensive response during pregnancy, for example, could kill the fetus.

Another common condition, anxiety, obviously originated as a defense in dangerous situations by promoting escape and avoidance. A 1992 study by Lee A. Dugatkin of the University of Louisville evaluated the benefits of fear in guppies. He grouped them as timid, ordinary or bold, depending on their reaction to the presence of smallmouth bass. The timid hid, the ordinary simply swam away, and the bold maintained their ground and eyed the bass. Each guppy group was then left alone in a tank with a bass. After 60 hours, 40 percent of the timid guppies had survived, as had only 15 percent of the ordinary fish. The entire complement of bold guppies, on the other hand, wound up aiding the transmission of bass genes rather than their own.

Selection for genes promoting anxious behaviors implies that there should be people who experience too much anxiety, and indeed there are. There should also be hypophobic individuals who have insufficient anxiety, either because of genetic tendencies or antianxiety drugs. The exact nature and frequency of such a syndrome is an open question, as few people come to psychiatrists complaining of insufficient apprehension. But if sought, the pathologically non-anxious may be found in emergency rooms, jails and unemployment lines.

The utility of common and unpleasant conditions such as diarrhea, fever and anxiety is not intuitive. If natural selection shapes the mechanisms that regulate defensive responses, how can people get away with using drugs to block these defenses without doing their bodies obvious harm? Part of the answer is that we do, in fact, sometimes do ourselves a disservice by disrupting defenses.

Herbert L. DuPont of the University of Texas at Houston and Richard B. Hornick of Orlando Regional Medical Center studied the diarrhea caused by *Shigella* infection and found that people who took antidiarrhea drugs stayed sick longer and were more likely to have complications than those who took a placebo. In another example, Eugene D. Weinberg of Indiana University has documented that well-intentioned attempts to correct perceived iron deficiencies have led to increases in infectious disease, especially amebiasis, in parts of Africa. Although the iron in most oral supplements is unlikely to make much difference in otherwise healthy people with everyday infections, it can severely harm those who are infected and malnourished. Such people cannot make enough protein to bind the iron, leaving it free for use by infectious agents.

On the morning-sickness front, an antinausea drug was recently blamed for birth defects. It appears that no consideration was given to the possibility that the drug itself might be harmless to the fetus but could still be associated with birth defects, by interfering with the mother's defensive nausea.

Another obstacle to perceiving the benefits of defenses arises from the observation that many individuals regularly experience seemingly worthless reactions of anxiety, pain, fever, diarrhea or nausea. The explanation requires an analysis of the regulation of defensive responses in terms of signal-detection theory. A circulating toxin may come from something in the stomach. An organism can expel it by vomiting, but only at a price. The cost of a false alarm—vomiting when no toxin is truly present—is only a few calories. But the penalty for a single missed authentic alarm—failure to vomit when confronted with a toxin—may be death.

Natural selection therefore tends to shape regulation mechanisms with hair triggers, following what we call the smoke-detector principle. A smoke alarm that will reliably wake a sleeping family in the event of any fire will necessarily give a false alarm every time the toast burns. The price of the human body's numerous "smoke alarms" is much suffering that is completely normal but in most instances unnecessary. This principle also explains why blocking defenses is so often free of tragic consequences. Because most defensive reactions occur in response to insignificant threats, interference is usually harmless; the vast majority of alarms that are stopped by removing the battery from the smoke alarm are false ones, so this strategy may seem reasonable. Until, that is, a real fire occurs.

CONFLICTS WITH OTHER ORGANISMS

Natural selection is unable to provide us with perfect protection against all pathogens, because they tend to evolve much faster than humans do. *E. coli*, for example, with its rapid rates of reproduction, has as much opportunity for mutation and selection in one day as humanity gets in a millennium. And our defenses, whether natural or artificial, make for potent selection forces. Pathogens either quickly evolve a counterdefense or become extinct. Amherst College biologist Paul W. Ewald has suggested classifying phenomena associated with infection according to whether they benefit the host, the pathogen, both or neither. Consider the runny nose associated with a cold. Nasal mucous secretion could expel intruders, speed the pathogen's transmission to new hosts or both [see "The Evolution of Virulence," by Paul W. Ewald; Scientific American, April 1993]. Answers could come from studies examining whether blocking nasal secretions shortens or prolongs illness, but few such studies have been done.

Humanity won huge battles in the war against pathogens with the development of antibiotics and vaccines. Our victories were so rapid and seemingly complete that in 1969 U.S. Surgeon General William H. Stewart said that it was "time to close the book on infectious disease." But the enemy, and the power or natural selection, had been underestimated. The sober reality is that pathogens apparently can adapt to every chemical researchers develop. ("The war has been won," one scientist more recently quipped. "By the other side.")

Antibiotic resistance is a classic demonstration of natural selection. Bacteria that happen to have genes that allow them to prosper despite the presence of an antibiotic reproduce faster than others, and so the genes that confer resistance spread quickly. As shown by Nobel laureate Joshua Lederberg of the Rockefeller University, they can even jump to different species of bacteria, borne on bits of infectious DNA. Today some strains of tuberculosis in New York City are resistant to all three main antibiotic treatments; patients with those strains have no better chance of surviving than did TB patients a century ago. Stephen S. Morse of Columbia University notes that the multidrug-resistant strain that has spread throughout the East Coast may have originated in a homeless shelter across the street from Columbia-Presbyterian Medical Center. Such a phenomenon would indeed be predicted in an environment where fierce selection pressure quickly weeds out less hardy strains. The surviving bacilli have been bred for resistance.

Many people, including some physicians and scientists, still believe the outdated theory that pathogens necessarily become benign after long association with hosts. Superficially, this makes sense. An organism that kills rapidly may never get to a new host, so natural selection would seem to favor lower virulence. Syphilis, for instance, was a highly virulent disease when it first arrived in Europe, but as the centuries passed it became steadily more mild. The virulence of a pathogen is, however, a life history trait that can increase as well as decrease, depending on which option is more advantageous to its genes.

For agents of disease that are spread directly from person to person, low virulence tends to be beneficial, as it allows the host to remain active and in contact with other potential hosts. But some diseases, like malaria, are transmitted just as well—or better—by the incapacitated. For such pathogens, which usually rely on intermediate vectors like mosquitoes, high virulence can give a selective advantage. This principle has direct implications for infection control in hospitals, where health care workers' hands can be vectors that lead to selection for more virulent strains.

In the case of cholera, public water supplies play the mosquitoes' role. When water for drinking and bathing is contaminated by waste from immobilized patients, selection tends to increase virulence, because more diarrhea enhances the spread of the organism even if individual hosts quickly die. But, as Ewald has shown, when sanitation improves, selection acts against classical *Vibrio cholerae* bacteria in favor of the more benign El Tor biotype. Under these conditions, a dead host is a dead end. But a less ill and more mobile host, able to infect many others over a much longer time, is an effective vehicle for a pathogen of lower virulence. In another example, better sanitation leads to displacement of the aggressive *Shigella flexneri* by the more benign *S. sonnei*.

Such considerations may be relevant for public policy. Evolutionary theory predicts that clean needles and the encouragement of safe sex will do more than save numerous individuals from HIV infection. If humanity's behavior itself slows HIV transmission rates, strains that do not soon kill their hosts have the long-term survival advantage over the more virulent viruses that then die with their hosts, denied the opportunity to spread. Our collective choices can change the very nature of HIV.

Conflicts with other organisms are not limited to pathogens. In times past, humans were at great risk from predators looking for a meal. Except in a few places, large carnivores now pose no threat to humans. People are in more danger today from smaller organisms' defenses, such as the venoms of spiders and snakes. Ironically, our fears of small creatures, in the form of phobias, probably cause more harm than any interactions with those organisms do. Far more dangerous than predators or poisoners are other members of our own species. We attack each other not to get meat but to get mates, territory and other resources. Violent conflicts between individuals are overwhelmingly between young men in competition and give rise to organizations to advance these aims. Armies, again usually composed of young men, serve similar objectives, at huge cost.

Even the most intimate human relationships give rise to conflicts having medical implications. The reproductive interests of a mother and her infant, for instance, may seem congruent at first but soon diverge. As noted by biologist Robert L. Trivers in a now classic 1974 paper, when her child is a few years old, the mother's genetic interests may be best served by becoming pregnant again, whereas her offspring benefits from continuing to nurse. Even in the womb mere is contention. From the mother's vantage point, the optimal size of a fetus is a bit smaller than that which would best serve the fetus and the father. This discord, according to David Haig of Harvard University, gives rise to an arms race between fetus and mother over her levels of blood pressure and blood sugar, sometimes resulting in hypertension and diabetes during pregnancy.

COPING WITH NOVELTY

Making rounds in any modern hospital provides sad testimony to the prevalence of diseases humanity has brought on itself. Heart attacks, for example, result mainly from atherosclerosis, a problem that became widespread only in this century and that remains rare among hunter-gatherers. Epidemiological research furnishes the information that should help us prevent heart attacks: limit fat intake, eat lots of vegetables, and exercise hard each day. But hamburger chains proliferate, diet foods languish on the shelves, and exercise machines serve as expensive clothing hangers throughout the land. The proportion of overweight Americans is one third and rising. We all know what is good for us. Why do so many of us continue to make unhealthy choices?

Our poor decisions about diet and exercise are made by brains shaped to cope with an environment substantially different from the one our species now inhabits. On the African savanna, where the modern human design was fine-tuned, fat, salt and sugar were scarce and precious. Individuals who had a tendency to consume large amounts of fat when given the rare opportunity had a selective advantage. They were more likely to survive famines that killed their thinner companions. And we, their descendants, still carry those urges for foodstuffs that today are anything but scarce. These evolved desires—inflamed by advertisements from competing food corporations that themselves survive by selling us more of whatever we want to buy—easily defeat our intellect and willpower. How ironic that humanity worked for centuries to create environments that are almost literally flowing with milk and honey, only to see our success responsible for much modern disease and untimely death.

Increasingly, people also have easy access to many kinds of drugs, especially alcohol and tobacco, that are responsible for a huge proportion of disease, health care costs and premature death. Although individuals have always used psychoactive substances, widespread problems materialized only following another environmental novelty: the ready availability of concentrated drugs and new, direct routes of administration, especially injection. Most of these substances, including nicotine, cocaine and opium, are products of natural selection that evolved to protect plants from insects. Because humans share a common evolutionary heritage with insects, many of these substances also affect our nervous system.

This perspective suggests that it is not just defective individuals or disordered societies that are vulnerable to the dangers of psychoactive drugs, all of us are susceptible because drugs and our biochemistry have a long history of interaction. Understanding the details of that interaction, which is the focus of much current research from both a proximate and evolutionary perspective, may well lead to better treatments for addiction.

The relatively recent and rapid increase in breast cancer must be the result in large part of changing environments and ways of life, with only a few cases resulting solely from genetic abnormalities. Boyd Eaton and his colleagues at Emory University reported that the rate of breast cancer in today's "nonmodern" societies is only a tiny fraction of that in the U.S. They hypothesize that the amount of time between menarche and first pregnancy is a crucial risk factor, as is the related issue of total lifetime number of menstrual cycles. In hunter-gatherers, menarche occurs at about age 15 or later, followed within a few years by pregnancy and two or three years of nursing, then by another pregnancy soon after. Only between the end of nursing and the next pregnancy will the woman menstruate and thus experience the high levels of hormones that may adversely affect breast cells.

In modern societies, in contrast, menarche occurs at age 12 or 13—probably at least in part because of a fat intake sufficient to allow an extremely young woman to nourish a fetus—and the first pregnancy may be decades later or never. A female hunter-gatherer may have a total of 150 menstrual cycles, whereas the average woman in modern societies has 400 or more. Although few would suggest that women should become pregnant in their teens to prevent breast cancer later, early administration of a burst of hormones to simulate pregnancy may reduce the risk. Trials to test this idea are now under way at the University of California at San Diego.

TRADE-OFFS AND CONSTRAINTS

Compromise is inherent in every adaptation. Arm bones three times their current thickness would almost never break, but *Homo sapiens* would be lumbering creatures on a never-ending quest for calcium. More sensitive ears might sometimes be useful, but we would be distracted by the noise of air molecules banging into our eardrums.

Such trade-offs also exist at the genetic level. If a mutation offers a net reproductive advantage, it will tend to increase in frequency in a population even if it causes vulnerability to disease. People with two copies of the sickle cell gene, for example, suffer terrible pain and die young. People with two copies of the "normal" gene are at high risk of death from malaria. But individuals with one of each are protected from both malaria and sickle cell disease. Where malaria is prevalent, such people are fitter, in the Darwinian sense, than members of either other group. So even though the sickle cell gene causes disease, it is selected for where malaria persists. Which is the "healthy" allele in this environment? The question has no answer. There is no one normal human genome—there are only genes.

Many other genes that cause disease must also have offered benefits, at least in some environments, or they would not be so common. Because cystic fibrosis (CF) kills one out of 2,500 Caucasians, the responsible genes would appear to be at great risk of being eliminated from the gene pool. And yet they endure. For years, researchers mused that the CF gene, like the sickle cell gene, probably conferred some advantage. Recently a study by Gerald B. Pier of Harvard Medical School and his colleagues gave substance to this informed speculation: having one copy of the CF gene appears to decrease the chances of the bearer acquiring a typhoid fever infection, which once had a 15 percent mortality.

Aging may be the ultimate example of a genetic trade-off. In 1957 one of us (Williams) suggested that genes that cause aging and eventual death could nonetheless be selected for if they had other effects that gave an advantage in youth, when the force of selection is stronger. For instance, a hypothetical gene that governs calcium metabolism so that bones heal quickly but that also happens to cause the steady deposition of calcium in arterial walls might well be selected for even though it kills some older people. The influence of such pleiotropic genes (those having multiple effects) has been seen in fruit flies and flour beetles, but no specific example has yet been found in humans. Gout, however, is of particular interest, because it arises when a potent antioxidant, uric acid, forms crystals that precipitate out of fluid in joints. Antioxidants have antiaging effects, and plasma levels of uric acid in different species of primates are closely correlated with average adult life span. Perhaps high levels of uric acid benefit most humans by slowing tissue aging, while a few pay the price with gout.

Other examples are more likely to contribute to more rapid aging. For instance, strong immune defenses protect us from infection but also inflict continuous, low-level tissue damage. It is also possible, of course, that most genes that cause aging have no benefit at any age—they simply never decreased reproductive fitness enough in the natural environment to be selected against. Nevertheless, over the next decade research will surely identify specific genes that accelerate senescence, and researchers will soon thereafter gain the means to interfere with their actions or even change them. Before we tinker, however, we should determine whether these actions have benefits early in life.

Because evolution can take place only in the direction of time's arrow, an organism's design is constrained by structures already in place. As noted, the vertebrate eye is arranged backward. The squid eye, in contrast, is free from this defect, with vessels and nerves running on the outside, penetrating where necessary and pinning down the retina so it cannot detach. The human eye's flaw results from simple bad luck; hundreds of millions of years ago, the layer of cells that happened to become sensitive to light in our ancestors was positioned differently from the corresponding layer in ancestors of squids. The two designs evolved along separate tracks, and there is no going back.

Such path dependence also explains why the simple act of swallowing can be life-threatening. Our respiratory and food passages intersect because in an early lungfish ancestor the air opening for breathing at the surface was understandably located at the top of the snout and led into a common space shared by the food passage way. Because natural selection cannot start from scratch, humans are stuck with the possibility that food will clog the opening to our lungs.

The path of natural selection can even lead to a potentially fatal cul-de-sac, as in the case of the appendix, that vestige of a cavity that our ancestors employed in digestion. Because it no longer performs that function, and as it can kill when infected, the expectation might be that natural selection would have eliminated it. The reality is more complex. Appendicitis results when inflammation causes swelling, which compresses the artery supplying blood to the appendix. Blood flow protects against bacterial growth, so any reduction aids infection, which creates more swelling. If the blood supply is cut off completely, bacteria have free rein until the appendix bursts. A slender appendix is especially susceptible to this chain of events, so appendicitis may, paradoxically, apply the selective pressure that maintains a large appendix. Far from arguing that everything in the body is perfect, an evolutionary analysis reveals that we live with some very unfortunate legacies and that some vulnerabilities may even be actively maintained by the force of natural selection.

Selected Principles of Darwinian Medicine

A Darwinian approach to medical practice leads to a shift in perspective. The following principles provide a foundation for considering health and disease in an evolutionary context:

- DEFENSES and DEFECTS are two fundamentally different manifestations of disease

- BLOCKING defenses has costs as well as benefits

- Because natural selection shapes defense regulation according to the SMOKE-DETECTOR PRINCIPLE, much defensive expression and associated suffering are unnecessary in the individual instance

- Modern epidemics are most likely to arise from the mismatch between PHYSIOLOGICAL DESIGN of our bodies and NOVEL ASPECTS of our environment

- Our DESIRES, shaped in the ancestral environment to lead us to actions that tended to maximize reproductive success, now often lead us to disease and early death

- The body is a bundle of COMPROMISES

- There is no such thing as "the NORMAL body"

- There is no such thing as "the NORMAL human genome"

- Some GENES that cause disease may also have benefits, and others are quirks that cause disease only when they interact with novel environmental factors

- GENETIC SELF-INTEREST will drive an individual's actions, even at the expense of the health and longevity of the individual created by those genes

- VIRULENCE is a trait of the pathogen that can increase as well as decrease

- SYMPTOMS of infection can benefit the pathogen, the host, both or neither

- Disease is INEVITABLE because of the way that organisms are shaped by evolution

- Each disease needs a PROXIMATE EXPLANATION of why some people get it and others don't, as well as an EVOLUTIONARY EXPLANATION of why members of the species are vulnerable to it

- Diseases are not products of natural selection, but most of the VULNERABILITIES that lead to disease are shaped by the process of natural selection

- Aging is better viewed as a TRADE-OFF than a disease

- Specific clinical recommendations must be based on CLINICAL STUDIES; clinical interventions based only on theory are not scientifically grounded and may cause harm

EVOLUTION OF DARWINIAN MEDICINE

Despite the power of the Darwinian paradigm, evolutionary biology is just now being recognized as a basic science essential for medicine. Most diseases decrease fitness, so it would seem that natural selection could explain only health, not disease. A Darwinian approach makes sense only when the object of explanation is changed from diseases to the traits that make us vulnerable to diseases. The assumption that natural selection maximizes health also is incorrect—selection maximizes the reproductive success of genes. Those genes that make bodies having superior reproductive success will become more common, even if they compromise the individual's health in the end.

Finally, history and misunderstanding have presented obstacles to the acceptance of Darwinian medicine. An evolutionary approach to functional analysis can appear akin to naive teleology or vitalism, errors banished only recently, and with great effort, from medical thinking. And, of course, whenever evolution and medicine are mentioned together, the specter of eugenics arises. Discoveries made through a Darwinian view of how all human bodies are alike in their vulnerability to disease will offer great benefits for individuals, but such insights do not imply that we can or should make any attempt to improve the species. If anything, this approach cautions that apparent genetic defects may have unrecognized adaptive significance, that a single "normal" genome is nonexistent and that notions of "normality" tend to be simplistic.

The systematic application of evolutionary biology to medicine is a new enterprise. Like biochemistry at the beginning of this century, Darwinian medicine very likely will need to develop in several incubators before it can prove its power and utility. If it must progress only from the work of scholars without funding to gather data to test their ideas, it will take decades for the field to mature. Departments of evolutionary biology in medical schools would accelerate the process, but for the most part they do not yet exist. If funding agencies had review panels with evolutionary expertise, research would develop faster, but such panels remain to be created. We expect that they will.

The evolutionary viewpoint provides a deep connection between the states of disease and normal functioning and can integrate disparate avenues of medical research as well as suggest fresh and important areas of inquiry. Its utility and power will ultimately lead to recognition of evolutionary biology as a basic medical science.

FURTHER READING

EVOLUTION OF INFECTIOUS DISEASE. P. W. Ewald. Oxford University Press, 1994.
DARWINIAN PSYCHIATRY. M. T. McGuire and A. Troisi. Harvard University Press, 1998.
EVOLUTION IN HEALTH AND DISEASE. Edited by S. Stearns. Oxford University Press, 1998.
EVOLUTIONARY MEDICINE. W. R. Trevathan et al. Oxford University Press (in press).

VIRAL INFECTIONS

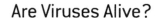

Are Viruses Alive?

LUIS P. VILLARREAL

ORIGINALLY PUBLISHED IN DECEMBER 2004

In an episode of the classic 1950s television comedy *The Honeymooners,* Brooklyn bus driver Ralph Kramden loudly explains to his wife, Alice, "You know that I know how easy you get the virus." Half a century ago even regular folks like the Kramdens had some knowledge of viruses—as microscopic bringers of disease. Yet it is almost certain that they did not know exactly what a virus was. They were, and are, not alone.

For about 100 years, the scientific community has repeatedly changed its collective mind over what viruses are. First seen as poisons, then as life-forms, then biological chemicals, viruses today are thought of as being in a gray area between living and nonliving: they cannot replicate on their own but can do so in truly living cells and can also affect the behavior of their hosts profoundly. The categorization of viruses as nonliving during much of the modern era of biological science has had an unintended consequence: it has led most researchers to ignore viruses in the study of evolution. Finally, however, scientists are beginning to appreciate viruses as fundamental players in the history of life.

COMING TO TERMS

It is easy to see why viruses have been difficult to pigeonhole. They seem to vary with each lens applied to examine them. The initial interest in viruses stemmed from their association with diseases—the word "virus" has its roots in the Latin term for "poison." In the late 19th century researchers realized that certain diseases, including rabies and foot-and-mouth, were caused by particles that seemed to behave like bacteria but were much smaller. Because they were clearly biological themselves and could be spread from one victim to another with obvious biological effects, viruses were then thought to be the simplest of all living, gene-bearing life-forms.

Their demotion to inert chemicals came after 1935, when Wendell M. Stanley and his colleagues, at what is now the Rockefeller University in New York City, crystallized a virus—tobacco mosaic virus—for the first time. They saw that it consisted of a package of complex biochemicals. But it lacked essential systems necessary for metabolic functions, the biochemical activity of life. Stanley shared the 1946 Nobel Prize—in chemistry, not in physiology or medicine—for this work.

Further research by Stanley and others established that a virus consists of nucleic acids (DNA or RNA) enclosed in a protein coat that may also shelter viral proteins involved in infection. By that description, a virus seems more like a chemistry set than an organism. But when a virus enters a cell (called a host after infection), it is far from inactive. It sheds its coat, bares its genes and induces the cell's own replication machinery to reproduce the intruder's DNA or RNA and manufacture more viral protein based on the instructions in the viral nucleic acid. The newly created viral bits assemble and, voilà, more virus arises, which also may infect other cells.

These behaviors are what led many to think of viruses as existing at the border between chemistry and life. More poetically, virologists Marc H. V. van Regenmortel of the University of Strasbourg in France and Brian W. J. Mahy of the Centers for Disease Control and Prevention have recently said that with their dependence on host cells, viruses lead "a kind of borrowed life." Interestingly, even though biologists long favored the view that viruses were mere boxes of chemicals, they took advantage of viral activity in host cells to determine how nucleic acids code for proteins: indeed, modern molecular biology rests on a foundation of information gained through viruses.

Molecular biologists went on to crystallize most of the essential components of cells and are today accustomed to thinking about cellular constituents—for example, ribosomes, mitochondria, membranes, DNA and proteins—as either chemical machinery or the stuff that the machinery uses or produces. This exposure to multiple complex chemical structures that carry out the processes of life is probably a reason that most molecular biologists do not spend a lot of time puzzling over whether viruses are alive. For them, that exercise might seem equivalent to pondering whether those individual subcellular constituents are alive on their own. This myopic view allows them to see only how viruses co-opt cells or cause disease. The more sweeping question of viral contributions to the history of life on earth, which I will address shortly, remains for the most part unanswered and even unasked.

TO BE OR NOT TO BE

The seemingly simple question of whether or not viruses are alive, which my students often ask, has probably defied a simple answer all these years because it raises a fundamental issue: What exactly defines "life?" A precise scientific definition of life is an elusive thing, but most observers would agree that life includes certain qualities in addition to an ability to replicate. For example, a living entity is in a state bounded by birth and death. Living organisms also are thought to require a degree of biochemical autonomy, carrying on the metabolic activities that produce the molecules and energy needed to sustain the organism. This level of autonomy is essential to most definitions.

Viruses, however, parasitize essentially all biomolecular aspects of life. That is, they depend on the host cell for the raw materials and energy necessary for nucleic acid synthesis, protein synthesis, processing and transport, and all other biochemical activities that allow the virus to multiply and spread. One might then conclude that even though these processes come under viral direction, viruses are simply nonliving parasites of living metabolic systems. But a spectrum may exist between what is certainly alive and what is not.

A rock is not alive. A metabolically active sack, devoid of genetic material and the potential for propagation, is also not alive. A bacterium, though, is alive. Although it is a single cell, it can generate energy and the molecules needed to sustain itself, and it can reproduce. But what about a seed ? A seed might not be considered alive. Yet it has a potential for life, and it may be destroyed. In this regard, viruses resemble seeds more than they do live cells. They have a certain potential, which can be snuffed out, but they do not attain the more autonomous state of life.

Another way to think about life is as an emergent property of a collection of certain nonliving things. Both life and conscious-ness are examples of emergent complex systems. They each require a critical level of complexity or interaction to achieve their respective states. A neuron by itself, or even in a network of nerves, is not conscious—whole brain complexity is needed. Yet even an intact human brain can be biologically alive but incapable of consciousness, or "brain-dead." Similarly, neither cellular nor viral individual genes or proteins are by themselves alive. The enucleated cell is akin to the state of being brain-dead, in that it lacks a full critical complexity. A virus, too, fails ro reach a critical complexity. So life itself is an emergent, complex state, but it is made from the same fundamental, physical building blocks that constitute a virus. Approached

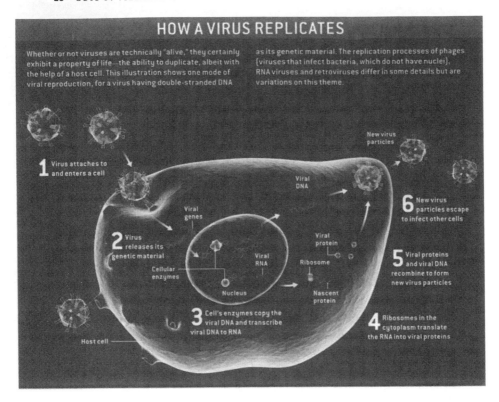

HOW A VIRUS REPLICATES

Whether or not viruses are technically "alive," they certainly exhibit a property of life—the ability to duplicate, albeit with the help of a host cell. This illustration shows one mode of viral reproduction, for a virus having double-stranded DNA as its genetic material. The replication processes of phages (viruses that infect bacteria, which do not have nuclei), RNA viruses and retroviruses differ in some details but are variations on this theme.

1 Virus attaches to and enters a cell

2 Virus releases its genetic material

3 Cell's enzymes copy the viral DNA and transcribe viral DNA to RNA

4 Ribosomes in the cytoplasm translate the RNA into viral proteins

5 Viral proteins and viral DNA recombine to form new virus particles

6 New virus particles escape to infect other cells

Viral DNA

Viral genes

Cellular enzymes

Viral RNA

Nucleus

Ribosome

Viral protein

Nascent protein

New virus particles

Host cell

from this perspective, viruses, though not fully alive, may be thought of as being more than inert matter: they verge on life.

In fact, in October, French researchers announced findings that illustrate a fresh just how close some viruses might come. Didier Raoult and his colleagues at the University of the Mediterranean in Marseille announced that they had sequenced the genome of the largest known virus, Mimi-virus, which was discovered in 1992. The virus, about the same size as a small bacterium, infects amoebae. Sequence analysis of the virus revealed numerous genes previously thought to exist only in cellular organisms. Some of these genes are involved in making the proteins encoded by the viral DNA and may make it easier for Mimivirus to co-opt host cell replication systems. As the research team noted in its report in the journal *Science,* the enormous complexity of the Mimivirus's genetic complement "challenges the established frontier between viruses and parasitic cellular organisms."

IMPACT ON EVOLUTION

Debates over whether to label viruses as living lead naturally to another question: Is pondering the status of viruses as living or nonliving more than a philosophical exercise, the basis of a lively and heated rhetorical debate but with little real consequence? I think the issue *is* important, because how scientists regard this question influences their thinking about the mechanisms of evolution.

Viruses have their own, ancient evolutionary history, dating to the very origin of cellular life. For example, some viral-repair enzymes—which excise and resynthesize damaged DNA, mend oxygen radical damage, and so on—are unique to certain viruses and have existed almost unchanged probably for billions of years.

Nevertheless, most evolutionary biologists hold that because viruses are not alive, they are unworthy of serious consideration when trying to understand evolution. They also look on viruses as coming from host genes that somehow escaped the host and acquired a protein coat. In this view, viruses are fugitive host genes that have degenerated into parasites. And with viruses thus dismissed from the web of life, important contributions they may have made to the origin of species and the maintenance of life may go unrecognized. (Indeed, only four of the 1,205 pages of the 2002 volume *The Encyclopedia of Evolution* are devoted to viruses.)

Of course, evolutionary biologists do not deny that viruses have had some role in evolution. But by viewing viruses as inanimate, these investigators place them in the same category of influences as, say, climate change. Such external influences select among individuals having varied, genetically controlled traits; those individuals most able to survive and thrive when faced with these challenges go on to reproduce most successfully and hence spread their genes to future generations.

But viruses directly exchange genetic information with living organisms—that is, within the web of life itself. A possible surprise to most physicians, and perhaps to most evolutionary biologists as well, is that most known viruses are persistent and innocuous, not pathogenic. They take up residence in cells, where they may remain dormant for long periods or take advantage of the cells' replication apparatus to reproduce at a slow and steady rate. These viruses have developed many clever ways to avoid detection by the host immune system—essentially every step in the immune process can be altered or controlled by various genes found in one virus or another.

Furthermore, a virus genome (the entire complement of DNA or RNA) can permanently colonize its host, adding viral genes to host lineages and ultimately becoming a critical part of the host species' genome. Viruses therefore surely have effects that are faster and more direct than those of external forces that simply select among more slowly generated, internal genetic variations. The huge population of viruses, combined with their rapid rates of replication and mutation, makes them the world's leading source of genetic innovation: they constantly "invent" new genes. And unique genes of viral origin may travel, finding their way into other organisms and contributing to evolutionary change.

Data published by the International Human Genome Sequencing Consortium indicate that somewhere between 113 and 223 genes present in bacteria and in the human genome are absent in well-studied organisms—such as the yeast *Saccharomyces cerevisiae*, the fruit fly *Drosophila melanogaster* and the nematode *Caenorhabditis elegans*—that lie in between those two evolutionary extremes. Some researchers thought that these organisms, which arose after bacteria but before vertebrates, simply lost the genes in question at some point in their evolutionary history. Others suggested that these genes had been transferred directly to the human lineage by invading bacteria.

My colleague Victor DeFilippis of the Vaccine and Gene Therapy Institute of the Oregon Health and Science University and I suggested a third alternative: viruses may originate genes, then colonize two different lineages—for example, bacteria and vertebrates. A gene apparently bestowed on humanity by bacteria may have been given to both by a virus.

In fact, along with other researchers, Philip Bell of Macquarie University in Sydney, Australia, and I contend that the cell nucleus itself is of viral origin. The advent of the nucleus—which differentiates eukaryotes (organisms whose cells contain a true nucleus), including humans, from prokaryotes, such as bacteria—cannot be satisfactorily explained solely by the gradual adaptation of prokaryotic cells until they became eukaryotic. Rather the nucleus may have evolved from a persisting large DNA virus that made a permanent home within prokaryotes. Some support for this idea comes from sequence data showing that the gene for a DNA polymerase (a DNA-copying enzyme) in the virus called T4, which infects bacteria, is closely related to other DNA polymerase genes in both eukaryotes and the viruses that infect them. Patrick Forterre of the University of Paris-Sud has also analyzed enzymes responsible for DNA replication and has concluded that the genes for such enzymes in eukaryotes probably have a viral origin.

From single-celled organisms to human populations, viruses affect all life on earth, often determining what will survive. But viruses themselves also evolve. New viruses, such as the AIDS-causing HIV-1, may be the only biological entities that researchers can actually witness come into being, providing a real-time example of evolution in action.

Viruses matter to life. They are the constantly changing boundary between the worlds of biology and biochemistry. As we continue to unravel the genomes of more and more organisms, the contributions from this dynamic and ancient gene pool should become apparent. Nobel laureate Salvador Luria mused about the viral influence on evolution in 1959. "May we not feel," he wrote, "that in the virus, in their merging with the cellular genome and reemerging from them, we observe the units and process which, in the course of evolution, have created the successful genetic patterns that underlie all living cells ?" Regardless of whether or not we consider viruses to be alive, it is time to acknowledge and study them in their natural context—within the web of life.

FURTHER READING

VIRAL QUASISPECIES. Manfred Eigen in *Scientific American*, Vol. 269, No. 1, pages 42–49; July 1993.

DNA VIRUS CONTRIBUTION TO HOST EVOLUTION. L. P. Villarreal in *Origin and Evolution of Viruses*. Edited by E. Domingo et al. Academic Press, 1999.

LATERAL GENE TRANSFER OR VIRAL COLONIZATION? Victor DeFilippis and Louis Villarreal in *Science*, Vol. 293, page 1048; August 10, 2001.

VIRUSES AND THE EVOLUTION OF LIFE. Luis Villarreal. ASM Press (in press). All the Virology on the WWW is at www.virology.net

New Hope For Defeating Rotavirus

ROGER I. GLASS

ORIGINALLY PUBLISHED IN APRIL 2006

The thought of a murderous virus often conjures images of patients suffering from Ebola virus in Africa, SARS in Asia or hantavirus in the U.S. Yet those evildoers have taken far fewer lives than rotavirus, whose name is virtually unknown. This virus infects nearly all children in their first few years of life. It causes vomiting followed by diarrhea. The diarrhea is often so severe that, if left untreated, it can lead to shock from dehydration and then death. Worldwide, rotavirus kills an estimated 610,000 children every year, accounting for about 5 percent of all deaths among those younger than five years. In the U.S., few children perish from the virus, but as many as 70,000 require hospitalization for it annually, and several million suffer quietly at home.

Scientists, though, are now about to break the grip of this devastating disease. In January—some three decades after investigators first identified the pathogen—researchers reported that two rotavirus vaccines had proved successful in massive clinical trials. The process of developing rotavirus vaccines has been more difficult and complicated than anyone imagined, full of setbacks and surprises. But today both the World Health Organization and the Global Alliance for Vaccines and Immunization consider rotavirus vaccine a top priority, and the final battle to get immunizations to the young children who so desperately need them has begun.

IDENTIFYING THE CONTAGION

Rotavirus was first identified as a cause of human disease in 1973 by Ruth Bishop, a young microbiologist working on gastrointestinal diseases at the Royal Children's Hospital in Melbourne, Australia. At the time, investigators were perplexed by diarrhea in children. Although the disorder was common and frequently severe, the causative agent was rarely identified. Searching for clues, Bishop's group looked through an electron microscope at biopsied tissue from the duodenum, or small intestine, of acutely sick children. What they saw astounded them: an infestation of wheel-shaped viruses in the epithelial cells that form the intestinal lining.

My own involvement with rotavirus began in 1979, when my wife and I moved to Bangladesh to work at the International Center for Diarrheal Disease Research. Young and idealistic, we were drawn by the prospect of helping children in a country where severe diarrhea was a leading cause of death. The center's hospital in Dhaka admitted so many patients with unspecified "intestinal" flu annually that some had to be cared for in hallways and in tents outside. Believing the cause of their diarrhea to be bacterial, we were surprised to find many of the children were suffering not from cholera, salmonella, shigella or *Escherichia coli* but from rotavirus, about which we knew little. With the help of a simple test, we determined that rotavirus was responsible for the admittance of between 25 and 40 percent of all children younger than five to our hospital for diarrhea.

Studies from around the globe yielded similar results. What is more, they revealed that rotavirus was not only widespread but a major cause of death in the poorest nations. By 1985 such data compelled the Institute of Medicine to put rotavirus infection atop a list of diseases for which vaccines were urgently needed in the developing world.

At the same time, surprisingly little was known about the incidence and distribution of rotavirus in the U.S. In 1986, when I returned to the U.S. Centers for Disease Control, the disease was rarely diagnosed and, in fact, was not even listed in the *International Classification of Diseases*. Having seen the impact of the disease overseas, my co-workers and I were intent on finding out whether it was affecting many people in the states.

But how does one assess the burden of a disease that is rarely diagnosed, is never listed as the cause of hospitalization in discharge records, and goes unrecognized by a majority of pediatricians who commonly treat it? My colleague, Mei-Shang Ho, began by looking at U.S. data on childhood hospitalizations. She found that diarrhea was a common cause of hospital stays, accounting for 12 percent of hospitalizations in children younger than five, and that most cases were coded as being of unknown etiology. Further studies revealed that a lion's share of the undiagnosed cases were attributable to rotavirus. Three other interesting facts about rotavirus in the U.S. emerged as well. First, infection follows a distinctly seasonal pattern, peaking from December to March; second, the vast majority of children hospitalized for this virus are younger than five years; and third, regardless of season, rotavirus causes most cases of severe diarrhea in young children.

Global Distribution of Deaths from Rotavirus

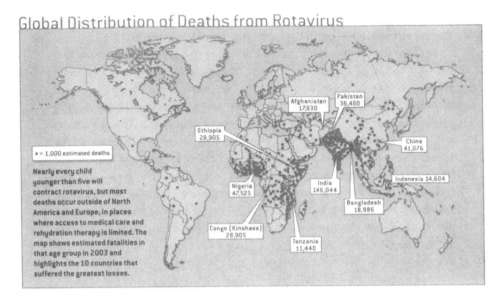

Afghanistan 17,930

Pakistan 36,450

Ethiopia 28,905

China 41,076

● = 1,000 estimated deaths

Nearly every child younger than five will contract rotavirus, but most deaths occur outside of North America and Europe, in places where access to medical care and rehydration therapy is limited. The map shows estimated fatalities in that age group in 2003 and highlights the 10 countries that suffered the greatest losses.

Nigeria 47,525

India 146,044

Indonesia 14,604

Bangladesh 18,986

Congo (Kinshasa) 28,905

Tanzania 11,440

Epidemiologists now know that rotavirus is far and away the leading cause of childhood diarrhea worldwide, infecting virtually all children between the ages of three months and five years. Unlike bacteria that spread via contaminated food and water and thus disproportionately affect people in poor regions, rotavirus shows no regard for geographic borders. Indeed, the very ubiquity of the pathogen—with Americans facing the same risk of infection as Bangladeshis—suggests the virus is highly contagious, spreading as easily as, say, a cold virus. And, as is true of cold viruses, sanitation and clean drinking water have little power to block transmission.

Molecular and clinical studies bear witness to its virulence. Just 10 virus particles can start trouble in a young child. A virus-laden droplet landing on a baby's thumb or toy is all it takes. Popped into the mouth, the virus makes its way to the epithelial cells lining the small intestine, where it replicates at astonishing speed: within 24 hours, 10 viruses become millions, filling and killing the cells with their proteins, toxins and newly made particles. Soon the gut epithelium sloughs, and a flood of fluids and electrolytes exits the body in diarrheal bursts. Without rehydration therapy, a child can lose as much as 10 percent of his or her body weight and go into shock in just one or two days.

Fortunately, children who survive their first infection suffer no long-term consequences, and few ever experience another bout of rotavirus diarrhea. They have natural immunity—that is, their immune system has become primed to quickly recognize and prevent replication of rotavirus when it next invades. But because so many children become severely ill with the first infection, scientists consider a vaccine that could mimic this natural immunity to be the best hope for saving lives.

QUEST FOR A VACCINE BEGINS

Vaccines are powerful weapons in the human arsenal against infectious disease and among the most effective interventions in public health. Made from either live or killed microorganisms or from their key proteins, vaccines trick a recipient's immune system into believing it is under attack. In response, the immune system produces antibodies against the vaccine (which poses no biological threat), just as it would against the virus itself. And as in natural immunity, should the disease-causing agent ever invade, the immune system is fully primed, ready to pump out antibodies to immobilize it.

Twenty years ago several pharmaceutical companies became interested in developing a vaccine against rotavirus. With a potential market both large in size and global in scope, the high costs of vaccine development appeared reasonable. In addition, distribution would be easy even in remote places: rotavirus vaccine could be added to the Universal Program for Childhood Immunization, which under the auspices of the WHO and UNICEF already delivers routine vaccines to about 80 percent of the world's children.

Although different approaches to vaccines have been considered—human versus animal strains, live versus killed viruses, whole virus or protein subunits—rotavirus researchers followed the lead of Albert Sabin, creator of the oral poliomyelitis vaccine. Sabin believed that live vaccines, which can replicate somewhat but are too weak to trigger disease, best mimic the protection acquired through natural infection. Also, in the case of rotavirus, oral vaccines would prompt an immune response where it is most desirable—in the gastrointestinal tract. Vaccine developers quickly focused on live but weakened, or attenuated, strains of rotavirus that could be administered by mouth, without needles.

In 1983 the first rotavirus vaccine was ready for testing. Francis Andre of Smith Kline-RIT (now GlaxoSmithKline Biologicals) in Rixensart, Belgium, and Timo Vesikari, a pediatrician at the University of Tampere in Finland, prepared and tested a vaccine derived from a rotavirus strain found in cows. They chose a bovine rotavirus because it grew well in culture and was thought to be naturally attenuated in humans.

From all vantages, the first trial, conducted in Finland, was a landmark success: the vaccine reduced the chances that a vaccinated child would get severe rotavirus by 88 percent, demonstrating that immunity could be induced with a live oral vaccine. Moreover, the vaccine had no troubling side effects. Encouraged, Smith Kline-RIT launched trials in other countries, and by the late 1980s the end of rotavirus-related deaths seemed at hand. But then results from trials in Africa and Peru proved inconsistent and disappointing. Lacking certainty about the reasons for the troubles—although poor health, untreated infections, malnutrition and parasites are known to affect a child's immune response to vaccines—the company put its rotavirus program on hold.

BACK TO THE DRAWING BOARD

Researchers at the National Institutes of Health and the Wistar Institute in Philadelphia sought to explain the failure of the RIT vaccine. Possibly, the bovine strain was overattenuated—that is, it was too weak to replicate and elicit a good immune response under challenging conditions. They began looking for new formulations. Albert Kapikian of the NIH, for example, identified a rhesus strain of virus, and Fred Clark and Stanley Plotkin of Wistar identified another bovine strain that might replicate more vigorously. The strains were prepared for clinical trials, but these, too, showed both success and failure. Several more years were needed to rethink the science.

Meanwhile other researchers were unraveling the virus's molecular structure. Though wheel like in cross section, rotavirus is actually a three-layered sphere containing 11 segments of double-stranded RNA, each of which consists of a single gene encoding a protein. The proteins fall into two basic types: ones that are structural (composing the virus) and ones that are nonstructural (made within infected cells). The structural viral proteins, or VPs, are numerically named: VP1, VP2 and so on, as are the nonstructural proteins, or NSPs, which participate in viral replication and in deranging intestinal function.

The outermost shell, important in eliciting the host's immune response, has been a focus of attention in vaccine development. VP7 fashions its lumpy surface, and the VP4 protein forms the spikes on the outside of the "wheel." VP6, the most abundant protein in the virus, sits underneath VP7 and participates in producing viral proteins in infected cells. A non-structural molecule called NSP4 is a toxin that may play a role in triggering profuse diarrhea.

The proteins come in several varieties, and separate strains sport different mixes of proteins. When two viral strains infect the same cell, their gene segments can reassort just like figures on a slot machine, forming

Rotavirus Up Close

Structural studies reveal that rotavirus, shown below in two cutaway views, consists of three protein layers that encase the genome. Its structural proteins—those present in particles that spread from person to person—are called VPs and are denoted by numbers.

VP7 forms the outer surface and is studded with VP4 spikes. These two proteins elicit a host's disease-fighting immune response and thus play a central role in vaccines. VP4 also facilitates viral entry into cells, as do VP5 and VP8 [not shown],

which result from cleavage of VP4 in a host's body. VP6 composes the middle layer and is required for gene transcription, a process essential to the synthesis of viral proteins in infected cells. VP2 makes up the inner shell, and VP1 and VP3 are enzymes involved in copying viral genes.

The genome comprises 11 segments of double-stranded RNA tightly coiled and packed together. These segments code for the VPs as well as for nonstructural proteins [NSPs], including a toxin called NSP4 that is made after the virus enters cells.

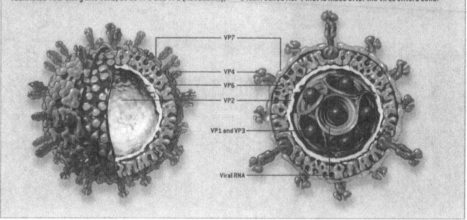

new combinations and thus novel versions of the virus. New reassortant viruses arise constantly, but as is true of most mutations, few offer—survival advantages to the virus. Consequently, of the 42 unique rotavirus strains identified to date based on their combinations of VP7 and VP4 varieties, only four or five account for more than 90 percent of rotavirus disease worldwide.

Exploiting the natural ability of rotavirus to reassort its genes, Kapikian and his NIH colleague Harry Greenberg developed a laboratory method to create reassortants that had features useful for vaccines but would not cause disease in humans. They began by making a reassortant virus that combined 10 genes from a monkey rotavirus—giving it the property of attenuation—with one gene encoding a surface protein, VP7, from a human strain. They made three such reassortants, each displaying a different human version of VP7, and one purely rhesus virus, displaying a fourth VP7 found in both monkey and human rotaviruses. They mixed all four into a cocktail called a tetravalent vaccine intended to offer protection against the four most prevalent human strains of rotavirus.

In 1991 the Food and Drug Administration granted the pharmaceutical company Wyeth Ayerst (later Wyeth Pharmaceuticals) permission to make

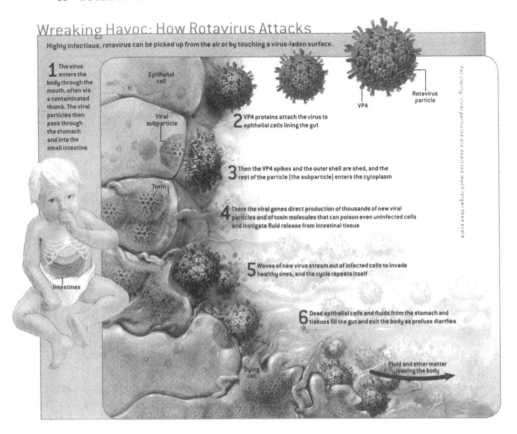

Wreaking Havoc: How Rotavirus Attacks

Highly infectious, rotavirus can be picked up from the air or by touching a virus-laden surface.

1 The virus enters the body through the mouth, often via a contaminated thumb. The viral particles then pass through the stomach and into the small intestine

Epithelial cell

Viral subparticle

Toxin

Intestines

Rotavirus particle

VP4

2 VP4 proteins attach the virus to epithelial cells lining the gut

3 Then the VP4 spikes and the outer shell are shed, and the rest of the particle (the subparticle) enters the cytoplasm

4 There the viral genes direct production of thousands of new viral particles and of toxin molecules that can poison even uninfected cells and instigate fluid release from intestinal tissue

5 Waves of new virus stream out of infected cells to invade healthy ones, and the cycle repeats itself

6 Dead epithelial cells and fluids from the stomach and tissues fill the gut and exit the body as profuse diarrhea

Dying cell

Fluid and other matter leaving the body

For clarity, viral particles are depicted much larger than at scale

and test this vaccine, which they named RotaShield. Over the next five years it launched large-scale clinical trials in the U.S., Finland and Venezuela, verifying RotaShield's safety, ability to induce a protective immune response, and lasting efficacy. In 1998 RotaShield was licensed by the FDA and recommended by the CDC's Advisory Committee on Immunization Practices and the American Academy of Pediatrics for routine immunization of all American children. Over the next nine months more than 600,000 children received an estimated 1.2 million doses of RotaShield.

These were heady times. The vaccine still had to be tested on undernourished children in developing nations, where live oral vaccines for other diseases—including polio and cholera—were known to be less effective than elsewhere. Also, the price per dose was still high for most developing nations. But for the first time, the world had a tool with which to combat rotavirus, and many of us were jubilant.

Then disaster struck. In 1999 several infants suffered a serious complication within two weeks of receiving the vaccine: a segment of the intestine folded into a nearby region (like a part of a telescope collapses into another), creating a blockage called intussusception. The condition can be excruciatingly painful and must be quickly reversed with either an air or fluid enema or fixed surgically. In rare cases, the intestine perforates and the infant dies. The CDC, which was monitoring experience with RotaShield, called for an immediate halt to the immunization program, thereby sinking a vaccine that had taken 15 years and several hundred million dollars to launch.

The agency initially estimated the risk to be one intussusception in 2,500 vaccine recipients, which was considered unacceptable. Later studies pegged the probability at only one in 11,000. Then Lone Simonsen of the NIH correlated risk with age: infants younger than three months were in less danger than older ones. If the vaccine were given only to young babies, the likelihood of intussusception could drop 10-fold, to perhaps one in 30,000.

The new data raised new questions. Was this risk acceptable in the U.S., where children are often hospitalized but rarely die of rotavirus? Were the odds more palatable in the developing world, where one child in 200 dies of rotavirus? If 150 lives could be saved for each complication from intussusception, might the risk be justified? Given these statistics, was it unethical, in fact, to withhold a vaccine that might save half a million lives a year? Or no matter what the risk-benefit analysis showed, was it unethical to market a vaccine in the developing world that had been withdrawn from use in the U.S.?

The CDC and the WHO called a meeting of policymakers from developing countries. After heated discussion, science bowed to politics. As a high-ranking Indian official said, "I know this vaccine would save 100,000 children in my country. But when the first case of intestinal blockage occurred, I would not be forgiven for allowing a vaccine that had been withdrawn in the United States to be used in my country."

BACK ON TRACK

Researchers continued to study the link between vaccination and intussusception. Children who contracted rotavirus naturally had no greater incidence of blockage than other children, so why should vaccination per se raise their risk? Some began to suspect the problem was specific to the rhesus strains, not an effect common to all live oral rotavirus vaccines.

Betting the intussusception problem could be overcome, two vaccine makers renewed their interest in rotavirus. GlaxoSmithKline dusted off its program and pressed forward with a new monovalent vaccine derived

entirely from a single attenuated human strain. Because natural rotavirus infection was not associated with intussusception, they reasoned
their vaccine would similarly not increase the risk of this complication.
In addition, the company would select for study only infants who were
six weeks to 13 weeks old, a stage when natural intussusception is rare. At
the same time, Merck developed a pentavalent vaccine derived from five
human-bovine reassortant strains that together would target the major
strains of rotavirus. Merck scientists knew bovine strains did not grow or
replicate as well as the rhesus strain and also did not cause the low-grade
fever many children developed after being immunized with the rhesus
vaccine. Also, the company would limit eligibility in its clinical trials exclusively to infants six to 12 weeks old.

Both companies conferred with the FDA on their plans to conduct
clinical trials. The FDA, wanting to ensure that the next generation of
rotavirus vaccines would be safer than RotaShield, insisted that the trials
be large enough to detect any risks, however small, that might be associated with the vaccine. An initial target of 60,000 participants per trial
was set, making these the largest and most expensive safety trials of any
vaccine ever tested before licensing. Not only would the trials be costly,
but the undertaking itself was risky—each one would instantly collapse
if the rate of intussusception among vaccinated babies exceeded that of
non-vaccinated ones. The developers pressed on with some trepidation.

Now, six years after the intussusception debacle, the rotavirus gamble
is paying off. GlaxoSmithKline and Merck have completed their clinical
trials, and the results for both vaccines are encouraging. They offer from
85 to 98 percent protection against severe rotavirus diarrhea. Moreover,
the vaccinated children showed no more cases of intussusception than
did nonvaccinated children.

The GlaxoSmithKline vaccine, Rotarix, was tested primarily in Latin
America. Since 2004, it has won approval from more than 20 countries
and, most recently, from the European Union; it is under review in the U.S.
Merck, in contrast, targeted the U.S. market first, wanting to prove that its
vaccine, RotaTeq, is safe here before introducing it elsewhere. The company has gained approval in Mexico and the U.S. and expects to have it for
Europe this year; such approvals are a prelude to introducing the vaccine
to many countries.

Vaccine manufacturers in the developing world are also interested in
rotavirus. Unlike those that require sophisticated bioengineering techniques, a rotavirus vaccine, like that for polio, can be made using traditional tissue culture methods and so is within the reach of smaller com-

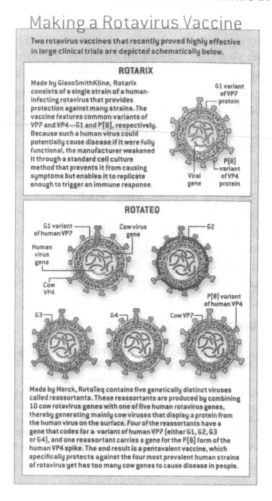

Making a Rotavirus Vaccine

Two rotavirus vaccines that recently proved highly effective in large clinical trials are depicted schematically below.

ROTARIX

Made by GlaxoSmithKline, Rotarix consists of a single strain of a human-infecting rotavirus that provides protection against many strains. The vaccine features common variants of VP7 and VP4—G1 and P[8], respectively. Because such a human virus could potentially cause disease if it were fully functional, the manufacturer weakened it through a standard cell culture method that prevents it from causing symptoms but enables it to replicate enough to trigger an immune response.

G1 variant of VP7 protein

P[8] variant of VP4 protein

Viral gene

ROTATEQ

G1 variant of human VP7

Cow virus gene

G2

Human virus gene

Cow VP4

P[8] variant of human VP4

G3

G4

Cow VP7

Made by Merck, RotaTeq contains five genetically distinct viruses called reassortants. These reassortants are produced by combining 10 cow rotavirus genes with one of five human rotavirus genes, thereby generating mainly cow viruses that display a protein from the human virus on the surface. Four of the reassortants have a gene that codes for a variant of human VP7 (either G1, G2, G3 or G4), and one reassortant carries a gene for the P[8] form of the human VP4 spike. The end result is a pentavalent vaccine, which specifically protects against the four most prevalent human strains of rotavirus yet has too many cow genes to cause disease in people.

panies. Today more than 10 makers in India, China, Indonesia and Brazil are preparing live oral rotavirus vaccines; a Chinese firm has already gained approval to sell its product.

FUTURE CHALLENGES

The prospect of new vaccines fuels hope that rotavirus's grip may soon be broken. Still, hurdles remain. Because many policymakers in—developing countries have not heard of rotavirus, they fail to understand its dire consequences. Surveillance efforts in more than 40 countries—being conducted by Joseph Bresee and Umesh D. Parashar of the CDC, with the

WHO and the Program for Appropriate Technology in Health—are just beginning to provide data that decision makers will need before welcoming the vaccines into their nations. In addition, confirmation that live oral vaccines are safe and effective in the poorest areas is still lacking. Moreover, the vaccines, which cost several hundred million dollars each to develop, must be affordable to those responsible for the 135 million children born worldwide every year.

Yet momentum is building, and many of us hope that within a decade, this major cause of diarrhea and principal killer of children in the developing world will be eliminated by the most cost-effective public health measure we have today: immunization. With help from a committed global community, rotavirus will soon join such microorganisms as polio, smallpox and diphtheria, which have been vanquished by vaccines and are now sidelined and obscure. Epidemiologists hope the anonymity that has historically characterized this disease will define it once again, its regained obscurity a true testament to power of vaccination.

FURTHER READING

Global Illness and Deaths Caused by Rotavirus Disease in Children. U. D. Parashar et al. in *Emerging Infectious Diseases,* Vol. 9, No. 5, pages 565–572; May 2003.

The Future of Rotavirus Vaccines: A Major Setback Leads to New Opportunities. Viewpoint. R. Glass et al. in *Lancet,* Vol. 363, Issue 9420, pages 154–550; May 2004.

Safety and Efficacy of an Attenuated Vaccine against Severe Rotavirus Gastroenteritis. G. Ruiz-Palacios et al. in *New England Journal of Medicine,* Vol. 354, pages 11–22; January 5, 2006.

Safety and Efficacy of a Pentavalent Human-Bovine (WC3) Reassortant Rotavirus Vaccine. T. Vesikari et al. in *New England Journal of Medicine,* Vol. 354, pages 23–32; January 5, 2006.

Capturing a Killer Flu Virus

JEFFERY K. TAUBENBERGER, ANN H. REID
AND THOMAS G. FANNING

ORIGINALLY PUBLISHED IN JANUARY 2005

On September 7, 1918, at the height of World War I, a soldier at an army training camp outside Boston came to sick call with a high fever. Doctors diagnosed him with meningitis but changed their minds the next day when a dozen more soldiers were hospitalized with respiratory symptoms. Thirty-six new cases of this unknown illness appeared on the 16th. Incredibly, by September 23rd, 12,604 cases had been reported in the camp of 45,000 soldiers. By the end of the outbreak, one third of the camp's population would come down with this severe disease, and nearly 800 of them would die. The soldiers who perished often developed a bluish skin color and struggled horribly before succumbing to death by suffocation. Many died less than 48 hours after their symptoms appeared, and at autopsy their lungs were filled with fluid or blood.

Because this unusual suite of symptoms did not fit any known malady, a distinguished pathologist of the era, William Henry Welch, speculated that "this must be some new kind of infection or plague." Yet the disease was neither plague nor even new. It was just influenza. Still, this particularly virulent and infectious strain of the flu virus is thought to have killed as many as 40 million people around the world between 1918 and 1919.

This most lethal flu outbreak in modern history disappeared almost as quickly as it emerged, and its cause was long believed lost to time. No one had preserved samples of the pathogen for later study because influenza would not be identified as a virus until the 1930s. But thanks to incredible foresight by the U.S. Army Medical Museum, the persistence of a pathologist named Johan Hultin, and advances in genetic analysis of old tissue samples, we have been able to retrieve parts of the 1918 virus and study their features. Now, more than 80 years after the horrible natural disaster of 1918–1919, tissues recovered from a handful of victims are answering fundamental questions both about the nature of this pandemic strain and about the workings of influenza viruses in general.

The effort is not motivated merely by historical curiosity. Because influenza viruses continually evolve, new influenza strains continually threaten human populations. Pandemic human flu viruses have emerged twice since 1918—in 1957 and 1968. And flu strains that usually infect only animals have also periodically caused disease in humans, as seen in the recent outbreak of avian influenza in Asia. Our two principal goals are determining what made the 1918 influenza so virulent, to guide development of influenza treatments and preventive measures, and establishing the origin of the pandemic virus, to better target possible sources of future pandemic strains.

HUNTING THE 1918 VIRUS

In many respects, the 1918 influenza pandemic was similar to others before it and since. Whenever a new flu strain emerges with features that have never been encountered by most people's immune systems, widespread flu outbreaks are likely. But certain unique characteristics of the 1918 pandemic have long remained enigmatic.

For instance, it was exceptional in both its breadth and depth. Outbreaks swept across Europe and North America, spreading as far as the Alaskan wilderness and the most remote islands of the Pacific. Ultimately, one third of the world's population may have been infected. The disease was also unusually severe, with death rates of 2.5 to 5 percent—up to 50 times the mortality seen in other influenza outbreaks.

By the fall of 1918 everyone in Europe was calling the disease the "Spanish" influenza, probably because neutral Spain did not impose the wartime censorship of news about the outbreak prevalent in combatant countries. The name stuck, although the first outbreaks, or spring wave, of the pandemic seemingly arose in and around military camps in the U.S. in March 1918. The second, main wave of the global pandemic occurred from September to November 1918, and in many places yet another severe wave of influenza hit in early 1919.

Antibiotics had yet to be discovered, and most of the people who died during the pandemic succumbed to pneumonia caused by opportunistic bacteria that infected those already weakened by the flu. But a subset of influenza victims died just days after the onset of their symptoms from a more severe viral pneumonia—caused by the flu itself—that left their lungs either massively hemorrhaged or filled with fluid. Furthermore, most deaths occurred among young adults between 15 and 35 years old, a group that rarely dies from influenza. Strikingly, people younger than

65 years accounted for more than 99 percent of all "excess" influenza deaths (those above normal annual averages) in 1918–1919.

Efforts to understand the cause of the 1918 pandemic and its unusual features began almost as soon as it was over, but the culprit virus itself remained hidden for nearly eight decades. In 1951 scientists from the University of Iowa, including a graduate student recently arrived from— Sweden named Johan Hultin, went as far as the Seward Peninsula of Alaska seeking the 1918 strain. In November 1918 flu spread through an Inuit fishing village now called Brevig Mission in five days, killing 72 people—about 85 percent of the adult population. Their bodies had since been buried in permafrost, and the 1951 expedition members hoped to find the 1918 virus preserved in the victims' lungs. Unfortunately, all attempts to culture live influenza virus from these specimens were unsuccessful.

In 1995 our group initiated an attempt to find the 1918 virus using a different source of tissue: archival autopsy specimens stored at the Armed Forces Institute of Pathology (AFIP). For several years, we had been— developing expertise in extracting fragile viral genetic material from damaged or decayed tissue for diagnostic purposes. In 1994, for instance, we were able to use our new techniques to help an AFIP marine mammal pathologist investigate a mass dolphin die-off that had been blamed on red tide. Although the available dolphin tissue samples were badly decayed, we extracted enough pieces of RNA from them to identify a new virus, similar to the one that causes canine distemper, which proved to be the real cause of the dolphin deaths. Soon we began to wonder if there were any older medical mysteries we might solve with our institute's resources.

A descendant of the U.S. Army Medical Museum founded in 1862, the AFIP has grown along with the medical specialty of pathology and now has a collection of three million specimens. When we realized that these included autopsy samples from 1918 flu victims, we decided to go after the pandemic virus. Our initial study examined 78 tissue samples from victims of the deadly fall wave of 1918, focusing on those with the severe lung damage characteristic of patients who died rapidly. Because the influenza virus normally clears the lungs just days after infection, we had the greatest chance of finding virus remnants in these victims.

The standard practice of the era was to preserve autopsy specimens in formaldehyde and then embed them in paraffin, so fishing out tiny genetic fragments of the virus from these 80-year-old "fixed" tissues pushed the very limits of the techniques we had developed. After an agonizing year of negative results, we found the first influenza-positive sample in

1996, a lung specimen from a soldier who died in September 1918 at Fort Jackson, S.C. We were able to determine the sequence of nucleotides in small fragments of five influenza genes from this sample.

But to confirm that the sequences belonged to the lethal 1918 virus, we kept looking for more positive cases and identified another one in 1997. This soldier also died in September 1918, at Camp Upton, N.Y. Having a second sample allowed us to confirm the gene sequences we had, but the tiny quantity of tissue remaining from these autopsies made us worry that we would never be able to generate a complete virus sequence.

A solution to our problem came from an unexpected source in 1997: Johan Hultin, by then a 73-year-old retired pathologist, had read about our initial results. He offered to return to Brevig Mission to try another exhumation of 1918 flu victims interred in permafrost. Fortysix years after his first attempt, with permission from the Brevig Mission Council, he obtained frozen lung biopsies of four flu victims. In one of these samples, from a woman of unknown age, we found influenza RNA that provided the key to sequencing the entire genome of the 1918 virus.

More recently, our group, in collaboration with British colleagues, has also been surveying autopsy tissue samples from 1918 influenza victims from the Royal London Hospital. We have been able to analyze flu virus genes from two of these cases and have found that they were nearly identical to the North American samples, confirming the rapid worldwide spread of a uniform virus. But what can the sequences tell us about the virulence and origin of the 1918 strain? Answering those questions requires a bit of background about how influenza viruses function and cause disease in different hosts.

FLU'S CHANGING FACE

Each of the three novel influenza strains that caused pandemics in the past 100 years belonged to the type A group of flu viruses. Flu comes in three main forms, designated A, B and C. The latter two infect only humans and have never caused pandemics. Type A influenza viruses, on the other hand, have been found to infect a wide variety of animals, including poultry, swine, horses, humans and other mammals. Aquatic birds, such as ducks, serve as the natural "reservoir" for all the known subtypes of influenza A, meaning that the virus infects the bird's gut without causing symptoms. But these wild avian strains can mutate over time or exchange genetic material with other influenza strains, producing novel viruses that are able to spread among mammals and domestic poultry.

FLU HIJACKS HOSTS TO REPLICATE AND EVOLVE

Influenza is a small and simple virus—just a hollow lipid ball studded with a few proteins and bearing only eight gene segments (below). But that is all it needs to induce the cells of living hosts to make more viruses (bottom). One especially important protein on influenza's surface, hemagglutinin (HA), allows the virus to enter cells. Its shape determines which hosts a flu virus strain can infect. Another protein, neuraminidase (NA), cuts newly formed

viruses loose from an infected cell, influencing how efficiently the virus can spread. Slight changes in these and other flu proteins can help the virus infect new kinds of hosts and evade immune attack. The alterations can arise through mistakes that occur while viral genes are being copied. Or they can be acquired in trade when the genes of two different flu viruses infecting the same cell intermingle (right).

INFLUENZA VIRUS

The two major surface proteins, HA and NA, protrude from a lipid bilayer. Inside (cutaway), eight separate RNA segments specify additional proteins that determine all aspects of the virus's function.

— HA
— NA
Lipid bilayer

INFECTION AND REPLICATION

A flu virus's HA protein links to sialic acid on the surface of a host organism's cell (a), allowing the virus to slip inside (b), where it releases its RNA (c), which enters the cell's nucleus (d). There the viral RNA is copied and is genetic instructions are "read," prompting cellular machinery to produce overviral proteins (e). The new viral RNA and proteins then assemble into viruses (f) that bud from the cell membrane (f). At first, their surfaces are coated with sialic acid. To prevent viruses (from binding to one another's hemagglutinin proteins and to the host cell surfaces, neuraminidase clips the sialic acid (g), freeing the viruses to infect other cells.

Sialic acid
— viral RNA
Nucleus
Viral RNA copies

a b c d e f g

REASSORTMENT

New flu strains can result when two different viruses infect the same cell (above). Copies of their RNA can mix and produce progeny with combinations of genes from both parent viruses. In this manner, a bird or animal flu strain can gain genes conferring the ability to spread more easily among humans.

Nucleus
Host cell
Strain 1
Strain 2
Reassorted viruses

The life cycle and genomic structure of influenza A virus allow it to evolve and exchange genes easily. The virus's genetic material consists of eight separate RNA segments encased in a lipid membrane studded with proteins [see illustration]. To reproduce, the virus binds to and then enters a living cell, where it commandeers cellular machinery, inducing it to manufacture new viral proteins and additional copies of viral RNA. These pieces then assemble themselves into new viruses that escape the host cell, proceeding to infect other cells. No proofreading mechanism ensures that the RNA copies are accurate, so mistakes leading to new mutations are common. What is more, should two different influenza virus strains infect the same cell, their RNA segments can mix freely there, producing progeny viruses that contain a combination of genes from both the original viruses. This "reassortment" of viral genes is an important mechanism for generating diverse new strains.

Different circulating influenza A viruses are identified by referring to two signature proteins on their surfaces. One is hemagglutinin (HA), which has at least 15 known variants, or subtypes. Another is neuraminidase (NA), which has nine subtypes. Exposure to these proteins produces distinctive antibodies in a host, thus the 1918 strain was the first to be named, "H1N1," based on antibodies found in the bloodstream of pandemic survivors. Indeed, less virulent descendants of H1N1 were the predominant circulating flu strains until 1957, when an H2N2 virus emerged, causing a pandemic. Since 1968, the H3N2 subtype, which provoked the pandemic that year, has predominated.

The HA and NA protein subtypes present on a given influenza A virus are more than just identifiers; they are essential for viral reproduction and are primary targets of an infected host's immune system. The HA molecule initiates infection by binding to receptors on the surface of certain host cells. These tend to be respiratory lining cells in mammals and intestinal lining cells in birds. The NA protein enables new virus copies to escape the host cell so they can go on to infect other cells.

After a host's first exposure to an HA subtype, antibodies will block receptor binding in the future and are thus very effective at preventing reinfection with the same strain. Yet flu viruses with HA subtypes that are new to humans periodically appear, most likely through reassortment with the extensive pool of influenza viruses infecting wild birds. Normally, influenza HAs that are adapted to avian hosts bind poorly to the cell-surface receptors prevalent in the human respiratory tract, so an avian virus's HA binding affinity must be somewhat modified before the virus can replicate and spread efficiently in humans. Until recently, existing evidence suggested that a wholly avian influenza virus probably could not directly infect humans, but 18 people were infected with an avian H5N1 influenza virus in Hong Kong in 1997, and six died.

Outbreaks of an even more pathogenic version of that H5N1 strain became widespread in Asian poultry in 2003 and 2004, and more than 30 people infected with this virus have died in Vietnam and Thailand.

The virulence of an influenza virus once it infects a host is determined by a complex set of factors, including how readily the virus enters different tissues, how quickly it replicates, and the violence of the host's immune response to the intruder. Thus, understanding exactly what made the 1918 pandemic influenza strain so infectious and so virulent could yield great insight into what makes any influenza strain more or less of a threat.

A KILLER'S FACE

With the 1918 RNA we have retrieved, we have used the virus's own genes as recipes for manufacturing its component parts—essentially re-creating pieces of the killer virus itself. The first of these we were eager to examine was the hemagglutinin protein, to look for features that might explain the exceptional virulence of the 1918 strain.

We could see, for example, that the part of the 1918 HA that binds with a host cell is nearly identical to the binding site of a wholly avian influenza HA. In two of the 1918 isolates, this receptor-binding site differs from an avian form by only one amino acid building block. In the other three isolates, a second amino acid is also altered. These seemingly subtle mutations may represent the minimal change necessary to allow an avian-type HA to bind to mammalian-type receptors.

But while gaining a new binding affinity is a critical step that allows a virus to infect a new type of host, it does not necessarily explain why the 1918 strain was so lethal. We turned to the gene sequences themselves, looking for features that could be directly related to virulence, including two known mutations in other flu viruses. One involves the HA gene: to become active in a cell, the HA protein must be cleaved into two pieces by a gut-specific protein-cutting enzyme, or protease, supplied by the host. Some avian H5 and H7 subtype viruses acquire a gene mutation that adds one or more basic amino acids to the cleavage site, allowing HA to be activated by ubiquitous proteases. In chickens and other birds, infection by such a virus causes disease in multiple organs and even the central nervous system, with a very high mortality rate. This mutation has been observed in the H5N1 viruses currently circulating in Asia. We did not, however, find it in the 1918 virus.

The other mutation with a significant effect on virulence has been seen in the NA gene of two influenza virus strains that infect mice. Again, mutations at a single amino acid appear to allow the virus to replicate in many different body tissues, and these flu strains are typically lethal in laboratory mice. But we did not see this mutation in the NA of the 1918 virus either.

Because analysis of the 1918 virus's genes was not revealing any characteristics that would explain its extreme virulence, we initiated a collaborative effort with several other institutions to re-create parts of the 1918 virus itself so we could observe their effects in living tissues.

A new technique called plasmid-based reverse genetics allows us to copy 1918 viral genes and then combine them with the genes of an existing influenza strain, producing a hybrid virus. Thus, we can take an influenza strain adapted to mice, for example, and give it different combinations of 1918 viral genes. Then, by infecting a live animal or a human tissue culture with this engineered virus, we can see which components of the pandemic strain might have been key to its pathogenicity.

For instance, the 1918 virus's distinctive ability to produce rapid and extensive damage to both upper and lower respiratory tissues suggests that it replicated to high numbers and spread quickly from cell to cell. The viral protein NS1 is known to prevent production of type I interferon (IFN)—an "early warning" system that cells use to initiate an immune response against a viral infection. When we tested recombinant viruses in a tissue culture of human lung cells, we found that a virus with the 1918 NSI gene was indeed more effective at blocking the host's type I IFN system.

To date, we have produced recombinant influenza viruses containing between one and five of the 1918 genes. Interestingly, we found that any of the recombinant viruses possessing both the 1918 HA and NA genes were lethal in mice, causing severe lung damage similar to that seen in some of the pandemic fatalities. When we analyzed these lung tissues, we found signatures of gene activation involved in common inflammatory responses. But we also found higher than normal activation of genes associated with the immune system's offensive soldiers, T cells and macrophages, as well as genes related to tissue injury, oxidative damage, and apoptosis, or cell suicide.

More recently, Yoshihiro Kawaoka of the University of Wisconsin—Madison reported similar experiments with 1918 flu genes in mice, with similar results. But when he tested the HA and NA genes separately, he found that only the 1918 HA produced the intensive immune response, suggesting that for reasons as yet unclear, this protein may have played a key role in the 1918 strain's virulence.

These ongoing experiments are providing a window to the past, helping scientists understand the unusual characteristics of the 1918 pandemic. Similarly, these techniques will be used to study what types of changes to the current H5N1 avian influenza strain might give that extremely lethal virus the potential to become pandemic in humans. An equally compelling question is how such virulent strains emerge in the first place, so our group has also been analyzing the 1918 virus's genes for clues about where it might have originated.

SEEKING THE SOURCE

The best approach to analyzing the relationships among influenza viruses is phylogenetics, whereby hypothetical family trees are constructed using viral gene sequences and knowledge of how often genes typically mutate. Because the genome of an influenza virus consists of eight discrete RNA segments that can move independently by reassortment, these evolutionary studies must be performed separately for each gene segment.

We have completed analyses of five of the 1918 virus's eight RNA segments, and so far our comparisons of the 1918 flu genes with those of numerous human, swine and avian influenza viruses always place the 1918 virus within the human and swine families, outside the avian virus group [see box on page 44]. The 1918 viral genes do have some avian features, however, so it is probable that the virus originally emerged from an avian reservoir sometime before 1918. Clearly by 1918, though, the virus had acquired enough adaptations to mammals to function as a human pandemic virus. The question is, where?

When we analyzed the 1918 hemagglutinin gene, we found that the sequence has many more differences from avian sequences than do the 1957 H2 and 1968 H3 subtypes. Thus, we concluded, either the 1918 HA gene spent some length of time in an intermediate host where it accumulated many changes from the original avian sequence, or the gene came directly from an avian virus, but one that was markedly different from known avian H1 sequences.

To investigate the latter possibility that avian H1 genes might have changed substantially in the eight decades since the 1918 pandemic, we collaborated with scientists from the Smithsonian Institution's Museum of Natural History and Ohio State University. After examining many preserved birds from the era, our group isolated an avian subtype H1—

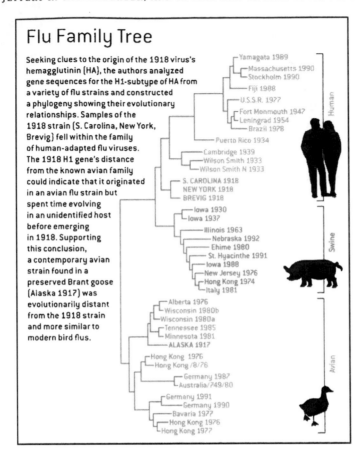

Flu Family Tree

Seeking clues to the origin of the 1918 virus's hemagglutinin (HA), the authors analyzed gene sequences for the H1-subtype of HA from a variety of flu strains and constructed a phylogeny showing their evolutionary relationships. Samples of the 1918 strain (S. Carolina, New York, Brevig) fell within the family of human-adapted flu viruses. The 1918 H1 gene's distance from the known avian family could indicate that it originated in an avian flu strain but spent time evolving in an unidentified host before emerging in 1918. Supporting this conclusion, a contemporary avian strain found in a preserved Brant goose (Alaska 1917) was evolutionarily distant from the 1918 strain and more similar to modern bird flus.

Yamagata 1989
Massachusetts 1990
Stockholm 1990
Fiji 1988
U.S.S.R. 1977
Fort Monmouth 1947
Leningrad 1954
Brazil 1978
Puerto Rico 1934
Cambridge 1939
Wilson Smith 1933
Wilson Smith N 1933
S. CAROLINA 1918
NEW YORK 1918
BREVIG 1918
Iowa 1930
Iowa 1937
Illinois 1963
Nebraska 1992
Ehime 1980
St. Hyacinthe 1991
Iowa 1988
New Jersey 1976
Hong Kong 1974
Italy 1981
Alberta 1976
Wisconsin 1980b
Wisconsin 1980a
Tennessee 1985
Minnesota 1981
ALASKA 1917
Hong Kong 1976
Hong Kong /8/76
Germany 1987
Australia/749/80
Germany 1991
Germany 1990
Bavaria 1977
Hong Kong 1976
Hong Kong 1977

Human

Swine

Avian

influenza strain from a Brant goose collected in 1917 and stored in ethanol in the Smithsonian's bird collections. As it turned out, the 1917 avian H1 sequence was closely related to modern avian North American H1 strains, suggesting that avian H1 sequences have changed little over the past 80 years. Extensive sequencing of additional wild bird H1 strains may yet identify a strain more similar to the 1918 HA, but it may be that no avian H1 will be found resembling the 1918 strain because, in fact, the HA did not reassort directly from a bird strain.

In that case, it must have had some intermediate host. Pigs are a widely suggested possibility because they are known to be susceptible to both human and avian viruses. Indeed, simultaneous outbreaks of influenza were seen in humans and swine during the 1918 pandemic, but we believe that the direction of transmission was most probably from humans to pigs. There are numerous examples of human influenza A virus strains infecting swine since 1918, but swine influenza strains have been isolated only sporadically from humans. Nevertheless, to explore the possibility that the 1918 HA may have started as an avian form that gradually adapted to mammalian hosts in swine, we looked at a current example of how avian viruses evolve in pigs—an avian H1N1 influenza lineage that has become established in European swine over the past 25 years. We found that even 20 years of evolution in swine has not resulted in the number of changes from avian sequences exhibited by the 1918 pandemic strain.

When we applied these types of analyses to four other 1918 virus genes, we came to the same conclusion: the virus that sparked the 1918 pandemic could well have been an avian strain that was evolutionarily isolated from the typical wild waterfowl influenza gene pool for some time—one that, like the SARS coronavirus, emerged into circulation among humans from an as yet unknown animal host.

FUTURE INVESTIGATIONS

Our analyses of five RNA segments from the 1918 virus have shed some light on its origin and strongly suggest that the pandemic virus was the common ancestor of both subsequent human and swine H1N1 lineages, rather than having emerged from swine. To date, analyzing the viral genes has offered no definitive clue to the exceptional virulence of the 1918 virus strain. But experiments with engineered viruses—containing 1918 genes indicate that certain of the 1918 viral proteins could promote rapid virus replication and provoke an intensely destructive host immune response.

In future work, we hope that the 1918 pandemic virus strain can be placed in the context of influenza viruses that immediately preceded and followed it. The direct precursor of the pandemic virus, the first or spring wave virus strain, lacked the autumn wave's exceptional virulence and seemed to spread less easily. At present, we are seeking influenza RNA samples from victims of the spring wave to identify any genetic differences between the two strains that might help elucidate why the autumn wave was more severe. Similarly, finding pre-1918 human influenza RNA samples would clarify which gene segments in the 1918 virus were completely novel to humans. The unusual mortality among young people during the 1918 pandemic might be explained if the virus shared features with earlier circulating strains to which older people had some immunity. And finding samples of H1N1 from the 1920s and later would help us understand the 1918 virus's subsequent evolution into less virulent forms.

We must remember that the mechanisms by which pandemic flu strains originate are not yet fully understood. Because the 1957 and

1968 pandemic strains had avian-like HA proteins, it seems most likely that they originated in the direct reassortment of avian and human virus strains. The actual circumstances of those reassortment events have never been identified, however, so no one knows how long it took for the novel strains to develop into human pandemics.

The 1918 pandemic strain is even more puzzling, because its gene sequences are consistent neither with direct reassortment from a known avian strain nor with adaptation of an avian strain in swine. If the 1918 virus should prove to have acquired novel genes through a different mechanism than subsequent pandemic strains, this could have important public health implications. An alternative origin might even have contributed to the 1918 strain's exceptional virulence. Sequencing of many more avian influenza viruses and research into alternative intermediate hosts other than swine, such as poultry, wild birds or horses, may provide more clues to the 1918 pandemic's source. Until the origins of such strains are better understood, detection and prevention efforts may overlook the beginning of the next pandemic.

FURTHER READING

DEVIL'S FLU: THE WORLD'S DEADLIEST INFLUENZA EPIDEMIC AND THE SCIENTIFIC HUNT FOR THE VIRUS THAT CAUSED IT. Pete Davies. Henry Holt and Co., 2000.

AMERICA'S FORGOTTEN PANDEMIC: THE INFLUENZA OF 1918. Second edition. Alfred W. Crosby. Cambridge University Press, 2003.

THE ORIGIN OF THE 1918 PANDEMIC INFLUENZA VIRUS: A CONTINUING ENIGMA. Ann H. Reid and Jeffery K. Taubenberger in Journal of General Virology, Vol. 84, Part 9, pages 2285–2292; September 2003.

GLOBAL HOST IMMUNE RESPONSE: PATHOGENESIS AND TRANSCRIPTIONAL PROFILING OF TYPE A INFLUENZA VIRUSES EXPRESSING THE HEMAGGLUTININ AND NEURAMINIDASE GENES FROM THE 1918 PANDEMIC VIRUS. J. C. Kash, C. F. Basler, A. Garcia-Sastre, V. Carter, R. Billharz, D. E. Swayne, R. M. Przygodzki, J. K. Taubenberger, M. G. Katze and T. M. Tumpey in Journal of Virology, Vol. 78, No. 17, pages 9499–9511; September 2004.

The Unmet Challenges of Hepatitis C

ADRIAN M. DI BISCEGLIE AND BRUCE R. BACON

ORIGINALLY PUBLISHED IN OCTOBER 1999

As recently as the late 1980s few people other than physicians had heard of hepatitis C, a slowly progressing viral infection that over a couple of decades can lead to liver failure or liver cancer. Today the condition is widely recognized as a huge public health concern. Some 1.8 percent of the U.S. adult population, almost four million people, are infected with the hepatitis C virus, most of them without knowing it. The virus is one of the major causes of chronic liver disease, probably accounting for even more cases than excessive alcohol use, and is the most common reason for liver transplants. Some 9,000 people die each year in the U.S. from complications of the infection, a number that is expected to triple by 2010. Information about the incidence of hepatitis C in other countries is less reliable, but it is clear that the virus is a major public health problem throughout the world.

Physicians, historians and military leaders have long recognized hepatitis—inflammation of the liver—as a cause of jaundice. This yellow discoloration of the whites of the eyes and skin occurs when the liver fails to excrete a pigment called bilirubin, which then accumulates in the body. In recent decades, however, the diagnosis of hepatitis has progressively improved, and physicians can now distinguish several distinct forms. At least five different viruses can cause the condition, as can drugs and toxins such as alcohol.

Researchers first studied viral hepatitis in the 1930s and 1940s in settings where jaundice was common, such as prisons and mental institutions. They identified two distinct forms with different patterns of transmission. One was transmitted by contact with feces of infected individuals and was called infectious hepatitis, or hepatitis A. The other appeared to be passed only through blood and was termed serum hepatitis, or hepatitis B.

An important development occurred in the 1950s, when researchers devised tests for liver injury based on certain enzymes in blood serum.

When liver cells—known as hepatocytes—die, they release these enzymes into the circulation, where their concentrations can be easily measured. Elevated serum levels of alanine aminotransferase (ALT) and, especially, aspartate aminotransferase (AST) became recognized as more reliable signs of liver trouble than jaundice. (In addition to hepatitis, some uncommon inherited metabolic diseases can cause elevated Liver enzymes.)

There things stood until Baruch Blumberg, working at the National Institutes of Health, made a breakthrough in the mid-1960s. Blumberg identified the signature of a viral agent, now known as hepatitis B virus, in the blood of patients with that disease. Blumberg's discovery won him a Nobel Prize and allowed researchers to develop reliable blood tests for the virus. A decade later Stephen M. Feinstone, a researcher at the same institution, identified a different viral agent in the stool of patients with hepatitis A. This work led quickly to the development of tests that accurately detect antibodies to hepatitis A virus in the blood of those infected.

Hepatitis had long been a significant risk for recipients of blood transfusions and blood products. As many as 30 percent of patients receiving a blood transfusion in the 1960s developed elevated levels of ALT and AST, or even jaundice, some weeks later. Workers had suspected an infectious agent was responsible. When the new tests for hepatitis A and B became available in the 1970s, researchers soon found that a substantial proportion of cases of post-transfusion hepatitis were caused by neither of these two viruses. The new disease was labeled "non-A, non-B" hepatitis.

Most investigators expected that the agent responsible for these cases would soon be discovered. In reality, it took nearly 15 years before Michael Houghton and his colleagues at Chiron Corporation, a biotechnology company in Emeryville, Calif., finally identified the hepatitis C virus, using samples of serum from infected chimpanzees provided by Daniel W. Bradley of the Centers for Disease Control and Prevention. Hepatitis C accounts for most cases of viral hepatitis that are not types A or B, although a few result from other, rarer viruses.

THE NEEDLE IN AN RNA HAYSTACK

Hepatitis C virus proved difficult to identify because it cannot be reliably grown in cell cultures, and chimpanzees and tamarins appear to be the only nonhuman animals that can be infected. Because both species are very expensive to use in research, only small numbers of animals can be employed. These obstacles, which still impede the study of the virus, explain why it was the first infectious agent discovered entirely by cloning nucleic acid.

The Chiron researchers first extracted RNA from serum samples strongly suspected to contain the unknown viral agent. A chemical variant of DNA, RNA is used by many viruses as their genetic material. RNA is also found in healthy cells, so the problem was to identify the tiny fraction corresponding to the unknown viral genome.

The Chiron workers used an enzyme to copy multiple fragments of DNA from the RNA, so that each carried some part of its genetic sequence. Next, they inserted this "complementary DNA" into viruslike entities that infect *Escherichia coli* bacteria, which induced some bacteria to manufacture protein fragments that the DNA encoded. The researchers grew the bacteria to form colonies, or clones, that were then tested for their ability to cause a visible reaction with serum from chimpanzees and a human with non-A, non-B hepatitis.

The hope was that antibodies in the serum would bind to any clones producing protein from the infectious agent. Out of a million bacterial clones tested, just one was found that reacted with serum from chimpanzees with the disease but not with serum from the same chimpanzees before they had been infected. The result indicated that this clone contained genetic sequences of the disease agent. Using the clone as a toehold, investigators subsequently characterized the remainder of the virus's genetic material and developed the first diagnostic assay, a test that detects antibodies to hepatitis C in blood. Since 1990 that test and subsequent versions have allowed authorities to screen all blood donated to blood banks for signs of infection.

The antibody test soon showed hepatitis C to be a much bigger threat to public health than had generally been recognized. A remarkable feature—one that sets it apart from most other viruses—is its propensity to cause chronic disease. Most other viruses are self-limited: infection with hepatitis A, for example, usually lasts for only a few weeks. In contrast, nearly 90 percent of people with hepatitis C have it for years or decades.

Few patients know the source of their virus, but on direct questioning many recall having a blood transfusion, an episode of injection drug use or an injury from a hypodermic needle containing blood from an infected individual. About 40 percent of patients have none of these clear risk factors but fall into one of several categories identified in epidemiologic studies. These include having had sexual contact with someone with hepatitis, having had more than one sexual partner in the past year, and being of low socioeconomic status.

Whether hepatitis C is sexually transmitted is controversial. Instances of transmission between partners in stable, monogamous relationships are rarely identified, and the rate of infection in promiscuous gay men

HOW THE HEPATITIS C VIRUS WAS DISCOVERED

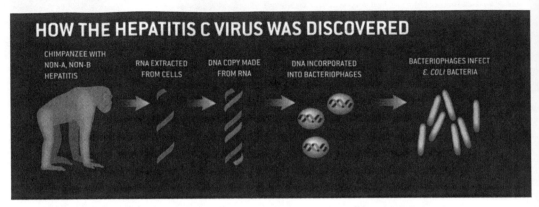

| CHIMPANZEE WITH NON-A, NON-B HEPATITIS | RNA EXTRACTED FROM CELLS | DNA COPY MADE FROM RNA | DNA INCORPORATED INTO BACTERIOPHAGES | BACTERIOPHAGES INFECT *E. COLI* BACTERIA |

Researchers identified the hepatitis C virus by making DNA copies of RNA from the cells of infected chimpanzees. They cloned the DNA by using bacteriophages to carry it into bacteria. Colonies were then tested with serum from infected chimps. One produced an immune reaction, indicating it carried viral genetic sequences.

is no higher than in the population in general. These observations suggest that sexual transmission is uncommon, but they are hard to reconcile with the epidemiologic findings. The paradox has not been resolved. Some patients who deny injection drug use may be unwilling or unable to recall it. Others might have been infected from unsterile razors or tattooing instruments. Shared straws put into the nose and used to snort street drugs might also transmit the virus via minute amounts of blood.

SLOW PROGRESS

The discovery of hepatitis C virus and the development of an accurate test for it mark an important victory for public health. The formerly substantial risk of infection from a blood transfusion has been virtually eliminated. Moreover, the rate of infection appears to be dropping among injection drug users, although this may be because anti-AIDS campaigns have discouraged sharing of needles. Yet hepatitis C still presents numerous challenges, and the prospects for eradicating the virus altogether appear dismal. Attempts to develop a vaccine have been hampered because even animals that successfully clear the virus from their bodies acquire no immunity to subsequent infection. Moreover, millions of people who are chronically infected are at risk of developing severe liver disease.

The mechanism of damage is known in outline. Viral infections can cause injury either because the virus kills cells directly or because the immune system attacks infected cells. Hepatitis C virus causes disease through the second mechanism. The immune system has two operating

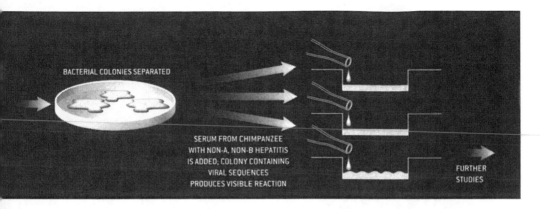

BACTERIAL COLONIES SEPARATED

SERUM FROM CHIMPANZEE
WITH NON-A, NON-B HEPATITIS
IS ADDED, COLONY CONTAINING
VIRAL SEQUENCES
PRODUCES VISIBLE REACTION

FURTHER
STUDIES

divisions. The humoral arm, which is responsible for producing anti-bodies, appears to be largely ineffective against hepatitis C virus. Although it produces antibodies to various viral components, the antibodies fail to neutralize the invader, and their presence does not indicate immunity, as is the case with hepatitis B.

It seems likely that hepatitis C virus evades this defense through its high mutation rate, particularly in regions of its genome responsible for the manufacture of proteins on the outside of the virus to which anti-bodies might bind. Two such hypervariable regions have been identified within the so-called envelope regions of the genome. As many as six distinct genotypes and many more subtypes of the virus have been identified; numerous variants exist even within a single patient.

In contrast to the humoral arm, the cellular arm of the immune system, which specializes in viral infections, mounts a vigorous defense against hepatitis C. It appears to be responsible for most of the liver injury. Cytotoxic T lymphocytes primed to recognize hepatitis C proteins are found in the circulation and in the liver of chronically infected individuals and are thought to kill hepatocytes that display viral proteins. Fortunately, liver tissue can regenerate well, but that from hepatitis patients often contains numerous dead or dying hepatocytes, as well as chronic inflammatory cells such as lymphocytes and monocytes.

LONG-TERM CONSEQUENCES

If hepatitis persists for long enough—typically some years—the condition escalates, and normally quiescent cells adjacent to hepatocytes, called hepatic stellate cells, become abnormally activated. These cells then secrete collagen and other proteins, which disrupt the fine-scale structure of the

HOW THE HEPATITIS C VIRUS REPRODUCES ITSELF

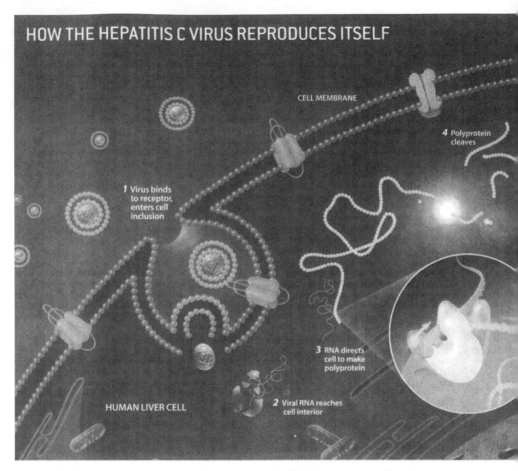

CELL MEMBRANE

4 Polyprotein cleaves

1 Virus binds to receptor, enters cell inclusion

3 RNA directs cell to make polyprotein

HUMAN LIVER CELL

2 Viral RNA reaches cell interior

Hepatitis C infection starts when viral particles in the circulation find their way to susceptible cells, particularly hepatocytes. A viral protein called E2 appears to facilitate entry by latching onto a specific receptor. On entering, the virus loses its lipid coat and its protein envelope, freeing the RNA cargo. Enzymes in the cell then use this RNA as a template to make a large viral protein, the polyprotein. It is cleaved into a variety of small proteins that go on to form new viral particles and help to copy the viral RNA.

The original RNA is copied to yield a "negative-stranded" RNA that carries the inverse, or complement, of the original sequence. This serves as a template to make multiple copies of the original RNA, which are incorporated into new viral particles, along with structural proteins, at a body called the Golgi complex. Complete viral particles are eventually released from the infected cell, after acquiring a lipid surface layer. Recent studies suggest that a patient produces as many as 1,000 billion, copies of hepatitis C virus a day, most of them from the liver.

liver and slowly impair its ability to process materials. This pathology is known as fibrosis. Stellate cells are similar in origin and function to the fibrosis-producing cells found in other organs, such as fibroblasts in the skin and mesangial cells in the kidney. They store vitamin A as well as produce the liver's extracellular matrix, or framework. It is likely that

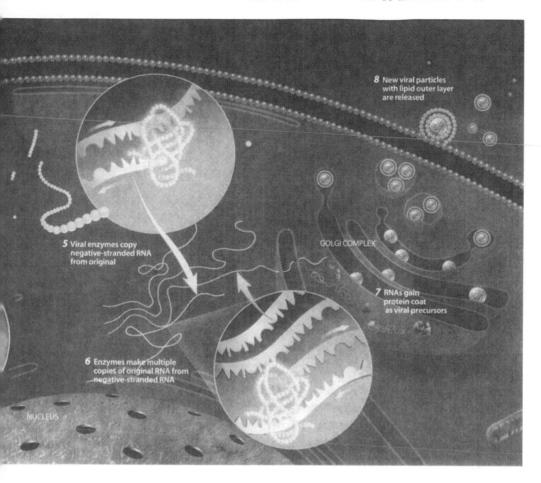

8 New viral particles
with lipid outer layer
are released

5 Viral enzymes copy
negative-stranded RNA
from original

GOLGI COMPLEX

7 RNAs gain
protein coat
as viral precursors

6 Enzymes make multiple
copies of original RNA from
negative-stranded RNA

NUCLEUS

many of the processes that initiate the fibrotic response in the liver occur
in these other tissues as well.

If fibrosis progresses far enough, it results in cirrhosis, which is characterized by bands of fibrosis enclosing nodules of regenerating hepatocytes. Progression is faster in people over age 50 at the time of infection,
in those who consume more than 50 grams of alcohol a day, and in men,
but cirrhosis can result even in patients who never drink alcohol. Fibrosis
and cirrhosis are generally considered irreversible, although recent findings cast some doubt on that conclusion.

About 20 percent of patients develop cirrhosis over the first 20 years of
infection. Thereafter some individuals may reach a state of equilibrium
without further liver damage, whereas others may continue to experience
very slow but progressive fibrosis. End-stage liver disease often manifests

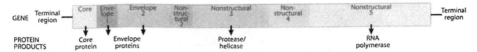

Hepatitis C virus genome consists of a single RNA gene plus two terminal regions. The gene encodes a polyprotein, which subsequently cleaves to form a variety of smaller proteins. Some of these are used to make new virus particles; others are enzymes that help to replicate the viral RNA for inclusion into new viruses.

itself as jaundice, ascites (accumulation of fluid within the abdomen), bleeding from varicose veins within the esophagus, and confusion. Hepatitis C infection has also come to be recognized as a major indirect cause of primary liver cancer. The virus itself seems not to put people at increased risk, but cirrhosis induced by the virus does.

Cirrhosis is responsible for almost all the illness caused by the hepatitis C virus. Although a small proportion of patients recollect an episode of jaundice when they probably acquired their infection, chronic hepatitis C is often asymptomatic. When symptoms do occur, they are nonspecific: patients sometimes complain of vague feelings of fatigue, nausea or general unwellness. The insidious nature of the condition is probably another reason why hepatitis C remained undiscovered for as long as it did. The disease plays out over decades. An aspect confounding investigators is that not all infected individuals react in the same way. Some may carry the virus for decades without significant injury; others experience serious damage within only a few years.

Liver transplantation can save some end-stage patients, but the supply of human livers available for transplant is woefully inadequate. Researchers are therefore working intensively to develop treatments that will eradicate the virus in patients.

The first therapeutic agent shown to be effective was alpha interferon, a protein that occurs naturally in the body. Interferon appears to have a nonspecific antiviral action and may also enhance immune system activity. The drug is generally given by subcutaneous injection three times a week for 12 months. Only 15 to 20 percent of patients, however, exhibit a sustained response, as defined by the return of ALT and AST to normal levels and the absence of detectable hepatitis C RNA in serum for at least six months after stopping treatment. Why treatment fails in most patients is essentially unknown, although some viral genotypes seem to be more susceptible to interferon than others.

Last year the Food and Drug Administration approved another drug, ribavirin, to treat hepatitis C in conjunction with interferon. Ribavirin, which can be swallowed in pill form, inhibits many viruses. Interestingly, though, it appears to have no effect against the hepatitis C virus by itself and is thought somehow to enhance interferon's effects on the immune system. Interferon and ribavirin given together for six to 12 months can expunge the virus in about 40 percent of patients, and clinical workers are now studying how to maximize the benefits from these two agents. Long-acting forms of interferon that require administration only once a week are one focus of interest.

A new drug is now being tested in small numbers of patients. Vertex Pharmaceuticals in Cambridge, Mass., is investigating a compound that inhibits a human enzyme called ionosine mono-phosphate—dehydrogenase. The hepatitis C virus relies on this enzyme to generate constituents of RNA. No results from these trials are yet available.

In the absence of medications capable of dependably eliminating the virus, the NIH recently embarked on a study to determine whether long-term administration of alpha interferon can slow liver damage in patients who fail to clear the virus. And we and other researchers are studying the simple expedient of taking a pint of blood from patients on a regular basis. This treatment reduces the amount of iron in the body, a manipulation that can reduce serum ALT and AST levels. Whether it slows liver damage is still uncertain.

TARGETING THE VIRUS

The best prospects for future treatment for hepatitis C appear to be agents targeted specifically against the virus, just as successful treatments for HIV target that agent. With that goal in mind, researchers have elucidated the structure of the hepatitis C virus in detail. Its genetic material, or genome, consists of a single strand of RNA. In size and organization the genome is similar to that of yellow fever and dengue fever viruses; hepatitis C virus has therefore been classified with them as a member of the family Flaviviridae. Enzymes in an infected cell use the viral RNA as a template to produce a single large protein called a polyprotein, which then cleaves to yield a variety of separate proteins with different functions. Some are structural proteins that go to form new viral particles; others are enzymes that replicate the original infecting RNA. At either end of the genome are short stretches of RNA that are not translated into protein. One of these terminal regions seems to prompt infected cells to manufacture the viral polyprotein; it is an important target for diagnostic assays. The other appears to play a role in initiating the replication of viral RNA.

The structural proteins include the core protein, which encloses the

RNA in a viral particle within a structure known as the nucleocapsid, and two envelope proteins that coat the nucleocapsid. The nonstructural proteins include a viral protease responsible for cleaving the polyprotein, as well as other enzymes responsible for chemically readying the components of viral RNA (triphosphatase), for copying the RNA (polymerase) and for unwinding the newly manufactured copy (helicase).

The protease and helicase enzymes have been well characterized and their detailed three-dimensional structure elucidated through x-ray crystallography, necessary first steps for designing drugs to inhibit an enzyme. Several drug companies, including Schering-Plough, Agouron Pharmaceuticals, and Eli Lilly and Vertex Pharmaceuticals, are now studying potential hepatitis C protease or helicase inhibitors. Clinical trials are probably only a few years away. Another viral enzyme, the polymerase, is also a possible target. Whether the virus will evolve resistance to such agents remains to be seen.

Developing anti-hepatitis C therapies may be about to get easier. Three months ago Ralf Bartenschlager and his colleagues at Johannes Gutenberg University in Mainz, Germany, published details of an RNA genetic construct that includes the regions coding for the virus's enzymes and reproduces itself in liver cancer cell lines. This construct may prove valuable for testing drugs targeted at these enzymes.

Another possible therapeutic avenue being investigated is disruption of the process that activates hepatic stellate cells and causes them to instigate fibrosis. This mechanism is known to involve cytokines, or signaling chemicals, that cells in the liver called Kupffer cells release when they are stimulated by lymphocytes. Turning this process off once it has started should prevent most of the untoward consequences of hepatitis C infection.

Some workers are trying to develop therapeutics aimed at the short terminal regions of the virus's genome. One idea, being pursued by Ribozyme Pharmaceuticals, is to develop therapeutic molecules that can cut specific constant sequences there. Ribozymes, short lengths of RNA or a chemical close relative, can accomplish this feat. The main challenge may be getting enough ribozymes into infected cells. Delivering adequate quantities of a therapeutic agent is also a problem for some other innovative treatment concepts, such as gene therapy to make liver cells resistant to infection, "antisense" RNA that can inhibit specified genes, and engineered proteins that activate a cell's self-destruct mechanism when they are cleaved by the hepatitis C protease.

All these attempts to counter hepatitis C are hampered by a serious shortage of funds tor research. The amount or federal support, considering the threat to millions of patients, is relatively small. We are confident that much improved therapies, and possibly a vaccine, will in time be available. An expanded research program could ensure that these developments come soon enough to help patients and those at risk.

FURTHER READING

THE CRYSTAL STRUCTURE OF HEPATITIS C VIRUS NS3 PROTEINASE REVEALS A TRYPSIN-LIKE FOLD AND A STRUCTURAL ZINC BINDING SITE. Robert A. Love et al. in *Cell*, Vol. 87, No. 2, pages 331–342; October 18,1996.

MANAGEMENT OF HEPATITIS C. NATIONAL INSTITUTES OF HEALTH CONSENSUS DEVELOPMENT CONFERENCE PANEL STATEMENT. In *Hepatology*, Vol. 26, Supplement No. 1, pages 2S–10S; 1997.

INTERFERON ALFA-2B ALONE OR IN COMBINATION WITH RIBAVIRIN AS INITIAL TREATMENT FOR CHRONIC HEPATITIS C. John G. McHutchison et al. in *New England Journal of Medicine*, Vol. 339, No. 21, pages 1485–1492; November 19,1998.

MOLECULAR CHARACTERIZATION OF HEPATITIS C VIRUS. Second edition. Karen E. Reed and Charles M. Rice in *Hepatitis C Virus*. Edited by H. W. Reesink. Karger, Basel, 1998.

REPLICATION OF SUBGENOMIC HEPATITIS C VIRUS RNAS IN A HEPATOMA CELL LINE. V. Lohmann, F. Körner, J.-O. Koch, U. Herian, L. Theilmann and R. Bartenschlager in *Science*, Vol. 285, pages 110–113; July 2,1999.

Hope in a Vial

CAROL EZZELL

ORIGINALLY PUBLISHED IN JUNE 2002

It wasn't supposed to be this hard. When HIV, the virus responsible for AIDS, was first identified in 1984, Margaret M. Heckler, then secretary of the U.S. Department of Health and Human Services, predicted that a vaccine to protect against the scourge would be available within two years. Would that it had been so straightforward.

Roughly 20 years into the pandemic, 40 million people on the planet are infected with HIV, and three million died from it last year (20,000 in North America). Although several potential AIDS vaccines are in clinical tests, so far none has lived up to its early promise. Time and again researchers have obtained tantalizing preliminary results only to run up against a brick wall later. As recently as two years ago, AIDS researchers were saying privately that they doubted whether even a partially protective vaccine would be available in their lifetime.

No stunning breakthroughs have occurred since that time, but a trickle of encouraging data is prompting hope to spring anew in the breasts of even jaded AIDS vaccine hunters. After traveling down blind alleys for more than a decade, they are emerging battered but not beaten, ready to strike out in new directions. "It's an interesting time for AIDS vaccine research," observes Gregg Gonsalves, director of treatment and prevention advocacy for Gay Men's Health Crisis in New York City. "I feel like it's Act Two now."

In the theater, Act One serves to introduce the characters and set the scene; in Act Two, conflict deepens and the real action begins. Act One of AIDS vaccine research debuted HIV, one of the first so-called retroviruses to cause a serious human disease. Unlike most other viruses, retroviruses insinuate their genetic material into that of the body cells they invade, causing the viral genes to become a permanent fixture in the infected cells and in the offspring of those cells. Retroviruses also reproduce rapidly and sloppily, providing ample opportunity for the emergence of mutations that allow HIV to shift its identity and thereby give the immune system or antiretroviral drugs the slip.

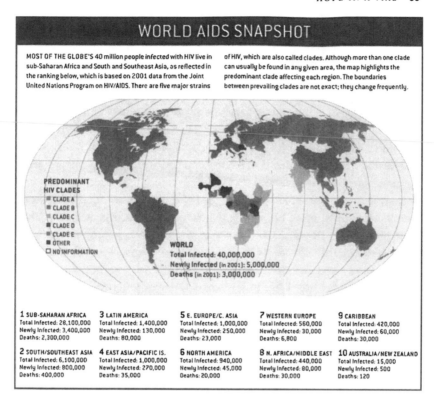

WORLD AIDS SNAPSHOT

MOST OF THE GLOBE'S 40 million people infected with HIV live in sub-Saharan Africa and South and Southeast Asia, as reflected in the ranking below, which is based on 2001 data from the Joint United Nations Program on HIV/AIDS. There are five major strains of HIV, which are also called clades. Although more than one clade can usually be found in any given area, the map highlights the predominant clade affecting each region. The boundaries between prevailing clades are not exact; they change frequently.

PREDOMINANT HIV CLADES
- CLADE A
- CLADE B
- CLADE C
- CLADE D
- CLADE E
- OTHER
- NO INFORMATION

WORLD
Total Infected: 40,000,000
Newly Infected (in 2001): 5,000,000
Deaths (in 2001): 3,000,000

1 SUB-SAHARAN AFRICA
Total Infected: 28,100,000
Newly Infected: 3,400,000
Deaths: 2,300,000

2 SOUTH/SOUTHEAST ASIA
Total Infected: 6,100,000
Newly Infected: 800,000
Deaths: 400,000

3 LATIN AMERICA
Total Infected: 1,400,000
Newly Infected: 130,000
Deaths: 80,000

4 EAST ASIA/PACIFIC IS.
Total Infected: 1,000,000
Newly Infected: 270,000
Deaths: 35,000

5 E. EUROPE/C. ASIA
Total Infected: 1,000,000
Newly Infected: 250,000
Deaths: 23,000

6 NORTH AMERICA
Total Infected: 940,000
Newly Infected: 45,000
Deaths: 20,000

7 WESTERN EUROPE
Total Infected: 560,000
Newly Infected: 30,000
Deaths: 6,800

8 N. AFRICA/MIDDLE EAST
Total Infected: 440,000
Newly Infected: 80,000
Deaths: 30,000

9 CARIBBEAN
Total Infected: 420,000
Newly Infected: 60,000
Deaths: 30,000

10 AUSTRALIA/NEW ZEALAND
Total Infected: 15,000
Newly Infected: 500
Deaths: 120

Act One also spotlighted HIV's opposition—the body's immune response—which consists of antibodies (Y-shaped molecules that stick to and tag invaders such as viruses for destruction) and cytotoxic, or killer, T cells (white blood cells charged with destroying virus-infected cells). For years after infection, the immune system battles mightily against HIV, pitting millions of new cytotoxic T cells against the billions of virus particles hatched from infected cells every day. In addition, the immune system deploys armies of antibodies targeted at HIV, at least early in the course of HIV infection, although the antibodies prove relatively ineffectual against this particular foe.

As the curtain rises for Act Two, HIV still has the stage. Results from the first large-scale trial of an AIDS vaccine should become available at the end of this year, but few scientists are optimistic about it: a preliminary analysis suggests that it works poorly. Meanwhile controversy surrounds a giant, U.S.-government-sponsored trial of another potential vaccine slated to begin this September in Thailand. But waiting in the wings are several approaches that are causing the AIDS research community to sit up and take notice. The strategies are reviving the debate about whether, to be useful, a vaccine must elicit immune responses that totally prevent HIV from colonizing a person's cells or whether a vaccine that falls somewhat short of that mark could be acceptable. Some scientists see potential value in vaccines that would elicit the kinds of immune responses that kick in soon after a virus establishes a foothold in cells. By constraining viral replication more effectively than the body's natural responses would, such vaccines, they argue, might at least help prolong the lives of HIV-infected people and delay the onset of the symptomatic, AIDS phase of the disease.

In the early 1990s scientists thought they could figure out the best vaccine strategy for preventing AIDS by studying long-term non-progressors, people who appeared to have harbored HIV for a decade or more but who hadn't yet fallen ill with AIDS. Sadly, many of the non-progressors have become ill after all. The key to their relative longevity seems to have been "a weakened virus and/or a strengthened immune system," says John P. Moore of Weill Medical College of Cornell University. In other words, they were lucky enough to have encountered a slow-growing form of HIV at a time when their bodies had the ammunition to keep it at bay.

NOT FOUND IN NATURE?

AIDS vaccine developers have struggled for decades to find the "correlates of immunity" for HIV—the magic combination of immune responses that, once induced by a vaccine, would protect someone against infection. But they keep coming up empty-handed, which leaves them with no road map to guide them in the search for an AIDS vaccine. "We're trying to elicit an immune response not found in nature," admits Max Essex of the Harvard School of Public Health. As a result, the quest for an AIDS vaccine has been a bit scattershot.

To be proved useful, a candidate AIDS vaccine must successfully pass through three stages of human testing. In phase I, researchers administer the vaccine to dozens of people to assess its safety and to establish an appropriate dose. Phase II involves hundreds of people and looks more closely at the vaccine's immunogenicity, its ability to prompt an immune response. In phase III, the potential vaccine is given to thousands of volunteers who are followed for a long time to see whether it protects them from infection. Phase III trials for any drug tend to be costly and difficult to administer. And the AIDS trials are especially challenging because of an ironic requirement: subjects who receive the vaccine must be counseled extensively on how to reduce their chances of infection. They are told, for instance, to use condoms or, in the case of intravenous drug users, clean needles because HIV is spread through sex or blood-to-blood contact. Yet the study will yield results only if some people don't heed the counseling and become exposed anyway.

The first potential vaccine to have reached phase III consists of gp120, a protein that studs the outer envelope of HIV and that the virus uses to latch onto and infect cells. In theory, at least, the presence of gp120 in the bloodstream should activate the recipient's immune system, causing it to quickly mount an attack targeted to gp120 if HIV later finds its way into the body.

This vaccine, which is produced by VaxGen in Brisbane, Calif.—a spin-off of biotech juggernaut Genentech in South San Francisco—is being tested in more than 5,400 people (mostly homosexual men) in North America and Europe and in roughly 2,500 intravenous drug users in Southeast Asia. The results from the North American/European trial, which began in 1998, are expected to be announced near the end of this year.

Many AIDS researchers are skeptical of VaxGen's approach because gp120 normally occurs in clumps of three on the surface of the virus, and the company's vaccine employs the molecule in its monomeric, or single-molecule, form. Moreover, vaccines made of just protein generally elicit only an antibody, or humoral, response, without greatly stimulating the cellular arm of the immune system, the part that includes activity by cytotoxic T cells. A growing contingent of investigators suspect that an antibody response alone is not sufficient; a strong cellular response must also be elicited to prevent AIDS.

Indeed, the early findings do not seem encouraging. Last October an independent data-monitoring panel did a preliminary analysis of the results of the North American/European data. Although the panel conducted the analysis primarily to ascertain that the vaccine was causing no dangerous side effects in the volunteers, the reviewers were empowered to recommend halting the trial early if the vaccine appeared to be working. They did not.

For its part, VaxGen asserts that it will seek U.S. Food and Drug Administration approval to sell the vaccine even if the phase III trials show that it reduces a person's likelihood of infection by as little as 30 percent. Company president and co-founder Donald P. Francis points out that the first polio vaccine, developed by Jonas Salk in 1954, was only 60 percent effective, yet it slashed the incidence of polio in the U.S. quickly and dramatically.

This approach could backfire, though, if people who receive a partially effective AIDS vaccine believe they are then protected from infection and can engage in risky behaviors. Karen M. Kuntz and Elizabeth Bogard of the Harvard School of Public Health have constructed a computer model simulating the effects of such a vaccine in a group of injection drug users in Thailand. According to their model, a 30 percent effective vaccine would not slow the spread of AIDS in a community if 90 percent of the people who received it went back to sharing needles or using dirty needles. They found that such reversion to risky behavior would not wash out the public health benefit if a vaccine were at least 75 percent effective.

The controversial study set to begin in Thailand is also a large-scale phase III trail, involving nearly 16,000 people. It combines the VaxGen vaccine with a canarypox virus into which scientists have stitched genes that encode gp120 as well as two other proteins—one that makes up the HIV core and one that allows it to reproduce. Because this genetically engineered canarypox virus (made by Aventis Pasteur, headquartered in Lyons, France) enters cells and causes them to display fragments of HIV on their surface, it stimulates the cellular arm of the immune system.

Political wrangling and questions over its scientific value have slowed widespread testing of the gp120/canarypox vaccine. Initially the National Institute of Allergy and Infectious Diseases (NIAID) and the U.S. Department of Defense were scheduled to conduct essentially duplicate trials of the vaccine. But NIAID pulled the plug on its trial after an examination of the data from a phase II study showed that fewer than 30 percent of the volunteers generated cytotoxic T cells against HIV. And in a bureaucratic twist, this past January the White House transferred the budget for the Defense Department trial over to NIAID as part of an effort to streamline AIDS research.

Peggy Johnston, assistant director of AIDS vaccines for NIAID, says she expects there will be a trial of the vaccine but emphasizes that "it will be a Thai trial; we won't have any [NIAID] people there on the ground running things."

Critics cite these machinations as a case study of politics getting in the way of progress against AIDS. "There's little science involved" in the trial, claims one skeptic, who wonders why the Thais aren't asking," 'If it's not good enough for America, how come it's good enough for us?'" Others point out that the trial, which was conceived by the Defense Department, will answer only the question of whether the vaccine works; it won't collect any data that scientists could use to explain its potential failure.

PARTIAL PROTECTION

Into this scene comes Merck, which is completing separate phase I trials of two different vaccine candidates that it has begun to test together. In February, Emilio A. Emini, Merck's senior vice president for vaccine research, wowed scientists attending the Ninth Conference on Retroviruses and Opportunistic Infections in Seattle with the company's initial data from the two trials.

The first trial is investigating a potential vaccine composed of only the HIV *gag* gene, which encodes the virus's core protein. It is administered as a so-called naked DNA vaccine, consisting solely of DNA. Cells take up the gene and use it as a blueprint for making the viral protein, which in turn stimulates a mild (and probably unhelpful) humoral response and a more robust cellular response [*see illustration*]. Emini and his colleagues reported that 42 percent of volunteers who received the highest dose of the naked DNA vaccine raised cytotoxic T cells capable of attacking HIV-infected cells.

The second trial employs the HIV *gag* gene spliced into a crippled adenovirus, the class responsible for many common colds. This altered adenovirus ferries the *gag* gene into cells, which then make the HIV core protein and elicit an immune response targeted to that protein. Emini told the conference that between 44 and 67 percent of people who received injections of the adenovirus-based vaccine generated a cellular immune response that varied in intensity according to the size of the dose the subjects received and how long ago they got their shots.

Merck is now beginning to test a combination of the DNA and adenovirus approaches because Emini predicts that the vaccines will work best when administered as part of the same regimen. "The concept," he says, "is not that the DNA vaccine will be a good vaccine on its own, but that it may work as a primer of the immune system," to be followed months later by a booster shot of the adenovirus vaccine. A possible stumbling block is that most people have had colds caused by adenoviruses. Accordingly, the immune systems of such individuals would already have an arsenal in place that could wipe out the adenovirus vaccine before it had a chance to deliver its payload of HIV genes and stimulate AIDS immunity. Increasing the dose of the adenovirus vaccine could get around this obstacle.

Emini says he and his co-workers are emphasizing cellular immunity in part because of the disappointing results so far with vaccines designed to engender humoral responses. "Antibodies continue to be a problem,"

One AIDS Vaccine Strategy

A VACCINE APPROACH being pioneered by Merck involves an initial injection of a naked DNA vaccine followed months later by a booster shot of crippled, genetically altered adenovirus particles. Both are designed to elicit an immune response targeted to the HIV core protein, Gag, and to primarily arouse the cellular arm of the immune system—the one that uses cytotoxic T cells to destroy virus-infected cells. The naked DNA vaccine also results in the production of antibody molecules against Gag, but such antibodies are not very useful in fighting HIV.

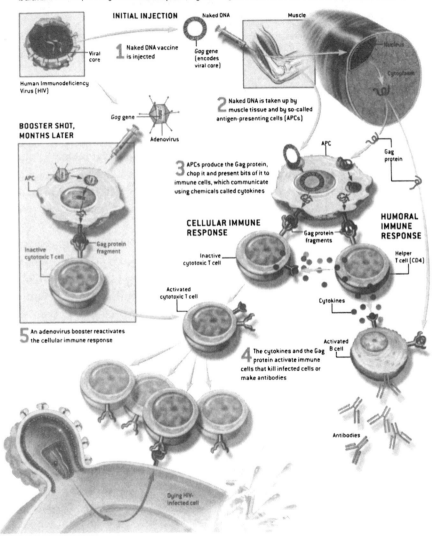

INITIAL INJECTION Naked DNA Muscle

Viral core

Human Immunodeficiency Virus (HIV)

1 Naked DNA vaccine is injected

Gag gene (encodes viral core)

Nucleus

Cytoplasm

Gag gene

Adenovirus

BOOSTER SHOT, MONTHS LATER

2 Naked DNA is taken up by muscle tissue and by so-called antigen-presenting cells (APCs)

APC

Gag protein

APC

3 APCs produce the Gag protein, chop it and present bits of it to immune cells, which communicate using chemicals called cytokines

Inactive cytotoxic T cell

Gag protein fragment

CELLULAR IMMUNE RESPONSE

Gag protein fragments

HUMORAL IMMUNE RESPONSE

Inactive cytotoxic T cell

Helper T cell (CD4)

Activated cytotoxic T cell

Cytokines

5 An adenovirus booster reactivates the cellular immune response

Activated B cell

4 The cytokines and the Gag protein activate immune cells that kill infected cells or make antibodies

Antibodies

Dying HIV-infected cell

he admits. "There are a handful of reasonably potent antibodies isolated from HIV-infected people, but we haven't figured out how to raise those antibodies using a vaccine."

Lawrence Corey of the Fred Hutchinson Cancer Research Center in Seattle agrees: "You'd like to have both [a cellular and an antibody response], but the greatest progress has been in eliciting a cellular response," says Corey, who is also principal investigator of the federally funded HIV Vaccine Trials Network.

Antibodies are important, too, because they are the immune system's first line of defense and are thought to be the key to preventing viruses from ever contacting the cells they infect. Corey says that vaccines that are designed primarily to evoke cellular immunity (as are Merck's) are not likely to prevent infection but should give someone a head start in combating the virus if he or she does become infected. "Instead of progressing to AIDS in eight years, you progress in 25 years," he predicts. But, Corey adds, it is unclear whether a vaccine that only slowed disease progression would stem the AIDS pandemic, because people would still be able to spread the infection to others despite having less virus in their bloodstream.

Finding a way to induce the production of antibodies able to neutralize HIV has been hard slogging for several reasons. For one, the virus's shape-shifting ways allow it to stay one step ahead of the immune response. "The thing that distinguishes HIV from all other human viruses is its ability to mutate so fast," Essex says. "By the time you make a neutralizing antibody [against HIV], it is only against the virus that was in you a month ago."

According to many scientists, vaccines using a logical molecule, gp120—the protein the virus uses to invade immune cells, as discussed above—haven't worked, probably because the antibodies that such vaccines elicit bind to the wrong part of the molecule. Gp120 shields the precise binding site it uses to latch onto CD4, its docking site on immune cells, until the last nanosecond, when it snaps open like a jackknife. One way to get around this problem, suggested in a paper published in *Science* three years ago by Jack H. Nunberg of the University of Montana and his colleagues, would be to make vaccines of gp120 molecules that have previously been exposed to CD4 and therefore have already sprung open. But those results have been "difficult to replicate," according to Corey, making researchers pessimistic about the approach.

Another possible hurdle to getting an AIDS vaccine that elicits effective anti-HIV antibodies is the variety of HIV subtypes, or clades, that—affect different areas of the world. There are five major clades, designated A through E [see illustration on page 59]. Although clade B is the predominant strain in North America and Europe, most of sub-Saharan Africa—the hardest-hit region of the globe—has clade C. The ones primarily responsible for AIDS in South and Southeast Asia—the second biggest AIDS hot spot—are clades B, C and E.

Several studies indicate that antibodies that recognize AIDS viruses from one clade might not bind to viruses from other clades, suggesting that a vaccine made from the strain found in the U.S. might not protect people in South Africa, for example. But scientists disagree about the significance of clade differences and whether only strains that match the most prevalent clade in a given area can be tested in countries there. Essex, who is gearing up to lead phase I tests of a clade C based vaccine in Botswana later this year, argues that unless researchers are sure that a vaccine designed against one clade can cross-react with viruses from another, they must stick to testing vaccines that use the clade prevalent in the populations being studied. Cross-reactivity could occur under ideal circumstances, but, he says, "unless we know that, it's important for us to use subtype-specific vaccines."

Using the corresponding clade also avoids the appearance that people in developing countries are being used as guinea pigs for testing a vaccine that is designed to work only in the U.S. or Europe. VaxGen's tests in Thailand are based on a combination of clades B and E, and in April the International AIDS Vaccine Initiative expanded tests of a clade A derived vaccine in Kenya, where clade A is found.

But in January, Malegapuru William Makgoba and Nandipha Solomon of the Medical Research Council of South Africa, together with Timothy Johan Paul Tucker of the South African AIDS Vaccine Initiative, wrote in the *British Medical Journal* that the relevance of HIV subtypes "remains unresolved." They assert that clades "have assumed a political and national importance, which could interfere with important international trials of efficacy."

Early data from the Merck vaccine trials suggest that clade differences blur when it comes to cellular immunity. At the retrovirus conference in February, Emini reported that killer cells from 10 of 13 people who received a vaccine based on clade B also reacted in laboratory tests to viral proteins from clade A or C viruses. "There is a potential for a substantial cross-clade response" in cellular immunity, he says, "but that's not going to hold true for antibodies." Corey concurs that clade variation "is likely to play much, much less of a role" for killer cells than for antibodies— because most cytotoxic T cells recognize parts of HIV that are the same from clade to clade.

Johnston of NIAID theorizes that one answer would be to use all five major clades in every vaccine. Chiron in Emeryville, Calif., is developing a multiclade vaccine, which is in early clinical trials. Such an approach could be overkill, however, Johnston says. It could be that proteins from only one clade would be recognized "and the other proteins would be wasted," she warns.

Whatever the outcome on the clade question, Moore of Weill Medical College says he and fellow researchers are more hopeful than they were a few years ago about their eventual ability to devise an AIDS vaccine that would elicit both killer cells and antibodies. "The problem is not impossible," he says, "just extremely difficult."

FURTHER READING

HIV VACCINE EFFORTS INCH FORWARD. Brian Vastag in *Journal of the Amercian Medical Association,* Vol. 286, No.15, pages 1826–1828; October 17, 2001.

For an overview of AIDS vaccine research, including the status of U.S.-funded AIDS clinical trials, visit www.niaid.nih.gov/daids/vaccine/default.htm

A global perspective on the AIDS pandemic and the need for a vaccine can be found at the International AIDS Vaccine Initiative Web site: www.iavi.org

Joint United Nations Program on HIV/AIDS: www.unaids.org

A New Assault on HIV

GARY STIX

ORIGINALLY PUBLISHED IN JUNE 2006

The field of virology spends a substantial chunk of its resources inspecting every minute step of the HIV life cycle—from the binding and entry of the virus into an immune cell to its replication and release of a new virus from the host cell and, finally, the seeking of a new cell on which to prey. The last major new class of anti-HIV drugs emerged about a decade ago with the introduction of the protease inhibitors, which curb the action of an enzyme that is critical to a late stage of viral replication.

At the time, a few members of the HIV research community wondered whether protease inhibitors could provide the basis for a cure. The ingenuity of the virus has proved the hollowness of that hope. As many as half of HIV-positive patients under treatment in the U.S. were found in one study to be infected with viruses that have developed resistance to at least one of the drugs in their regimen. Clinicians can choose from more than 20 pharmaceuticals among protease inhibitors and two classes of drug that prevent the invading virus from copying its RNA into DNA, thereby sabotaging viral replication. Combinations of these agents are administered to counteract the virus's inherent mutability, but that strategy does not always ward off resistance to the medicines, including the protease inhibitors. "Given increasing resistance to protease inhibitors, it's of paramount importance to identify new ways to interfere with the virus replication cycle," asserts Eric Freed, a researcher in the HIV drug resistance program at the National Institutes of Health.

Drugs that interrupt the beginning, middle and end of viral processing within the host are now in various stages of development. Academic researchers and Panacos, a small biotechnology outfit based in Watertown, Mass., are taking inspiration from the success of protease inhibitors by developing drug candidates known as maturation inhibitors that block protease activity in a novel way. Protease inhibitors mount a direct attack on the HIV protease, preventing the enzyme from processing a viral protein called GAG. When GAG proteins are cut properly, pieces spliced out

A NOVEL TREATMENT STRATEGY FOR HIV

Maturation inhibitors constitute a new class of HIV drugs under study. They attack the virus at a late stage in its life cycle—when freshly made components of the virus are coming together into new infectious particles that are beginning to "bud" from one infected T cell so that they can move on to infect another cell.

NORMAL HIV MATURATION

As a full HIV particle emerges, enveloped by a coat made from the T cell membrane and displaying viral proteins on its surface—the virus's protease enzymes have to chop molecules of the GAG protein to form other, smaller proteins (detail or bottom left).

Capsid proteins unite to form a conical core that, along with the interior nucleocapsid, encases and properly positions the viral genes—two molecules of single-stranded RNA.

TREATED VIRUS

The drug candidate PA-457 works by attaching to the GAG protein and preventing the protease from separating the capsid protein from its neighbor in GAG—the SP1 protein (detail).

In consequence, the capsid-SP1 structure and its interior capsule adopt abnormal shapes that are thought to prevent the virus from replicating properly.

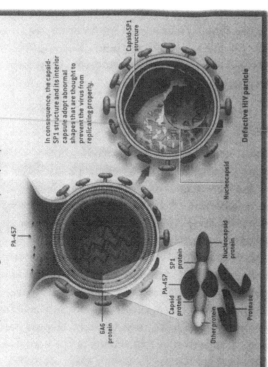

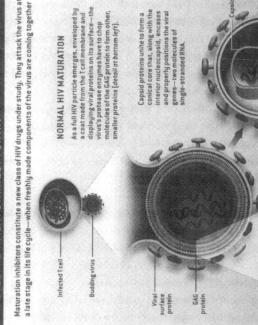

Infected T cell

Budding virus

Viral surface protein

GAG protein

Capsid

Viral RNA

Nucleocapsid

Normal HIV particle

Protease

SP1 protein

Nucleocapsid protein

Capsid protein

Other protein

PA-457

GAG protein

SP1 protein

Nucleocapsid protein

PA-457

Capsid protein

Protease

Other protein

Nucleocapsid

Capsid-SP1 structure

Defective HIV particle

of it form the conical protective core, or capsid, that encloses RNA. In contrast, the Panacos maturation inhibitor blocks a site on the GAG protein where the protease normally binds, keeping the protease from clipping GAG correctly. As a result, the capsid does not form appropriately and the virus cannot infect another cell.

LOOKING FOR LEADS

The path to the Panacos drug candidate began in the mid-1990s, when the company Boston Biomedica undertook a collaboration with a professor from the University of North Carolina at Chapel Hill to screen compounds from a collection of traditional Chinese herbs for biochemical activity against HIV. Kuo-Hsiung Lee's laboratory turned up a potential drug lead in a Taiwanese herb.

The substance, betulinic acid, had weak activity against HIV. After the lab separated the compound into its chemical constituents, the investigators found that one of these components, when chemically modified, exhibited a much stronger effect. "Betulinic acid had activity against HIV at the micromolar level," says Graham Allaway, Panacos's chief operating officer. "This derivative had activity at the nanomolar level."

Six years ago Boston Biomedica spun off its HIV research unit into Panacos, which began to investigate the compound, by then named PA-457. PA-457 was not just another Taxol, the anticancer drug that required the controversial felling of rare yew trees until a semisynthetic substitute was found. Panacos did not need a steady source of Taiwanese herbs. Betulinic acid could be extracted from ubiquitous plane and birch trees, and a subsequent processing step yielded the desired molecule.

Even though researchers understood that PA-457 seemed to have activity against all strains of HIV, they needed to find out how the betulinic acid derivative worked against the virus on the molecular level. The company wanted a new class of drug, not just another protease inhibitor. It contacted Freed's laboratory at the NIH, which studies the virus life cycle.

Freed's group and Panacos determined that the drug worked late in the viral replication process, apparently at the stage of capsid formation. The researchers already knew that the HIV capsid forms when newly made GAG attaches from inside the host T cell to the cell's membrane and is then chopped by the HIV protease into smaller pieces. They knew as well, from the development of protease inhibitors, that any disruption to the processing of GAG would cause the virus to become noninfectious.

So they began to study PA-457's interaction with GAG to see exactly how it bollixed up the cutting of GAG into its requisite parts.

CULTIVATION OF RESISTANCE

To understand how a compound works, scientists often begin by creating resistance, which lets them pin down the exact spot where the drug interacts with its target. To nurture resistance, Freed and his colleagues administered low doses of PA-457 to HIV-infected T cells in culture. The genome of the resistant viruses was sequenced and compared with that of viruses that succumbed to the drug. That analysis located the site on the newly produced viruses that changed in the resistant versions. It turned out to be a site on GAG where the protease binds, and this alteration prevented PA-457 from blocking the enzyme's activity.

Analyzing the resistant strains allowed the researchers to ascertain that PA-457 was not simply another protease inhibitor. Most drugs, not just HIV inhibitors, work by tinkering with enzymes. "Targeting the substrate [instead of the enzyme] was unknown and surprising," Allaway comments. "As a result, we believe we will have a fairly strong patent position."

Cultivating resistant strains does not necessarily mean that the drug will have a limited therapeutic life span. In fact, resistance to PA-457 may not develop quickly, because the site where it binds on the GAG protein does not readily change from one HIV strain to another through mutations.

PA-457 has already passed through a midphase clinical trial that checked for drug activity in patients who took it for 10 days while another group received a placebo. HIV replicates so rapidly that a short trial can be used to determine whether a drug is attacking the pathogen in the body. Viral levels averaged a drop of 92 percent at the highest dose of 200 milligrams. The study looked for a decrease in so-called viral load of at least 70 percent as a preliminary sign of the drug's effectiveness. Some patients, however, did not respond—and the company will determine in the next phase of testing whether it can give higher doses. "The main message is that this is an active drug and research should go forward," says Jeffrey M. Jacobson, chief of infectious diseases at Drexel University College of Medicine, who is the lead researcher in the clinical trials.

During the next round, investigators will be looking for interactions with other drugs, an essential test of any HIV drug prospect, because no treatment consists of a single drug therapy, given the threat of resistance.

The Food and Drug Administration is encouraging tests earlier during clinical trials these days. In developing new HIV drugs, researchers have at times detected these interactions only much later in the clinical trial process. If all goes according to plan, Panacos could file its final FDA approval application by 2008.

OTHER IMMATURITY PRESERVERS

PA-457 is not the only example of a maturation inhibitor, although it has progressed the furthest toward commercialization. At the University of Alabama and the University of Maryland, researchers working independently have identified small organic molecules that prevent the multitude of capsid subunits from joining up to form the finished casing. "We're trying to jam the parts so they don't fit together," says Peter Prevelige, a professor in the department of microbiology at the University of Alabama.

This strategy goes along with other approaches under development to sabotage the viral life cycle. Entry inhibitors, including one Panacos is working on, prevent the virus from entering the cell. (One injectable entry inhibitor has already received FDA approval, but the Panacos drug would be taken orally.) Among the other classes of drugs that have reached late-stage trials are integrase inhibitors, which undermine an enzyme that allows the viral-made DNA to integrate into the host DNA to produce new viral RNA. All these biological agents are needed—and more. Absent a vaccine—not a near-term prospect—the lowly virus, a nanometer-scale capsule of single-stranded RNA, will continue to outwit the best ideas that molecular biologists conjure.

FURTHER READING

PA-457: A POTENT HIV INHIBITOR THAT DISRUPTS CORE CONDENSATION BY TARGETING A LATE STEP IN GAG PROCESSING. F. Li et al. in *Proceedings of the National Academy of Sciences USA,* Vol. 100, No. 23, pages 13555–13560; November 11, 2003.

THE PREVALENCE OF ANTIRETROVIRAL DRUG RESISTANCE IN THE UNITED STATES. Douglas D. Richman et al. in *AIDS,* Vol. 18, No. 10, pages 1393–1401; July 2, 2004.

THE DISCOVERY OF A CLASS OF NOVEL HIV-1 MATURATION INHIBITORS AND THEIR POTENTIAL IN THE THERAPY OF HIV. Donglei Yu et al. in *Expert Opinion on Investigational Drugs,* Vol. 14, No. 6, pages 681–693; June 2005.

A Cure for Rabies?

RODNEY E. WILLOUGHBY, JR.

ORIGINALLY PUBLISHED IN APRIL 2007

Rabies is one of the oldest and most feared diseases. It attacks the brain, causing agitation, terror and convulsions. Victims suffer painful throat spasms when they try to drink or eat. Paralysis follows, yet people infected with rabies are intermittently alert until near death and can communicate their fear and suffering to family and caregivers. Although vaccines against the rabies virus can prevent the illness from developing, until recently doctors could hold out no hope for patients who failed to get immunized soon after being bitten by a rabid animal. Once the symptoms of rabies appeared—typically within two months of the bite—death was inevitable, usually in less than a week.

In 2004, however I was a member of a team of physicians at Children's Hospital of Wisconsin in Milwaukee that rescued a 15-year-old girl from a similar fate. Jeanna Giese of Fond du Lac, Wis., became the first known unimmunized survivor of rabies. (Five other people who had been immunized but developed rabies anyway have also survived.) Our novel treatment, dubbed the Milwaukee protocol, has stirred controversy among medical specialists; some claim that Jeanne's cure was a fluke. Although the few attempts to replicate the treatment have not saved the lives of any other rabies patients, I fervently hope that we are on the right track. At the very least, researchers should initiate animal studies to determine which of the elements in our protocol can help defeat rabies.

A cure for rabies would be a great boon for the developing world. The disease is quite rare in the U.S. and Europe because public health campaigns have almost eliminated the virus in domestic animals such as dogs, cats and cattle. Only two to three patients die of rabies in the U.S. every year, meaning that the chance that a person contracts the illness is only about one in 100 million. (In fact, the disease is so rare in the U.S. that it goes unrecognized in half its victims until after they die.) But the World Health Organization estimates that rabies kills 55,000 people annually in Asia, Africa and Latin America, with most of the victims infected by dog bites. If investigators can properly analyze the Milwaukee protocol and identify an inexpensive treatment that would provide similar benefits, the resulting therapy could save thousands of lives.

A BAT BITE

Rabies is an enveloped RNA virus, meaning that it has an outer membrane and uses ribonucleic acid as its genetic material rather than the deoxyribonucleic acid (DNA) used by humans and virtually all other forms of life. The bullet-shaped microbe invades human cells and forces them to churn out new viruses, inflicting all its damage by manufacturing just five proteins. Highly specialized for growth in the brain and nerves, the rabies virus is rarely found elsewhere in the body. After being transmitted by the bite of a rabid animal, which introduces infected saliva into a wound, the virus multiplies locally in muscle or skin. Because the virus is present at very low levels and does not travel through the bloodstream or lymph nodes, the body's immune system does not detect the microbe at this stage. This symptom-free incubation period typically lasts two to eight weeks but can be as long as several years. At some point the virus reaches a nerve and enters it, and then the game is over.

In the late 19th century French microbiologist Louis Pasteur discovered that injecting killed rabies virus stimulates the immune system into producing antibodies against the microbe. What is more, Pasteur realized that the time it takes the body to develop this immune response is shorter than the disease's incubation period. He injected killed virus from the spinal cords of infected rabbits into people bitten by rabid dogs, who survived by gaining immunity before they showed any symptoms. Rabies can still develop in the interval between immunization and immune response, so now doctors also inject patients with rabies-specific antibodies to bridge the gap. Properly cleaning a wound with soap and water (which kills the virus by removing its membrane) is also quite important. This postexposure prophylaxis—wound care, five shots of a very safe vaccine and one dose of antibody—has had no failures in the U.S. since its introduction in 1975.

Jeanna's battle with rabies began when a bat crashed into an interior window of her church during a service. As she picked it up by the wing tips to release it outside, the bat lunged at Jeanna's left hand, and she suffered a quarter-inch cut to her index finger. Bat teeth are small and razor-sharp, so their bites are often not felt and can be hard to find. For these reasons, health officials recommend rabies immunization for anyone

THE TERRIBLE COURSE OF RABIES

Vaccines against the rabies virus can prevent the illness from developing if given soon after the patient is bitten by a rabid animal. But the disease is deadly for those who fail to get immunized.

THE ROUTE TO THE BRAIN

The bite of a rabid animal introduces virus-filled saliva to the wound. After multiplying at the wound site for about a month, the virus enters a peripheral nerve. If sensory neurons are infected, the microbe travels to the sensory ganglia along the spinal cord, then to the thalamus and the sensory cortex. Patients may complain of numbness, pins and needles, burning or severe itching. If motor neurons are infected, the virus reaches the brain through long nerves connecting the spinal cord to the motor cortex. Symptoms include weakness, paralysis and muscle jerks that can progress to seizures.

Sensory cortex

Motor cortex

Thalamus

Sensory ganglia

Spinal cord

Rabies virus multiplying in finger

Sensory nerve

Motor nerve

RAVAGING THE BODY

Once inside the cortex, the virus spreads rapidly in the highly interconnected brain, showing a predilection for the cerebellum (which coordinates body movements), the hippocampus (involved in short-term memory) and the limbic system (which regulates emotions). Infected animals become irritable, agitated, aggressive and even prone to self-mutilation. To complete its life cycle, the rabies virus must make its way into the salivary and tear glands and from there proceed to its next victims. The microbe migrates from the brain through every peripheral nerve, including the autonomic nerves that connect the brain to the vital organs and to the salivary and tear glands. Death usually results from heart stoppage, stroke or respiratory failure.

Cerebellum

Hippocampus and limbic system

Tear gland

Salivary glands

Numbing of sensory nerve

Paralysis of motor nerve

coming into contact with a bat or sleeping in a room where a bat is discovered (unless the bat is captured and tests show that it is not rabid). Jeanna cleaned her scratch with hydrogen peroxide but did not seek immunization. Had she done so, she would likely have completed her sophomore year of high school without commotion.

Instead the rabies virus multiplied in Jeanna's finger for about a month, then entered a nerve and quickly traveled toward her brain, moving at a rate of about one centimeter an hour. Because of the exclusive targeting of the nervous system, where many types of immune cells do not operate, the body does not detect the virus until it has massively infiltrated the spinal cord and brain. The rabid patient eventually develops full paralysis from infected motor nerves and loses all sensation from infected sensory nerves. The mechanism behind this loss of neural activity is not known. Nor do researchers understand exactly how rabies kills a patient. Death can come in a variety of ways: stroke, heart stoppage, respiratory failure. The rabies virus seems to induce the brain to sabotage the vital organs, and it was this observation that inspired our treatment for Jeanna.

A gifted student and star of her high school volleyball team, Jeanna developed a flulike illness in October 2004, one month after being bitten. She then experienced numbness in her left hand, weakness in her left leg and double vision. She was admitted to her local hospital on a weekend and became lethargic and uncoordinated. These symptoms are typical of encephalitis, or inflammation of the brain, which is fairly common in medical practice, occurring in several thousand U.S. patients every year. Various types of viral and bacterial infections can cause encephalitis, but it can also be triggered by an immune response that goes awry and inflames the brain. Because the imaging of Jeanna's brain was normal, showing no sign of infection or stroke, her doctors assumed that she had this kind of postinfectious, autoimmune encephalitis. It appeared that she might progress to coma and need mechanical ventilation, so she was transferred to our hospital.

My colleagues and I would have almost certainly missed the diagnosis of rabies, but we got some help. Jeanna's local physician, Howard Dhonau, returned from his weekend off and did what beginning students are taught is one of the fundamentals of medicine: he took a repeat history of what had happened and heard about the bat. This history was critical to our success. I suggested that the team about to transport Jeanna to our hospital don protective equipment as a precaution. Although researchers have no proof that rabies can be transmitted from human to human, the tears and saliva of rabid animals are full of the virus, and contamination of wounds or mucous membranes (in the eyes, nose or mouth) is how rabies is perpetuated in nature. The medical staff caring for Jeanna wore visors, face masks, gowns and gloves during her first month of illness. The diagnosis of rabies required analysis of samples of her saliva, skin, blood and spinal fluid, which were shipped by air courier to the rabies section at the Centers for Disease Control and Prevention in Atlanta. This laboratory can provide first results in less than 24 hours.

Meanwhile I examined Jeanna. She was lethargic but performed simple commands. She could not maintain her balance and was weak in her left leg. Her reflexes were normal, allowing me to exclude the possibility of polio or West Nile virus. Also, her left arm was jerking intermittently. The pattern of numbness of her left hand and jerking of her left arm, correlating with the location of her bat bite, suggested rabies rather than the more common infections that cause encephalitis. But because doctors more often see an atypical presentation of a common disease than a truly rare disease, I reassured Jeanna's family, nurses and myself that it was almost impossible for her to have rabies. I was betting on another kind of encephalitis, most likely autoimmune, which is about 1,000 times more prevalent.

We had 24 hours to prepare a plan in case I was wrong. Between seeing other patients, I focused on another fundamental task of medicine: looking things up. I knew the conventional wisdom, which was that nothing works to treat rabies once symptoms appear; Jeanna was already dead if that was her diagnosis, and all we could do was limit her suffering. But I also knew that medicine is always advancing, so something new might be "out there." A search of the medical literature available online did not show any breakthroughs, but the lag between medical discovery and its publication can be as long as five years. I called Cathleen Hanlon, a rabies expert at the CDC, and received two depressing pieces of information. First, Jeanna's history and examination sounded like rabies to Hanlon, and second, nothing promising had emerged from recent scientific meetings or ongoing clinical trials.

Given the limited time, I decided on a different search strategy. Virtually nobody had survived rabies, so I avoided publications about how to treat the disease in humans. Research on treatments usually starts with the effects of drugs on viruses cultivated in test tubes. Although this is a necessary first step, drugs that show promise in early studies are often toxic or cannot be delivered in adequate amounts to the site of infection, so I avoided these publications as well. What did catch my eye, when looking at the remaining articles, was the mystery that rabies experts have puzzled over for nearly 30 years: rabies patients die with virtually nothing visibly wrong with their brains. Equally important, when patients die of rabies after several weeks of intensive care, the virus can no longer be found in their bodies. The human immune system can clear the virus over time, but the eradication happens too slowly to save the patient's life.

From these two facts, we improvised a strategy. The rabies virus was apparently capable of hijacking the brain into killing the body, but without directly damaging the brain tissue itself. If we could inactivate the malfunctioning brain through careful use of drugs, putting Jeanna into a state of prolonged unconsciousness, it would restrain the terrible havoc wrought on her body and perhaps keep her alive long enough to allow her immune system to catch up.

To choose which drugs made the most sense, I searched the medical literature for studies connecting rabies with neurotransmitters (the chemicals the brain uses to signal between cells) or with neuroprotection (the science of using drugs or other interventions to shield the brain from injury). This search uncovered an astonishing pair of papers by Henri Tsiang of the Pasteur Institute in Paris. In the early 1990s Tsiang and his colleagues reported that ketamine, an anesthetic, was capable of inhibiting the rabies virus in the cortical neurons of rats. The work was reassuring for three reasons. First, the research indicated that ketamine affected a key step in the life cycle of the virus, when it is transcribing its genetic information inside the neuron. Second, the drug hindered only the rabies virus and not other viruses, suggesting that its effect was not a result of general toxicity to the animal. Third, a similar but more toxic drug called MK-801 also inhibited rabies in rat neurons, so the benefit most likely applied to an entire class of compounds.

For more than 25 years, surgeons have used ketamine to induce and maintain unconsciousness in their patients, although it has largely been replaced because of its hallucinogenic side effects. (Illicit recreational users of the drug call it Special K.) Interestingly, ketamine's side effects appeared to offer another potential advantage for rabies patients. Ketamine acts as a neuroprotectant by blocking membrane proteins called NMDA glutamate receptors, which can kill neurons after becoming overactive following a stroke or other type of brain injury. Imagine my excitement to read about a drug that could contribute to suppressing the malfunctioning brain, while simultaneously clearing rabies virus and protecting the brain against further insult!

THE TREATMENT THAT SAVED JEANNA

On October 19, 2004, the day after Jeanna Giese arrived at Children's Hospital of Wisconsin in Milwaukee, doctors confirmed that she had rabies and used ketamine and midazolam to put her in a coma. Over the next week they also administered phenobarbital (a sedative), amantadine (an antiviral agent that helps to protect the brain) and ribavirin (a general antiviral drug). When Jeanna came out of the coma, her immune system was producing large amounts of antibodies

against rabies, particularly the neutralizing antibodies that stop the virus from invading new cells. But her recovery was slow until physicians gave her biopterin, a compound similar to folic acid.

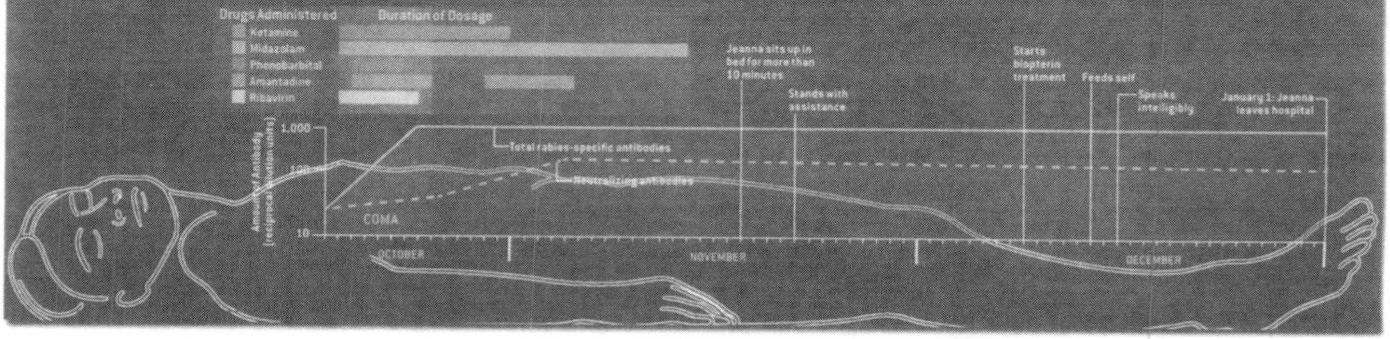

Drugs Administered

Ketamine
Midazolam
Phenobarbital
Amantadine
Ribavirin

Duration of Dosage

Amount of Antibody
(reciprocal dilution units)

1,000

10

Total rabies-specific antibodies

Neutralizing antibodies

COMA

Jeanna sits up in bed for more than 10 minutes

Stands with assistance

Starts biopterin treatment

Feeds self

Speaks intelligibly

January 1: Jeanna leaves hospital

OCTOBER

NOVEMBER

DECEMBER

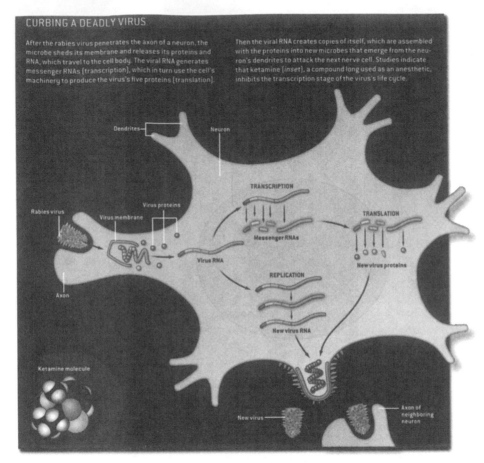

CURBING A DEADLY VIRUS

After the rabies virus penetrates the axon of a neuron, the microbe sheds its membrane and releases its proteins and RNA, which travel to the cell body. The viral RNA generates messenger RNAs (transcription), which in turn use the cell's machinery to produce the virus's five proteins (translation).

Then the viral RNA creates copies of itself, which are assembled with the proteins into new microbes that emerge from the neuron's dendrites to attack the next nerve cell. Studies indicate that ketamine (*inset*), a compound long used as an anesthetic, inhibits the transcription stage of the virus's life cycle.

A DESPERATE DECISION

As an infectious disease consultant, I did not have the skills to put Jeanna into a coma safely. I therefore applied another fundamental of medicine: seek help. I was new to Children's Hospital, so I asked Michael "Joe" Chusid, a senior infectious disease consultant, to help me find specialists in protecting the brain. Luckily, all the experts happened to be available that day. Kelly Tieves and Nancy Ghanayem, on call in the intensive care unit when Jeanna arrived, were experienced at minimizing injury to the brain following trauma and open-heart surgery, respectively. Our team also included Catherine Amlie-Lefond, a neurologist specializing in virus infections, and Michael Schwabe, an epilepsy expert who could provide us with continuous brain-wave monitoring to control the coma.

George Hoffman, the team's anesthesiologist, recognized that our plan for inducing the coma was a routine practice for other conditions.

The team members recommended other medications to moderate the side effects of ketamine, confer additional neuroprotection and achieve the coma that was our therapeutic goal. Amantadine, an antiviral agent, would also help shut down the neuronal NMDA receptors, binding to them at a different site than the one blocked by ketamine. Midazolam, a sedative in the benzodiazepine group, and phenobarbital would help suppress Jeanna's brain activity. Charles Rupprecht, a rabies expert at the CDC, later recommended that we administer a general antiviral drug, ribavirin. Although ribavirin had been tried on rabies patients before without success, we always listen to smart people.

Having redundant consultants in each specialty permitted us to critique the hypothesis and decide whether it was safe to proceed. Whenever new things are tried in medicine or biology, they usually fail and often cause harm. That is why treatments are supposed to progress from test tubes to animal studies to clinical trials. My hypothesis appeared too simple. It might result in a medical outcome even worse than death; four of the five immunized survivors of rabies ended up with severe disabilities. We convened for an hour after Rupprecht confirmed that Jeanna had rabies. We informed her parents of the diagnosis, outlined conventional options and then proposed our treatment. With Jeanna's death all but guaranteed, Ann and John Giese asked us to try something new so that more would be known for the next child with rabies.

We estimated that Jeanna's immune system would need about five to seven days to make antibodies targeting the rabies virus. From previous human cases of rabies, we knew that the natural immune response is vigorous once it is triggered. Given that Jeanna's brain was already full of rabies virus, adding more of it with a killed-virus vaccine was unlikely to help. It might actually cause harm by skewing the natural immune response away from the virus already in Jeanna's brain and toward variants overrepresented in the vaccine. For similar reasons, we chose to avoid boosting Jeanna's immune system by injecting her with rabies-specific antibodies or interferons (proteins that enhance immune activity). We elected to induce coma in Jeanna for a week and to analyze samples of her blood and spinal fluid over that period to confirm that she was producing her own antibodies.

Instead of the extreme commotion that characterizes the terminal care of rabies—including wild swings in heart rate and blood pressure— we encountered no major crises during Jeanna's coma. By the end of the week, her body had begun to produce large quantities of neutralizing antibodies that prevent the virus from invading new cells and perhaps clear the microbe by other unknown means. But the real test would come when we brought Jeanna back to consciousness. The day we removed her from coma was the worst day of my life: Jeanna was completely paralyzed and unresponsive. "We had no idea whether she was still in there or what would follow. But we knew that rabies patients could appear falsely brain-dead, so we did not lose hope. The next day Jeanna attempted to open her eyes and later developed reflexes in her legs. After six days she was fixing her gaze on her mother's face (in preference to mine) and open-ing her mouth to help her nurses clean it. By 12 days, she was sitting up in bed.

Full paralysis results in severe deconditioning of the body. Strength, stamina and coordination are all lost, as well as the ability to swallow and speak. Jeanna's recovery took a tremendous amount of hard work. In the first two months she ran into perplexing delays: for example, she made rapid progress in walking and exercise but still could not speak or swallow. She had other residual problems (such as a buildup of lactic acid in her body) that made me think of metabolism disorders. In consulta-tion with William Rhead, a geneticist at our hospital, we diagnosed that Jeanna had acquired a biopterin deficiency. Biopterin is chemically simi-lar to folic acid, the B vitamin essential to cell growth. The compound is present in limited amounts in the brain, where it is critical for mak-ing neurotransmitters such as dopamine, epinephrine, norepinephrine, serotonin and melatonin. Biopterin also controls how a brain enzyme, neuronal nitric oxide synthase, maintains the tone of blood vessels feed-ing the brain. In fact, we realized that low levels of biopterin could ex-plain most symptoms of rabies other than the late effects on peripheral nerves.

This was a stunning break, because biopterin is available as an oral supplement. After receiving biopterin, Jeanna quickly relearned how to speak and swallow. The speedy improvement allowed her to leave the hos-pital on January 1, 2005, three months earlier than expected. We have since confirmed biopterin deficiency in the only other rabies patient from whom samples were suitably preserved. We are testing whether low biop-terin levels occur in other animal species infected with rabies; if so, this deficiency may help explain how the virus ravages the body. Why rabies would reduce biopterin—most brain infections increase it—is not clear. We have prepared to test for and treat biopterin deficiency in future ra-bies patients who receive the Milwaukee protocol.

A CURABLE CONTROVERSY

By the first anniversary of her diagnosis with rabies, Jeanna attended an international meeting of rabies researchers in Canada as the guest of honor at their gala dinner, where she gave a speech. She rejoined her original classmates for her junior year of high school, received excellent grades and obtained a provisional driver's permit. The only lingering reminders of her bout with rabies are a small area of numbness on the bitten finger, an alteration in the tone of her left arm and a wider gait when she runs. She is graduating from high school this year and hopes to become a veterinarian.

But can the Milwaukee protocol save any other lives? Over the past two years the treatment has been tried six times, without success, in Germany, India, Thailand and the U.S. Unfortunately, several of the attempts violated key assumptions in our hypothesis or did not use most of the drugs in Jeanna's regimen. The medical community has been reluctant to repeat our therapy, with some experts publicly opposed. The resistance is understandable because Jeanna's survival appears to refute lab studies showing that the rabies virus kills brain cells. These studies may be misleading, however, because the strains of rabies used in the lab may be more prone to cause cell death than the viruses circulating in nature.

Other experts have argued that Jeanna survived because she was infected with an unusually weak variant of rabies. This argument is difficult to answer because we did not recover samples of the virus from Jeanna's body. (The CDC obtained rabies-specific antibodies from Jeanna but not the virus itself, which is difficult to isolate.) Analysis of the rabies virus carried by bats indicates that it is different from the strain in dogs, with a greater propensity for multiplying in skin rather than in muscle. But bat virus appears to be no less deadly than the canine version.

Perhaps, the best way to address these concerns is to apply our treatment to rabid animals to determine which parts of the protocol—the induced coma, the antiviral activity, the shutdown of NMDA receptors—are crucial in battling rabies. We have asked six veterinary schools to permit these studies, but school officials are wary about treating rabid animals in their intensive care units. The lack of follow-up research is a global loss, because we will not know whether our strategy actually works until others try it. If our success is replicated, investigators can systematically address which drugs are effective and at what doses and determine whether biopterin can significantly shorten convalescence. Furthermore, doctors must find ways to reduce the cost of treatment and rehabilitation—which totaled at least $800,000 in Jeanna's case—to make it a practical option in developing countries where rabies is most common. It may be too much to expect rabies to go from 100 percent fatal to 100 percent curable, but at least now we have a change to improve the odds.

FURTHER READING

INHIBITION OF RABIES VIRUS TRANSCRIPTION IN RAT CORTICAL NEURONS WITH THE DISSOCIATIVE ANESTHETIC KETAMINE. B. P. Lockhart, N. Tordo and H. Tsiang in *Antimicrobial Agents and Chemotherapy,* Vol. 36, No. 8, pages 1750–1755; August 1992.

PROPHYLAXIS AGAINST RABIES. C. E. Rupprecht and R. V. Gibbons in *New England Journal of Medicine,* Vol. 351, No. 25, pages 2626–2635; December 16, 2004.

SURVIVAL AFTER TREATMENT OF RABIES WITH INDUCTION OF COMA. R. E. Willoughby, Jr., K. S. Tieves, G. M. Hoffman, N. S. Ghanayem, C. M. Amlie-Lefond, M. J. Schwabe et al. in *New England Journal of Medicine,* Vol. 352, No. 24, pages 2508–2514; June 16, 2005.

More information about rabies and the Milwaukee protocol is available at www.mcw .edu/rabies

Emerging Viruses

BERNARD LE GUENNO

ORIGINALLY PUBLISHED IN OCTOBER 1995

In May 1993 a young couple in New Mexico died just a few days apart from acute respiratory distress. Both had suddenly developed a high fever, muscular cramps, headaches and a violent cough. Researchers promptly started looking into whether similar cases had been recorded elsewhere. Soon 24 were identified, occurring between December 1, 1992, and June 7, 1993, in New Mexico, Colorado and Nevada. Eleven of these patients had died.

Bacteriological, parasitological and virological tests conducted in the affected states were all negative. Samples were then sent to the Centers for Disease Control and Prevention (CDC) in Atlanta. Tests for all known viruses were conducted, and researchers eventually detected in the serum of several patients antibodies against a class known as hantaviruses. Studies using the techniques of molecular biology showed that the patients had been infected with a previously unknown type of hantavirus, now called Sin Nombre (Spanish for "no name").

New and more effective analytical techniques are identifying a growing number of infective agents. Most are viruses that 10 years ago would probably have passed unnoticed or been mistaken for other, known types. The Sin Nombre infections were not a unique occurrence. Last year a researcher at the Yale University School of Medicine was accidentally infected with Sabià, a virus first isolated in 1990 from an agricultural—engineer who died from a sudden illness in the state of São Paulo, Brazil.

Sabià and Sin Nombre both cause illnesses classified as hemorrhagic fevers. Patients initially develop a fever, followed by a general deterioration in health during which bleeding often occurs. Superficial bleeding reveals itself through skin signs, such as petechiae (tiny releases of blood from vessels under the skin surface), bruises or purpura (characteristic purplish discolorations). Other cardiovascular, digestive, renal and neurological complications can follow. In the most serious cases, the patient dies of massive hemorrhages or sometimes multiple organ failure.

Hemorrhagic fever viruses are divided into several families. The flaviviruses have been known for the longest. They include the Amaril virus that causes yellow fever and is transmitted by mosquitoes, as well as other viruses responsible for mosquito- and tick-borne diseases, such as dengue. Viruses that have come to light more recently belong to three other families: arenaviruses, bunyaviruses (a group that includes the hantaviruses) and filoviruses. They have names like Puumala, Guanarito and Ebola, taken from places where they first caused recognized outbreaks of disease.

All the arenaviruses and the bunyaviruses responsible for hemorrhagic fevers circulate naturally in various populations of animals. It is actually uncommon for them to spread directly from person to person. Epidemics are, rather, linked to the presence of animals that serve as reservoirs for the virus and sometimes as vectors that help to transfer it to people. Various species of rodent are excellent homes for these viruses, because the rodents show no signs when infected. Nevertheless, they shed viral particles throughout their lives in feces and, particularly, in urine. The filoviruses, for their part, are still a mystery: we do not know how they are transmitted.

Hemorrhagic fever viruses are among the most threatening examples of what are commonly termed emerging pathogens. They are not really new. Mutations or genetic recombinations between existing viruses can increase virulence, but what appear to be novel viruses are generally viruses that have existed for millions of years and merely come to light when environmental conditions change. The changes allow the virus to multiply and spread in host organisms. New illnesses may then sometimes become apparent.

IMPROVEMENTS IN DIAGNOSIS

The seeming emergence of new viruses is also helped along by rapid advances in the techniques for virological identification. The first person diagnosed with Sabià in São Paulo (called the index case) was originally thought to be suffering from yellow fever. The agent actually responsible was identified only because a sample was sent to a laboratory equipped for the isolation of viruses. That rarely happens, because most hemorrhagic fever viruses circulate in tropical regions, where hospitals generally have inadequate diagnostic equipment and where many sick people are not hospitalized. Even so, the rapid identification of Sin Nombre was possible only because of several years of work previously accumulated on hantaviruses.

Global Reach of Hemorrhagic Fever Viruses

Hantavirus Puumala causes frequent illness in northwest Europe; the infection is believed to result from inhalation of contaminated dust when handling wood.

Hantaviruses have caused illness with renal syndrome for more than 1,000 years.

Between 1951 and 1953, 2,000 United Nations troops are infected with Hantaan.

Dengue fever, caused by a flavivirus, is spreading from its home territory in Southeast Asia.

Seven laboratory workers preparing cell cultures from the blood of vervet monkeys die from Marburg virus in 1967.

Rift Valley fever infects 200,000 following construction of the Aswan Dam in 1970 and causes 600 deaths. A further outbreak occurs during the 1980s.

In 1976 and again in 1979, Ebola spreads widly through N'zara and Maridi in Sudan's southern grasslands.

Ebola, a filovirus, kills about 300 around a hospital in Yambuku, Zaire, in 1976.

More than 190 die from an Ebola outbreak in Kikwit, Zaire, in the spring of 1995.

A researcher handling samples from wild chimpanzees being decimated by an epidemic in Ivory Coast is infected with a type of Ebola.

In 1970, 25 hospital workers and patients suffer from Lassa fever, caused by an arenavirus, in Lassa, Nigeria.

Rift Valley fever outbreak in 1987 follows damming of the Senegal River in Mauritania.

In 1994 a researcher at Yale University is accidentally infected with Sabiá but survives.

Federal officials are put into a panic in 1989 when monkeys housed in a quarantine facility in Reston, Va., start dying from an Ebola-type filovirus.

Hantavirus Sin Nombre strikes 114 and kills 58 in New Mexico, Colorado and Nevada in 1993, after a rodent population grows rapidly.

More than 100 cases of illness are caused by Guanarito in 1989. The epidemic started in a rural community that had begun to clear a forest.

Machupo causes dozens of deaths in San Joaquin, Bolivia, during the 1950s; seven are infected in 1994.

In 1990 an agricultural engineer dies and a laboratory worker falls ill with the arenavirus Sabiá in the state of São Paulo, Brazil.

Junín kills many agricultural workers in the Argentinian pampas in the 1940s.

ARENAVIRUS

FILOVIRUS

ANIMAL FILOVIRUS OUTBREAK

FLAVIVIRUS

HANTAVIRUS

RIFT VALLEY FEVER (BUNYAVIRUS)

Hantaviruses typically cause an illness known as hemorrhagic fever with renal syndrome; it was described in a Chinese medical text 1,000 years ago. The West first became interested in this illness during the Korean War, when more than 2,000 United Nations troops suffered from it between 1951 and 1953. Despite the efforts of virologists, it was not until 1976 that the agent was identified in the lungs of its principal reservoir in Korea, a field mouse. It took more than four years to isolate the virus, to adapt it to a cell culture and to prepare a reagent that permitted a diagnostic serological test, essential steps in the study of a virus. It was named Hantaan, for a river in Korea. The virus also circulates in Japan and Russia, and a similar virus that produces an illness just as serious is found in the Balkans.

A nonfatal form exists in Europe. It was described in Sweden in 1934 as the "nephritic epidemic," but its agent was not identified until 1980, when it was detected in the lungs of the bank vole. Isolated in 1983 in Finland, the virus was named Puumala for a lake in that country. Outbreaks occur regularly in northwestern Europe. Since 1977, 505 cases have been recorded in northeastern France alone. The number of cases seems to be increasing, but this is probably because doctors are using more biological tests than formerly, and because the tests in recent years have become more sensitive.

Thus, it is only for about a decade that we have had the reagents necessary to identify hantaviruses. Thanks to these reagents and a research technique that spots antibodies marking recent infections, scientists at the CDC in 1993 were quickly on the track of the disease. The presence of specific antibodies is not always definite proof of an infection by the corresponding pathogen, however. False positive reactions and cross—reactions caused by the presence of antibodies shared by different viruses are possible. A more recent technology, based on the polymerase chain reaction, permits fragments of genes to be amplified (or duplicated) and sequenced. It provided confirmation that the patients were indeed infected with hantaviruses. The identification of Sin Nombre took no more than eight days.

THE INFECTIVE AGENTS

The primary cause of most outbreaks of hemorrhagic fever viruses is ecological disruption resulting from human activities. The expansion of the world population perturbs ecosystems that were stable a few decades ago and facilitates contacts with animals carrying viruses pathogenic to—humans. This was true of the arenavirus Guanarito, discovered in 1989 in an epidemic in Venezuela. The first 15 cases were found in a rural community that had started to clear a forested region in the center of the country. The animal reservoir is a species of cotton rat; workers had stirred up dust that had been contaminated with dried rat urine or excrement— one of the most frequent modes of transmission. Subsequently, more than 100 additional cases were diagnosed in the same area.

Other arenaviruses responsible for hemorrhagic fevers have been known for a long time—for example, Machupo, which appeared in Bolivia 1952, and Junín, identified in Argentina in 1958. Both those viruses can reside in species of rodents called vesper mice; the Bolivian species enters human dwellings. Until recently, an extermination campaign against the animals had prevented any human infections with Machupo since 1974. After a lull of 20 years, however, this virus has reappeared, in the same place: seven people, all from one family, were infected during the summer of 1994.

Junín causes Argentinian hemorrhagic fever, which appeared at the end of the 1940s in the pampas west of Buenos Aires. The cultivation of large areas of maize supported huge populations of the species of vesper mice that carry this virus and multiplied contacts between these rodents and agricultural workers. Today mechanization has put the operators of agricultural machinery on the front line: combine harvesters not only suspend clouds of infective dust, they also create an aerosol of infective blood when they accidentally crush the animals.

The arenavirus Sabià has, so far as is known, claimed only one life, but other cases have in all probability occurred in Brazil without being diagnosed. There is a real risk of an epidemic if agricultural practices bring the inhabitants of São Paulo into contact with rodent vectors. In Europe, the main reservoirs of the hantavirus Puumala—the bank vole and yellow-necked field mouse—are woodland animals. The most frequent route of contamination there is inhalation of contaminated dust while handling wood gathered in the forest or while working in sheds and barns.

Humans are not always the cause of dangerous environmental changes. The emergence of Sin Nombre in the U.S. resulted from heavier than usual rain and snow during spring 1993 in the mountains and deserts of New Mexico, Nevada and Colorado. The principal animal host of Sin Nombre is the deer mouse, which lives on pine kernels: the exceptional humidity favored a particularly abundant crop, and so the mice proliferated. The density of the animals multiplied 10-fold between 1992 and 1993.

TRANSMISSION BY MOSQUITOES

Some bunyaviruses are carried by mosquitoes rather than by rodents. Consequently, ecological perturbations such as the building of dams and the expansion of irrigation can encourage these agents. Dams raise the water table, which favors the multiplication of the insects and also brings humans and animals together in new population centers. These two factors probably explain two epidemics of Rift Valley fever in Africa: one in 1977 in Egypt and the other in 1987 in Mauritania.

The virus responsible was recognized as long ago as 1931 as the cause of several epizootics, or animal epidemics, among sheep in western and South Africa. Some breeders in contact with sick or dead animals became infected, but at the time the infection was not serious in humans. The situation became more grim in 1970. After the construction of the Aswan Dam, there were major losses of cattle; of the 200,000 people infected, 600 died. In 1987 a minor epidemic followed the damming of the Senegal River in Mauritania.

Rift Valley fever virus is found in several species of mosquitoes, notably those of the genus *Aedes*. The females transmit the virus to their eggs. Under dry conditions the mosquitoes' numbers are limited, but abundant rain or irrigation allows them to multiply rapidly. In the course of feeding on blood, they then transmit the virus to humans, with cattle acting as incubators.

CONTAMINATION BY ACCIDENT

Although important, ecological disturbances are not the only causes of the emergence of novel viruses. Poor medical hygiene can foster epidemics. In January 1969 in Lassa, Nigeria, a nun who worked as a nurse fell ill at work. She infected, before dying, two other nuns, one of whom died. A year later an epidemic broke out in the same hospital. An Inquiry found that 17 of the 25 persons infected had probably been in the room where the first victim had been hospitalized. Lassa is classed as an arenavirus.

Biological industries also present risks. Many vaccines are prepared from animal cells. If the cells are contaminated, there is a danger that an unidentified virus may be transmitted to those vaccinated. It was in this way that in 1967 a culture of contaminated blood cells allowed the discovery of a new hemorrhagic fever and a new family of viruses, the filoviruses.

The place was Marburg, Germany, where 25 people fell ill after preparing cell cultures from the blood of vervet monkeys. Seven died. Other cases were reported simultaneously in Frankfurt and in Yugoslavia, all in laboratories that had received monkeys from Uganda. The monkeys themselves also died, suggesting that they are not the natural reservoir of Marburg virus. Four cases of natural infection with Marburg have been reported in Africa, but neither the reservoir nor the natural modes of transmission have been discovered. What is clear is that Marburg can propagate in hospitals: secondary cases have occurred among medical personnel.

In 1976 two epidemics of fever caused by a different virus occurred two months apart in the south of Sudan and in northern Zaire. In Zaire, around Yambuku Hospital, by the Ebola River, 318 cases were counted, and 280 persons died. Eighty-five of them had received an injection in this hospital. The epidemic led to the identification of a new virus, Ebola.

The Marburg and Ebola viruses are classified as filoviruses, so called because under the electron microscope they can be seen as filamentous structures as much as 1,500 nanometers in length (the spherical particle of an arenavirus, for comparison, is about 300 nanometers in diameter). These two representatives of the filovirus family are exceedingly dangerous. In 1989 specialists at the CDC were put in a panic when they learned that crab-eating macaques from the Philippines housed in an animal quarantine facility in Reston, Va., were dying from an infection caused by an Ebola-type filovirus. The virus was also isolated from other animal facilities that had received monkeys from the Philippines. No human illnesses were recorded in the wake of this epizootic, however, which demonstrates that even closely related viruses can vary widely in their effects.

In January of this year we isolated a previously unknown type of Ebola from a patient who had infected herself handling samples from wild chimpanzees that were being decimated by a strange epidemic. That the chimpanzees, from Ivory Coast, succumbed is further evidence that primates are not filoviruses' natural reservoir, which has not yet been identified. Although Marburg has infected few people, Ebola surfaced again to cause a human epidemic in Zaire this past May.

A SHIFTING, HAZY TARGET

The extreme variability and speed of evolution found among hemorrhagic fever viruses are rooted in the nature of their genetic material. Hemorrhagic fever viruses, like many other types, generally have genes consisting of ribonucleic acid, or RNA, rather than the DNA employed by most living things. The RNA of these viruses is "negative stranded"—before it can be used to make viral proteins in an infected cell, it must be converted into a positive strand by an enzyme called RNA polymerase. RNA polymerases cause fairly frequent errors during this process. Because the errors are not corrected, an infected cell gives rise to a heterogeneous population of viruses resulting from the accumulating mutations. The existence of such "quasispecies" explains the rapid adaptation of these viruses to environmental changes. Some adapt to invertebrates and others to vertebrates, and they confound the immune systems of their hosts. Pathogenic variants can easily arise.

There is another source of heterogeneity, too. A characteristic common to arenaviruses and bunyaviruses is that they have segmented genomes. (The bunyaviruses have three segments of RNA, arenaviruses two.) When a cell is infected by two viruses of the same general class, they can then recombine so that segments from one become linked to segments from the other, giving rise to new viral types called reassortants.

Although we have a basic appreciation of the composition of these entities, we have only a poor understanding of how they cause disease. Far beyond the limited means of investigation in local tropical hospitals, many of these viruses are so hazardous they cannot be handled except in laboratories that conform to very strict safety requirements. There are only a few such facilities in the world, and not all of them have the required equipment. Although it is relatively straightforward to handle the agents safely in culture flasks, it is far more dangerous to handle infected monkeys: researchers risk infection from being scratched or bitten by sick animals. Yet the viruses cannot be studied in more common laboratory animals such as rats, because these creatures do not become ill when infected.

We do know that hemorrhagic fever viruses have characteristic effects on the body. They cause a diminution in the number of platelets, the principal cells of the blood-clotting system. But this diminution, called thrombocytopenia, is not sufficient to explain the hemorrhagic symptoms. Some hemorrhagic fever viruses destroy infected cells directly; others perturb the immune system and affect cells' functioning.

Among the first group, the cytolytic viruses, are the bunyaviruses that cause a disease called Crimean-Congo fever and Rift Valley fever; the filoviruses Marburg and Ebola; and the prototype of hemorrhagic fever viruses, the flavivirus Amaril. Their period of incubation is generally short, often less than a week. Serious cases are the result of an attack on several organs, notably the liver. When a large proportion of liver cells are destroyed, the body cannot produce enough coagulation factors, which partly explains the hemorrhagic symptoms. The viruses also modify the inner surfaces of blood vessels in such a way that platelets stick to them. This clotting inside vessels consumes additional coagulation factors. Moreover, the cells lining the vessels are forced apart, which can lead to the escape of plasma or to uncontrolled bleeding, causing edema, an accumulation of fluid in the tissue, or severely lowered blood pressure.

The arenaviruses fall into the noncytolytic group. Their period of incubation is longer, and although they invade most of the tissues in the body, they do not usually cause gross lesions. Rather the viruses inhibit the immune system, which delays the production of antibodies until perhaps a month after the first clinical signs of infection. Arenaviruses suppress the number of platelets only slightly, but they do inactivate them. Neurological complications are common.

Hantaviruses are like arenaviruses in that they do not destroy cells directly and also have a long period of incubation, from 12 to 21 days. They target cells lining capillary walls. Hantaan and Puumala viruses invade the cells of the capillary walls in the kidney, which results in edema and an inflammatory reaction caused by the organ's failure to work properly. Sin Nombre, in contrast, invades pulmonary capillaries and causes death by a different means: it leads to acute edema of the lung.

PROSPECTS FOR CONTROL

Several research groups are trying to establish international surveillance networks that will track all emerging infectious agents. The World Health Organization has established a network for tracking hemorrhagic fever viruses and other insect-borne viruses that is particularly vigilant.

Once a virus is detected, technology holds some promise for combating it. An antiviral medication, ribavirin, proved effective during an epidemic of hantavirus in China. A huge effort is under way in Argentina to develop a vaccine to protect people against Junin. Indeed, vaccines against the Rift Valley fever in animals, and against yellow fever in humans, are already approved for use. Yet despite the existence of yellow fever vaccine, that disease is now raging in Africa, where few are vaccinated.

Other approaches are constrained because it is difficult or impossible to control animals that are natural reservoirs and vectors for the viruses or to predict ecological modifications that favor outbreaks of disease. There was an effective campaign against rodent vectors during the Lassa and Machupo arenavirus outbreaks, but it is not usually possible to sustain such programs in rural regions for long periods.

Precautions can be taken in laboratories and hospitals, which have ironically served as amplifiers in several epidemics. In the laboratory, viruses responsible for hemorrhagic fevers must be handled in maximum confinement conditions (known in the jargon as biosafety level 4). The laboratory must be kept at lowered pressure, so that no potentially infectious particle can escape; the viruses themselves should be confined in sealed systems at still lower pressure. In hospitals, the risk of infection from a patient is high for some viruses, so strict safety measures must be followed: hospital personnel must wear masks, gloves and protective clothing; wastes must be decontaminated. A room with lowered pressure is an additional precaution.

Since penicillin has been in widespread use, many people had started to believe that epidemics were no longer a threat. The global pandemic of HIV, the virus that causes AIDS, has shown that view to be complacent. Hemorrhagic fever viruses are indeed a cause for worry, and the avenues to reduce their toll are still limited.

FURTHER READING

Genetic Identification of a Hantavirus Associated with an Outbreak of Acute Respiratory Illness. Stuart T. Nichol et al. in *Science*, Vol. 262, pages 914–917; November 5, 1993.

Hantavirus Epidemic in Europe, 1993. B. Le Guenno, M. A. Camprasse, J. C. Guilbaut, Pascale Lanoux and Bruno Hoen In *Lancet*, Vol. 343, No. 8889, pages 114–415; January 8,1994.

New Arenavirus Isolated in Brazil. Terezinha Lisieux M. Coimbra et al. in *Lancet*, Vol. 343, No. 8894, pages 391–392; February 12,1994.

Filoviruses as Emerging Pathogens. C. J. Peters et al. in *Seminars in Virology*, Vol. 5, No. 2, pages 147–154; April 1994.

Isolation and Partial Characterisation of a New Strain of Ebola Virus. Bernard Le Guenno, Pierre Formentry, Monique Wyers, Pierre Gounon, Francine Walker and Christophe Boesch in *Lancet*, Vol. 345, No. 8960, pages 1271–1274; May 20,1995.

Beyond Chicken Soup

WILLIAM A. HASELTINE

ORIGINALLY PUBLISHED IN NOVEMBER 2001

Back In the mid-1980s, when scientists first learned that a virus caused a relentless new disease named AIDS, pharmacy shelves were loaded with drugs able to treat bacterial infections. For viral diseases, though, medicine had little to offer beyond chicken soup and a cluster of vaccines. The story is dramatically different today. Dozens of antiviral therapies, including several new vaccines, are available, and hundreds more are in development. If the 1950s were the golden age of antibiotics, we are now in the early years of the golden age of antivirals.

This richness springs from various sources. Pharmaceutical companies would certainly point to the advent in the past 15 years of sophisticated techniques for discovering all manner of drugs. At the same time, frantic efforts to find lifesaving therapies for HIV, the cause of AIDS, have suggested creative ways to fight not only HIV but other viruses, too.

A little-recognized but more important force has also been at work: viral genomics, which deciphers the sequence of "letters," or nucleic acids, in a virus's genetic "text." This sequence includes the letters in all the virus's genes, which form the blueprints for viral proteins; these proteins, in turn, serve as the structural elements and the working parts of the virus and thus control its behavior. With a full or even a partial genome sequence in hand, scientists can quickly learn many details of how a virus causes disease—and which stages of the process might be particularly vulnerable to attack. In 2001 the full genome of any virus can be sequenced within days, making it possible to spot that virus's weaknesses with unprecedented speed.

The majority of antivirals on sale these days take aim at HIV, herpes viruses (responsible for a range of ills, from cold sores to encephalitis), and hepatitis B and C viruses (both of which can cause liver cancer). HIV and these forms of hepatitis will surely remain a main focus of investigation for some time; together they cause more than 250,000 cases of disease in the U.S. every year and millions in other countries. Biologists, however, are working aggressively to combat other viral illnesses as well. I cannot begin to describe all the classes of antivirals on the market and under study, but I do hope this article will offer a sense of the extraordinary advances that genomics and other sophisticated technologies have made possible in recent years.

DRUG-SEARCH STRATEGIES

The earliest antivirals (mainly against herpes) were introduced in the 1960s and emerged from traditional drug-discovery methods. Viruses are structurally simple, essentially consisting of genes and perhaps some enzymes (biological catalysts) encased in a protein capsule and sometimes also in a lipid envelope. Because this design requires viruses to replicate inside cells, investigators infected cells, grew them in culture and exposed the cultures to chemicals that might plausibly inhibit viral activities known at the time. Chemicals that reduced the amount of virus in the culture were considered for in-depth investigation. Beyond being a rather hit-or-miss process, such screening left scientists with few clues to other viral activities worth attacking. This handicap hampered efforts to develop drugs that were more effective or had fewer side effects.

Genomics has been a springboard for discovering fresh targets for attack and has thus opened the way to development of whole new classes of antiviral drugs. Most viral targets selected since the 1980s have been identified with the help of genomics, even though the term itself was only coined in the late 1980s, well after some of the currently available antiviral drugs were developed.

After investigators decipher the sequence of code letters in a given virus, they can enlist computers to compare that sequence with those already identified in other organisms, including other viruses, and thereby learn how the sequence is segmented into genes. Strings of code letters that closely resemble known genes in other organisms are likely to constitute genes in the virus as well and to give rise to proteins that have similar structures. Having located a virus's genes, scientists can study the functions of the corresponding proteins and thus build a comprehensive picture of the molecular steps by which the virus of interest gains a foothold and thrives in the body.

That picture, in turn, can highlight the proteins—and the domains within those proteins—that would be good to disable. In general, investigators favor targets whose disruption would impair viral activity most. They also like to focus on protein domains that bear little resemblance to those in humans, to avoid harming healthy cells and causing intolerable side effects. They take aim, too, at protein domains that are basically identical in all major strains of the virus, so that the drug will be useful against the broadest possible range of viral variants.

After researchers identify a viral target, they can enlist various techniques to find drugs that are able to perturb it. Drug sleuths can, for example, take advantage of standard genetic engineering (introduced in the 1970s) to produce pure copies of a selected protein for use in drug development. They insert the corresponding gene into bacteria or other types of cells, which synthesize endless copies of the encoded protein. The resulting protein molecules can then form the basis of rapid screening tests: only substances that bind to them are pursued further.

Alternatively, investigators might analyze the three-dimensional structure of a protein domain and then design drugs that bind tightly to that region. For instance, they might construct a compound that inhibits the active site of an enzyme crucial to viral reproduction. Drugmakers can also combine old-fashioned screening methods with the newer methods based on structures.

Advanced approaches to drug discovery have generated ideas for thwarting viruses at all stages of their life cycles. Viral species vary in the fine details of their reproductive strategies. In general, though, the stages of viral replication include attachment to the cells of a host, release of viral genes into the cells' interiors, replication of all viral genes and proteins (with help from the cells' own protein-making machinery), joining of the components into hordes of viral particles, and escape of those particles to begin the cycle again in other cells.

The ideal time to ambush a virus is in the earliest stage of an infection, before it has had time to spread throughout the body and cause symptoms. Vaccines prove their worth at that point, because they prime a person's immune system to specifically destroy a chosen disease-causing agent, or pathogen, almost as soon as it enters the body. Historically vaccines have achieved this priming by exposing a person to a killed or weakened version of the infectious agent that cannot make enough copies of itself to cause disease. So-called subunit vaccines are the most common alternative to these. They contain mere fragments of a pathogen; fragments alone have no way to produce an infection but, if selected carefully, can evoke a protective immune response.

An early subunit vaccine, for hepatitis B, was made by isolating the virus from the plasma (the fluid component of blood) of people who were infected and then purifying the desired proteins. Today a subunit hepatitis B vaccine is made by genetic engineering. Scientists use the gene for a specific hepatitis B protein to manufacture pure copies of the protein. Additional vaccines developed with the help of genomics are in development for other important viral diseases, among them dengue fever, genital herpes and the often fatal hemorrhagic fever caused by the Ebola virus.

Several vaccines are being investigated for preventing or treating HIV. But HIV's genes mutate rapidly, giving rise to many viral strains; hence, a vaccine that induces a reaction against certain strains might have no effect against others. By comparing the genomes of the various HIV strains, researchers can find sequences that are present in most of them and then use those sequences to produce purified viral protein fragments. These can be tested for their ability to induce immune protection against strains found worldwide. Or vaccines might be tailored to the HIV variants prominent in particular regions.

BAR ENTRY

Treatments become important when a vaccine is not available or not effective. Antiviral treatments effect cures for some patients, but so far most of them tend to reduce the severity or duration of a viral infection. One group of therapies limits viral activity by interfering with entry into a favored cell type.

The term "entry" actually covers a few steps, beginning with the binding of the virus to some docking site, or receptor, on a host cell and ending with "uncoating" inside the cell; during uncoating, the protein capsule (capsid) breaks up, releasing the virus's genes. Entry for enveloped viruses requires an extra step. Before uncoating can occur, these microorganisms must fuse their envelope with the cell membrane or with the membrane of a vesicle that draws the virus into the cell's interior.

Several entry-inhibiting drugs in development attempt to block HIV from penetrating cells. Close examination of the way HIV interacts with its favorite hosts (white blood cells called helper T cells) has indicated that it docks with molecules on those cells called CD4 and CCR5. Although blocking CD4 has failed to prevent HIV from entering cells, blocking CCR5 may yet do so.

Amantidine and rimantidine, the first two (of four) influenza drugs to be introduced, interrupt other parts of the entry process. Drugmakers found the compounds by screening likely chemicals for their overall ability to interfere with viral replication, but they have since learned more specifically that the compounds probably act by inhibiting fusion and uncoating. Fusion inhibitors discovered with the aid of genomic information are also being pursued against respiratory syncytial virus (a cause of lung disease in infants born prematurely), hepatitis B and C, and HIV.

Many colds could soon be controlled by another entry blocker, pleconaril, which is reportedly close to receiving federal approval. Genomic and structural comparisons have shown that a pocket on the surface of rhinoviruses (responsible for most colds) is similar in most variants. Pleconaril binds to this pocket in a way that inhibits the uncoating of the virus. The drug also appears to be active against enteroviruses, which can cause diarrhea, meningitis, conjunctivitis and encephalitis.

JAM THE COPIER

A number of antivirals on sale and under study operate after uncoating, when the viral genome, which can take the form of DNA or RNA, is freed for copying and directing the production of viral proteins. Several of the agents that inhibit genome replication are nucleoside or nucleotide analogues, which resemble the building blocks of genes. The enzymes that copy viral DNA or RNA incorporate these mimics into the nascent strands. Then the mimics prevent the enzyme from adding any further building blocks, effectively aborting viral replication.

Acyclovir, the earliest antiviral proved to be both effective and relatively nontoxic, is a nucleoside analogue that was discovered by screening selected compounds for their ability to interfere with the replication of herpes simplex virus. It is prescribed mainly for genital herpes, but chemical relatives have value against other herpesvirus infections, such as shingles caused by varicella zoster and inflammation of the retina caused by cytomegalovirus.

The first drug approved for use against HIV, zidovudine (AZT), is a nucleoside analogue as well. Initially developed as an anticancer drug, it was shown to interfere with the activity of reverse transcriptase, an enzyme that HIV uses to copy its RNA genome into DNA. If this copying step is successful, other HIV enzymes splice the DNA into the chromosomes of an invaded cell, where the integrated DNA directs viral reproduction.

AZT can cause severe side effects, such as anemia. But studies of reverse transcriptase, informed by knowledge of the enzyme's gene sequence, have enabled drug developers to introduce less toxic nucleoside analogues. One of these, lamivudine, has also been approved for hepatitis B, which uses reverse transcriptase to convert RNA copies of its

Antiviral Drugs Today

Sampling of antiviral drugs on the market appears below. Many owe their existence, at least in part, to viral genomics. About 30 other viral drugs based on an understanding of viral genomics are in human tests.

DRUG NAMES	SPECIFIC ROLES	MAIN VIRAL DISEASES TARGETED
DISRUPTORS OF GENOME		
abacavir, didanosine, stavudine, zalcitabine, zidovudine	Nucleoside analogue inhibitors of reverse transcriptase	HIV infection
acyclovir, ganciclovir, penciclovir	Nucleoside analogue inhibitors of the enzyme that duplicates viral DNA	Herpes infections; retinal inflammation caused by cytomegalovirus
cidofovir	Nucleotide analogue inhibitor of the enzyme that duplicates viral DNA	Retinal inflammation caused by cytomegalovirus
delavardine, efavirenz	Nonnucleoside, nonnucleotide inhibitors of reverse transcriptase	HIV infection
lamivudine	Nucleoside analogue inhibitor of reverse transcriptase	HIV, hepatitis B infections
ribavirin	Synthetic nucleoside that induces mutations in viral genes	Hepatitis C infection
DISRUPTORS OF PROTEIN SYNTHESIS		
amprenavir, indinavir, lopinavir, nelfinavir, ritonavir, saquinavir	Inhibitors of HIV protease	HIV infection
fomivirsen	Antisense molecule that blocks translation of viral RNA	Retinal inflammation caused by cytomegalovirus
interferon alpha	Activator of intracellular immune defenses that block viral protein synthesis	Hepatitis B and C infections
BLOCKERS OF VIRAL SPREAD FROM CELL TO CELL		
oseltamivir, zanamivir	Inhibitors of viral release	Influenza
palivizumab	Humanized monoclonal antibody that marks virus for destruction	Respiratory syncytial infection

DNA genome back into DNA. Intense analyses of HIV reverse transcriptase have led as well to improved versions of a class of reverse transcriptase inhibitors that do not resemble nucleosides.

Genomics has uncovered additional targets that could be hit to interrupt replication of the HIV genome. Among these is RNase H, a part of reverse transcriptase that separates freshly minted HIV DNA from RNA. Another is the active site of integrase, an enzyme that splices DNA into the chromosomal DNA of the infected cell. An integrase inhibitor is now being tested in HIV-infected volunteers.

IMPEDE PROTEIN PRODUCTION

All viruses must at some point in their life cycle transcribe genes into mobile strands of messenger RNA, which the host cell then "translates," or uses as a guide for making the encoded proteins. Several drugs in—development interfere with the transcription stage by preventing proteins known as transcription factors from attaching to viral DNA and switching on the production of messenger RNA.

Genomics helped to identify the targets for many of these agents. It also made possible a novel kind of drug: the antisense molecule. If genomic research shows that a particular protein is needed by a virus, workers can halt the protein's production by masking part of the corresponding RNA template with a custom-designed DNA fragment able to bind firmly to the selected RNA sequence. An antisense drug, fomivirsen, is already used to treat eye infections caused by cytomegalovirus in AIDS patients. And antisense agents are in development for other viral diseases; one of them blocks production of the HIV protein Tat, which is needed for the transcription of other HIV genes.

Drugmakers have also used their knowledge of viral genomes to identify sites in viral RNA that are susceptible to cutting by ribozymes—enzymatic forms of RNA. A ribozyme is being tested in patients with hepatitis C, and ribozymes for HIV are in earlier stages of development. Some such projects employ gene therapy: specially designed genes are introduced into cells, which then produce the needed ribozymes. Other types of HIV gene therapy under study give rise to specialized antibodies that seek targets inside infected cells or to other proteins that latch onto certain viral gene sequences within those cells.

Some viruses produce a protein chain in a cell that must be spliced to yield functional proteins. HIV is among them, and an enzyme known as a protease performs this cutting. When analyses of the HIV genome pinpointed this activity, scientists began to consider the protease a drug target. With enormous help from computer-assisted structure-based research, potent protease inhibitors became available in the 1990s, and more are in development. The inhibitors that are available so far can cause disturbing side effects, such as the accumulation of fat in unusual places, but they nonetheless prolong overall health and life in many people when taken in combination with other HIV antivirals. A new generation of protease inhibitors is in the research pipeline.

STOP TRAFFIC

Even if viral genomes and proteins are reproduced in a cell, they will be harmless unless they form new viral particles able to escape from the cell and migrate to other cells. The most recent influenza drugs, zanamivir and oseltamivir, act at this stage. A molecule called neuraminidase, which is found on the surface of both major types of influenza (A and B), has long been known to play a role in helping viral particles escape from the cells that produced them. Genomic comparisons revealed that the active site of neuraminidase is similar among various influenza strains, and structural studies enabled researchers to design compounds able to plug that site. The other flu drugs act only against type A.

Drugs can prevent the cell-to-cell spread of viruses in a different way—by augmenting a patient's immune responses. Some of these responses are nonspecific: the drugs may restrain the spread through the body of various kinds of invaders rather than homing in on a particular pathogen. Molecules called interferons take part in this type of immunity, inhibiting protein synthesis and other aspects of viral replication in infected cells. For that reason, one form of human interferon, interferon alpha, has been a mainstay of therapy for hepatitis B and C. (For hepatitis C, it is used with an older drug, ribavirin.) Other interferons are under study, too.

More specific immune responses include the production of standard antibodies, which recognize some fragment of a protein on the surface of a viral invader, bind to that protein and mark the virus for destruction by other parts of the immune system. Once researches have the gene

Deciphered Viruses

Some medically important viruses whose genomes have been sequenced are listed below. Frederick Sanger of the University of Cambridge and his colleagues determined the DNA sequence of the first viral genome—from a virus that infects bacteria—in 1977.

VIRUS	DISEASE	YEAR SEQUENCED
Human poliovirus	Poliomyelitis	1981
Influenza A virus	Influenza	1981
Hepatitis B virus	Hepatitis B	1984
Human rhinovirus type 14	Common cold	1984
HIV-1	AIDS	1985
Human papillomavirus type 16	Cervical cancer	1985
Dengue virus type 1	Dengue fever	1987
Hepatitis A virus	Hepatitis A	1987
Herpes simplex virus type 1	Cold sores	1988
Hepatitis C virus	Hepatitis C	1990
Cytomegalovirus	Retinal infections in HIV-infected people	1991
Variola virus	Smallpox	1992
Ebola virus	Ebola hemorrhagic fever	1993
Respiratory syncytial virus	Childhood respiratory infections	1996
Human parainfluenzavirus 3	Childhood respiratory infections	1998

sequence encoding a viral surface protein, they can generate pure, or "monoclonal," antibodies to selected regions of the protein. One monoclonal is on the market for preventing respiratory syncytial virus in babies at risk for this infection; another is being tested in patients suffering from hepatitis B.

Comparisons of viral and human genomes have suggested yet another antiviral strategy. A number of viruses, it turns out, produce proteins that resemble molecules involved in the immune response. Moreover, certain of those viral mimics disrupt the immune onslaught and thus help the virus to evade destruction. Drugs able to intercept such evasion-enabling proteins may preserve full immune re sponses and speed the organism's recovery from numerous viral diseases. The hunt for such agents is under way.

THE RESISTANCE DEMON

The pace of antiviral drug discovery is nothing short of breathtaking, but at the same time, drugmakers have to confront a hard reality: viruses are very likely to develop resistance, or insensitivity, to, many drugs. Resistance is especially probable when the compounds are used for long periods, as they are in such chronic diseases as HIV and in quite a few cases of hepatitis B and C. Indeed, for every HIV drug in the present arsenal, some viral strain exists that is resistant to it and, often, to additional drugs. This resistance stems from the tendency of viruses—especially RNA viruses and most especially HIV—to mutate rapidly. When a mutation enables a viral strain to overcome some obstacle to reproduction (such as a drug), that strain will thrive in the face of the obstacle.

To keep the resistance demon at bay until effective vaccines are found, pharmaceutical companies will have to develop more drugs. When mutants resistant to a particular drug arise, reading their genetic text can indicate where the mutation lies in the viral genome and suggest how that mutation might alter the interaction between the affected viral protein and the drug. Armed with that information, researchers can begin structure-based or other studies designed to keep the drug working despite the mutation.

Pharmaceutical developers are also selecting novel drugs based on their ability to combat viral strains that are resistant to other drugs. Recently, for instance, DuPont Pharmaceuticals chose a new HIV nonnucleoside reverse transcriptase inhibitor, DPC 083, for development precisely because of its ability to overcome viral resistance to such inhibitors. The company's researchers first examined the mutations in the reverse transcriptase gene that conferred resistance. Next they turned to computer modeling to find drug designs likely to inhibit the reverse transcriptase enzyme in spite of those mutations. Then, using genetic engineering, they created viruses that produced the mutant enzymes and selected the compound best able to limit reproduction by those viruses. The drug is now being evaluated in HIV-infected patients.

It may be some time before virtually all serious viral infections are either preventable by vaccines or treatable by some effective drug therapy. But now that the sequence of the human genome is available in draft form, drug designers will identify a number of previously undiscovered proteins that stimulate the production of antiviral antibodies or that energize other parts of the immune system against viruses. I fully expect these discoveries to translate into yet more antivirals. The insights gleaned from the human genome, viral genomes and other advanced drug-discovery methods are sure to provide a flood of needed antivirals within the next 10 to 20 years.

FURTHER READING

VIRAL STRATEGIES OF IMMUNE EVASION. Hidde L. Ploegh in Science, Vol. 280, No. 5361, pages 248–253; April 10, 1998.

STRATEGIES FOR ANTIVIRAL DRUG DISCOVERY. Philip S. Jones in Antiviral Chemistry and Chemotherapy, Vol. 9, No. 4, pages 283–302; July 1998.

NEW TECHNOLOGIES FOR MAKING VACCINES. Ronald W. Ellis in Vaccine, Vol. 17, No. 13–14, pages 1596–1604; March 26, 1999.

PROTEIN DESIGN OF AN HIV-1 ENTRY INHIBITOR. Michael J. Root, Michael S. Kay and Peter S. Kim in Science, Vol. 291, No. 5505, pages 884–888; February 2, 2001.

ANTIVIRAL CHEMOTHERAPY: GENERAL OVERVIEW. Jack M. Bernstein, Wright State University School of Medicine, Division of Infectious Diseases, 2000. Available at www.med.wright.edu/im/AntiviralChemotherapy.html

INFECTIOUS DISEASE

The Bacteria behind Ulcers

MARTIN J. BLASER

ORIGINALLY PUBLISHED IN FEBRUARY 1996

In 1979 J. Robin Warren, a pathologist at the Royal Perth Hospital in Australia, made a puzzling observation. As he examined tissue specimens from patients who had undergone stomach biopsies, he noticed that several samples had large numbers of curved and spiral-shaped bacteria. Ordinarily, stomach acid would destroy such organisms before they could settle in the stomach. But those Warren saw lay underneath the organ's thick mucus layer—a lining that coats the stomach's tissues and protects them from acid. Warren also noted that the bacteria were present only in tissue samples that were inflamed. Wondering whether the microbes might somehow be related to the irritation, he looked to the literature for clues and learned that German pathologists had witnessed similar organisms a century earlier. Because they could not grow the bacteria in culture, though, their findings had been ignored and then forgotten.

Warren, aided by an enthusiastic young trainee named Barry J. Marshall, also had difficulty growing the unknown bacteria in culture. He began his efforts in 1981. By April 1982 the two men had attempted to culture samples from 30-odd patients—all without success. Then the Easter holidays arrived. The hospital laboratory staff accidentally held some of the culture plates for five days instead of the usual two. On the fifth day, colonies emerged. The workers christened them *Campylobacter pyloridis* because they resembled pathogenic bacteria of the *Campylobacter* genus found in the intestinal tract. Early in 1983 Warren and Marshall published their first report, and within months scientists around the world had isolated the bacteria. They found that it did not, in fact, fit into the *Campylobacter* genus, and so a new genus, *Helicobacter,* was created. These researchers also confirmed Warren's initial finding, namely that *Helicobacter pylori* infection is strongly associated with persistent stomach inflammation, termed chronic superficial gastritis.

The link led to a new question: Did the inflamed tissue somehow invite *H. pylori* to colonize there, or did the organisms actually cause the inflammation? Research proved the second hypothesis correct. In one of the studies, two male volunteers—Marshall included—actually ingested the organisms. Both had healthy stomachs to start and subsequently developed gastritis. Similarly, when animals ingested *H. pylori*, gastritis ensued. In other investigations, antibiotics suppressed the infection and alleviated the irritation. If the organisms were eradicated, the inflammation went away, but if the infection recurred, so did the gastritis. We now know that virtually all people infected by *H. pylori* acquire chronic superficial gastritis. Left untreated, both the infection and the inflammation last for decades, even a lifetime. Moreover, this condition can lead to ulcers in the stomach and in the duodenum, the stretch of small intestine leading away from the stomach. *H. pylori* may be responsible for several forms of stomach cancer as well.

More than 40 years ago doctors recognized that most people with peptic ulcer disease also had chronic superficial gastritis. For a variety of reasons, though, when the link between *H. pylori* infection and gastritis was established, the medical profession did not guess that the bacteria might prompt peptic ulcer disease as well. Generations of medical students had learned instead that stress made the stomach produce more acid, which in turn brought on ulcers. The theory stemmed from work carried out by the German scientist K. Schwartz. In 1910, after noting that duodenal ulcers arose only in those individuals who had acid in their stomachs, he coined the phrase "No acid, no ulcer." Although gastric acidity is necessary for ulcers to form, it is not sufficient to explain their occurrence—most patients with ulcers have normal amounts of stomach acid, and some people who have high acid levels never acquire ulcers.

Nevertheless, the stress-acid theory of ulcers gained further credibility in the 1970s, when safe and effective agents to reduce gastric acid were introduced. Many patients felt free of pain for the first time while taking these medications, called histamine 2-receptor blockers (H_2-receptor blockers). The drugs often healed ulcers outright. But when patients stopped taking them, their ulcers typically returned. Thus, patients were consigned to take H_2-receptor blockers for years. Given the prevalence of ulcer disease—5 to 10 percent of the world's population are affected at some point during their lifetime—it is not surprising that H_2-receptor blockers became the most lucrative pharmaceutical agents in the world. Major drug companies felt little incentive to explore or promote alternative models of peptic ulcer disease.

THE BACTERIA AND ULCER DISEASE

In fact, ulcers can result from medications called nonsteroidal anti-inflammatory agents, which include aspirin and are often used to treat chronic arthritis. But all the evidence now indicates that *H. pylori* cause almost all cases of ulcer disease that are not medication related. Indeed, nearly all patients having such ulcers are infected by *H. pylori,* versus some 30 percent of age-matched control subjects in the U.S., for example. Nearly all individuals with ulcers in the duodenum have *H. pylori* present there. Studies show that *H. pylori* infection and chronic gastritis increase from three to 12 times the risk of a peptic ulcer developing within 10 to 20 years of infection with the bacteria. Most important, antimicrobial medications can cure *H. pylori* infection and gastritis, thus markedly lowering the chances that a patient's ulcers will return. But few people can overcome *H. pylori* infection without specific antibiotic treatment.

When someone is exposed to *H. pylori* his or her immune system reacts by making antibodies, molecules that can bind to and incapacitate some of the invaders. These antibodies cannot eliminate the microbes, but a blood test readily reveals the presence of antibodies, and so it is simple to detect infection. Surveys consistently show that one third to one half of the world's population carry *H. pylori.* In the U.S. and western Europe, children rarely become infected, but more than half of all 60-year-olds have the bacteria. In contrast, 60 to 70 percent of the children in developing countries show positive test results by age 10, and the infection rate

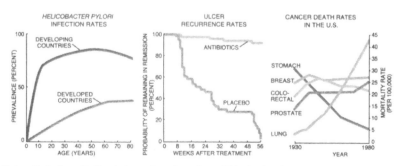

Rates of infection with *H. pylori* vary throughout the world. In developed countries, the infection is rare among children, but its prevalence rises with age. In developing countries, far more people are infected in all age groups (*left*). Supporting the fact that such infections cause ulcer disease, Enno Hentschel and his colleagues at Hanusch Hospital in Vienna found that antimicrobial therapy dramatically decreased the chance that a duodenal ulcer would recur (*center*). As infection rates have declined during the past century in the U.S., so, too, have the number of deaths from stomach cancer (*right*)—suggesting that *H. pylori* infection can, under some circumstances, cause that disease as well.

remains high for adults. H. *pylori* infection is also common among institutionalized children.

Although it is as yet unclear how the organisms pass from one person to another, poor sanitation and crowding clearly facilitate the process. As living conditions have improved in many parts of the world during the past century, the rate of H. *pylori* infection has decreased, and the average age at which the infection is acquired has risen. Gastric cancer has also become progressively less common during the past 80 years. At the start of the 20th century, it was the leading cause of death from cancer in the U.S. and many other developed countries. Now it is far down on the list. The causes for its decline are not well understood, but we have reason to believe that the drop in H. *pylori* infection rates deserves some credit.

A CONNECTION TO CANCER

In the 1970s Pelayo Correa, now at Louisiana State University Medical Center, proposed that gastric cancer resulted from a series of changes in the stomach taking place over a long period. In Correa's model, a normal stomach would initially succumb to chronic superficial gastritis for unknown reasons. We now know that H. *pylori* are to blame. In the second step—lasting for perhaps several decades—this gastritis would cause more serious harm in the form of a lesion, called atrophic gastritis. This lesion might then lead to further changes, among them intestinal metaplasia and dysplasia, conditions that typically precede cancer. The big mystery since finding H. *pylori* has been: Could the bacteria account for the second transition—from superficial gastritis to atrophic gastritis and possibly cancer—in Correa's model?

The first real evidence linking H. *pylori* and gastric cancer came in 1991 from three separate studies. All three had similar designs and reached the same conclusions, but I will outline the one in which I participated, working with Abraham Nomura of Kuakini Medical Center in Honolulu. I must first give some background. In 1942, a year after the bombing of Pearl Harbor, the selective service system registered young Japanese-American men in Hawaii for military service. In the mid-1960s medical investigators in Hawaii examined a large group of these men—those born between 1900 and 1919—to gain information on the epidemiology of heart disease, cancer and other ailments. By the late 1960s they had assembled a cohort of about 8,000 men, administered questionnaires and obtained and frozen blood samples. They then tracked and monitored these men for particular diseases they might develop.

For many reasons, by the time we began our study we had sufficient information on only 5,924 men from this original group. Among them, however, 137 men, or more than 2 percent, had acquired gastric cancer between 1968 and 1989. We then focused on 109 of these patients, each of whom was matched with a healthy member of the cohort. Next, we examined the blood samples frozen in the 1960s for antibodies to *H. pylori*. One strength of this study was that the samples had been taken from these men, on average, 13 years before they were diagnosed with cancer. With the results in hand, we asked the critical question: Was evidence of preexisting *H. pylori* infection associated with gastric cancer? The answer was a strong yes. Those men who had a prior infection had been six times more likely to acquire cancer during the 21-year follow-up period than had men showing no signs of infection. If we confined our analysis to cancers affecting the lower part of the stomach—an area where *H. pylori* often collect—the risk became 12 times as great.

The other two studies, led by Julie Parsonnet of Stanford University and by David Forman of the Imperial Cancer Research Fund in London, produced like findings but revealed slightly lower risks. Over the past five years, further epidemiological and pathological investigations have confirmed the association of *H. pylori* infection and gastric cancer. In June 1994 the International, Agency for Research in Cancer, an arm of the World Health Organization, declared that *H. pylori* a class-1 carcinogen— the most dangerous rank given to cancer-causing agents. An uncommon

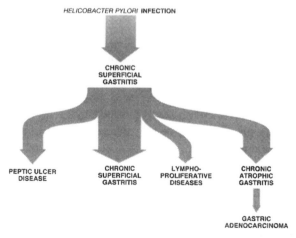

H. Pylori infection progresses to chronic superficial gastritis within months. Left untreated, the condition persists for life in most people. A small fraction, however, may develop peptic ulcer disease, lympho- proliferative diseases or severe chronic atrophic gastritis, leading to adenocarcinoma of the stomach.

cancer of the stomach, called gastric lymphoma, also appears to be largely caused by *H. pylori*. Recent evidence suggests that antimicrobial treatment to cure *H. pylori* infection may bring about regression in a subset of tumors of this kind, which is an exciting development in both clinical medicine and cancer biology.

HOW PERSISTENCE TAKES PLACE

Certainly most bacteria cannot survive in an acidic environment, but *H. pylori* are not the only exception. Since that bacteria's discovery, scientists have isolated 11 other organisms from the stomachs of other primates, dogs, cats, rodents, ferrets and even cheetahs. These bacteria, for now considered to be members of the *Helicobacter* family, seem to have a common ancestor. All are spiral-shaped and highly motile (they swim well)—properties enabling them to resist the muscle contractions that regularly empty the stomach. They grow best at oxygen levels of 5 percent, matching the level found in the stomach's mucus layer (ambient air is 21 percent oxygen). In addition, these microbes all manufacture large amounts of an enzyme called urease, which cleaves urea into ammonia and carbon dioxide. Fostering the production of ammonia may be one way helicobacters neutralize the acid in their local environment, further securing their survival.

An interesting puzzle involves what *H. pylori* eat. There are two obvious guesses: the mucus in which it lives and the food its human host ingests. But Denise Kirschner of Texas A&M University and I constructed a mathematical model showing that *H. pylori* would not be able to persist for years relying on those nutrient sources. In our model, the mathematics of persistence in the stomach requires some regulated interaction between the host cells and the bacteria. Inflammation provides one such interaction, and so I have proposed that *H. pylori* might trigger inflammation for the purpose of acquiring nutrients. An apparent paradox in *H. pylori* biology is that although the organisms do not invade the gastric tissue, they can cause irritation there. Rather, as we and others have found, the microbes release chemicals that the stomach tissue absorbs. These compounds attract phagocytic cells, such as leukocytes and macrophages, that induce gastritis.

The host is not entirely passive while *H. pylori* bombard it with noxious substances. Humans mount an immune response, primarily by making antibodies to the microbe. This response apparently does not function well, though, because the infection and the antibodies almost inevitably coexist for decades. In essence, faced with a pathogen that cannot be easily destroyed, humans had two evolutionary options: we could have evolved to fight *H. pylori* infection to its death, possibly involving the abrogation of normal gastric function, or we could have become tolerant and tried to ignore the organisms. I believe the choice was made long ago in favor of tolerance. The response to other persistent pathogens—such as the microbes responsible for malaria and leprosy—may follow the same paradigm, in which it is adaptive for the host to dampen its immune reaction.

Fortunately, it is not in *H. pylori's* best interest to take advantage of this passivity, growing to overwhelming numbers and ultimately killing its host. First, doing so would limit the infection's opportunity to spread. Second, even in a steady state, *H. pylori* reaches vast numbers (from 10^7 to 10^{10} cells) in the stomach. And third, further growth might exhaust the mechanisms keeping the immune system in check, leading to severe inflammation, atrophic gastritis and, eventually, a loss of gastric acidity. When low acidity occurs, bacteria from the intestines, such as *Escherichia coli*, are free to move upstream and colonize the stomach. Although *H. pylori* can easily live longer than *E. coli* in an acid environment *E. coli* crowds *H. pylori* but of more neutral surroundings. So to avoid any competition with intestinal bacteria, *H. pylori* must not cause too much inflammation, thereby upsetting the acid levels in the stomach.

Are *H. pylori* symbionts that have only recently evolved into disease-causing organisms? Or are they pathogens on the long and as yet incomplete road toward symbiosis? We do not yet know, but we can learn from the biology of *Mycobacterium tuberculosis*, the agent responsible for tuberculosis. It, too, infects about one third of the world's population. But as in *H. pylori* infection, only 10 percent of all infected people become sick at some point in their life; the other 90 percent experience no symptoms whatsoever. The possible explanations fall into several main categories. Differences among microbial strains or among hosts could explain why some infected people acquire certain diseases and others do not. Environmental cofactors, such as whether someone eats well or smokes, could influence the course of infection. And the age at which someone acquires an infection might alter the risks. Each of these categories affects the outcome of *H. pylori* infection, but I will describe in the next section the microbial differences.

NOT ALL BACTERIA ARE CREATED EQUAL

Given its abundance throughout the world, it is not surprising that *H. pylori* are highly diverse at the genetic level. The sundry strains share many structural, biochemical and physiological characteristics, but they are not all equally virulent. Differences among them are associated with variations in two genes. One encodes a large protein that 60 percent of all strains produce. Our group at Vanderbilt University, comprising Murali Tummura, Timothy L. Cover and myself, and a group at the company Biocine in Italy, led by Antonello Covacci and Rino Rappuoli, identified and cloned the gene nearly simultaneously in 1993 and by agreement called it *cagA*. Among patients suffering from chronic superficial gastritis alone, about 50 to 60 percent are infected by *H. pylori* strains having the *cagA* gene. In contrast, nearly all individuals with duodenal ulcers bear cagA strains. Recently we reexamined the results of the Hawaiian study and found that infection by a cagA strain was associated with a doubled risk of gastric cancer. Research done by Jean E. Crabtree of Leeds University in England and by the Vanderbilt group has shown that persons infected by cagA strains experience more severe inflammation and tissue injury than do those infected by strains lacking the *cagA* gene.

The other *H. pylori* gene that seems to influence disease encodes for a toxin. In 1988 Robert D. Leunk, working for Procter & Gamble—the makers of bismuth subsalicylate (Pepto-Bismol)—reported that a broth containing *H. pylori* could induce the formation of vacuoles, or small holes, in tissue cultures. In my group, Cover had clearly shown that a toxin caused this damage and that it was being made not only by *H. pylori* grown in the laboratory but also by those residing in human hosts. In 1991 we purified the toxin and confirmed Leunk's finding that only 50 to 60 percent of *H. pylori* strains produced it. Our paper was published in May 1992 and included a brief sequence of some of the amino acids that encode for the mature toxin. Based on that scanty information, within the next year four groups—two in the U.S., including our own, one in Italy and one in Germany—were able to clone the gene, which we all agreed to name *vacA*. The race to publish was on. Each of our four papers appeared in separate journals within a three-month period.

Lest this sounds like duplicated labor, I should point out that each team had in fact solved a different aspect of the problem. We learned, for example, that virtually all *H. pylori* strains possess *vacA*, whether or not they produce the toxin when grown in culture. We also discovered that there is an extraordinary amount of strain-to-strain variability in *vacA* itself. In addition, broth from toxin-producing strains inoculated directly into the stomach of mice brought about substantial injury. Strains that produce the toxin are some 30 to 40 percent overrepresented in ulcer patients compared with those having gastritis alone. And toxigenic strains usually but not always contain *cagA*, which is located far away from *vacA* on the chromosome.

Which Treatment Strategy Should You Choose?

	OLD MODEL	NEW MODEL
CAUSE	**Excess stomach acid** eats through tissues and causes inflammation	*Helicobacter pylori* bacteria secrete toxins and cause inflammation in the stomach, bringing about damage
TREATMENTS	**Bland diet,** including dairy products every hour, small meals, no citrus or spicy foods and no alcohol or caffeine **H$_2$-receptor blockers** lessen blood levels of histamine, which increases the production of stomach acid **Surgery** to remove ulcers that do not respond to medication or that bleed uncontrollably. In the 1970s, it was the most common operation a surgical resident learned. Now it is increasingly rare	**Antibiotic regimen.** In February 1994 an NIH panel endorsed a two-week course of antibiotics for treating ulcer disease: amoxicillin or tetracycline, metronidazole (Flagyl) and bismuth subsalicylate (Pepto-Bismol). In December 1995 an FDA advisory committee recommended approval of two new four-week treatments, involving clarithromycin (Biaxin) with either omeprazole (Prilosec) or ranitidine bismuth citrate (Tritec). One-week therapies are also highly effective
SUCCESS	Patients who stop taking H$_2$-receptor blockers face a **50 percent chance** that their ulcers will recur within six months and a 95 percent chance that they will reappear within two years	**No recurrence** after the underlying bacterial infection is eliminated
COST	H$_2$-receptor blockers cost from $60 to $100 per month, adding up to **thousands of dollars** over decades of care. Surgery can cost as much as $18,000	**Less than $200** for a standard one-week therapy

SLOW-ACTING BACTERIA AND DISEASE

Over the past 15 years, researchers and physicians have learned a good deal about H. pylori. This knowledge has revolutionized our understanding of gastritis, formerly thought to represent the aging stomach, and of peptic ulcer disease and gastric cancer. It has made possible new treatments and screening methods. In addition, a new field of study has emerged—the microbiology and immunology of the human stomach—that will undoubtedly reveal more about persistent infections within mucosal surfaces.

But let us extrapolate from these findings. Consider that slow-acting bacteria, H. pylori, cause a chronic inflammatory process, peptic ulcer disease, that was heretofore considered metabolic. And also keep in mind that this infection greatly enhances the risk of neoplasms developing, such adenocarcinomas and lymphomas. It seems reasonable, then, to suggest that persistent microbes may be involved in the etiology of other chronic inflammatory diseases of unknown origin, such as ulcerative colitis, Crohn's disease, sarcoidosis, Wegener's granulomatosis, systemic lupus erythematosus and psoriasis, as well as various neoplasms, including carcinomas of the colon, pancreas and prostate. I believe H. pylori are very likely the first in a class of slow-acting bacteria that may well account for a number of perplexing diseases that we are facing today.

FURTHER READING

UNIDENTIFIED CURVED BACILLI IN THE STOMACH OF PATIENTS WITH GASTRITIS AND PEPTIC ULCERATION. Barry J. Marshal and J. Robin Warren in *Lancet*, No. 8390, pages 1311–1315; June 16, 1984.

HELICOBACTER PYLORI INFECTION AND GASTRIC CARCINOMA AMONG JAPANESE AMERICANS IN HAWAII. A. Nomura, G. N. Stemmermnnn, P.-H. Chyou, I. Kato, G. I. Perez-Perez and M. J. Blasers in *New England Journal of Medicine*, Vol. 325, No. 16, pages 1132–1136; October 17, 1991.

HUMAN GASTRIC CARCINOGENESIS: A MULTISTEP AND MULTIFACTORIAL PROCESS. Pelayo Correa in *Cancer Research*, Vol. 52, No. 24, pages 6735–6740; December 15, 1992.

EFFECT OF RANITIDINE AND AMOXICILLIN PLUS METRONIDAZOLE ON THE ERADICATION OF HELICOBACTER PYLORI AND THE RECURRENCE OF DUODENAL ULCER. Enno Hentschel et al. in *New England Journal of Medicine*, Vol. 328, No. 5, pages 308–312; February 4, 1993.

REGRESSION OF PRIMARY LOW-GRADE B-CELL GASTRIC LYMPHOMA OF MUCOSA-ASSOCIATED LYMPHOID TISSUE TYPE AFTER ERADICATION OF HELICOBACTER PYLORI. A. C. Wotherspoon et al. in *Lancet*, Vol. 342, No. 8871, pages 575–577; September 4, 1993.

PARASITISM BY THE "SLOW" BACTERIUM HELICOBACTER PYLORI LEADS TO ALTERED GASTRIC HOMEOSTASIS AND NEOPLASIA. Martin J. Blaser and Julie Parsonnet in *Journal of Clinical Investigation*, Vol. 94, No. 1 pages 4–8; July 1994.

The Challenges of STDs

PHILIP E. ROSS

ORIGINALLY PUBLISHED IN FALL 2005

Chronic in the patient, persistent in the population, easy to prevent in principle but not in practice, sexually transmitted diseases (STDs) present some of the greatest challenges faced by applied biological science. STDs were in fact among the first diseases to be attacked by pharmacology and to attract the interest of public-health officials. While modern medicine has reduced the societal impact of many, such as syphilis, particularly in developed countries, they still present serious problems worldwide. And today's most notable STD, human immunodeficiency virus (HIV), the virus that causes AIDS, has become one of the greatest health threats to arise in modern times.

STDs have been a prime target of research from the early days of modern medicine, in large part because they fit so neatly into the foundation of such medicine: the germ theory of disease. They have remained important in part because they generally let their hosts live long enough to create a medical problem, a marketing opportunity and even a political lobby. Yet rational treatment of the diseases—and, still more, their rational prevention—have been complicated by nonmedical issues. More than any other type of disease, STDs have been associated in the popular mind with moral weakness and social decadence.

This association creates a potential conflict of interest between patients and their physicians, on the one hand, and public-health authorities, on the other. The doctor-patient relationship has been considered sacrosanct since the days of Hippocrates. Yet in the case of STDs, this respect for privacy and confidentiality must sometimes yield to wider concerns. Municipalities regularly quarantine people with dangerous, contagious illnesses, such as drug-resistant tuberculosis, but these quarantines are accompanied by special medical treatment, and can thus be portrayed as for the good of the patient as well as the community. In contrast, tracing a patient's prior sexual contacts often goes directly against his personal interests, given the embarrassment and shame often involved.

COMMON STDs

DISEASE	PATHOGEN TYPE	PATHOGEN NAME	EARLY SYMPTOMS	COMPLICATIONS	INCIDENCE IN THE U.S.	TREATMENT	PREVENTION
Genital herpes	Virus	Herpes simplex virus (HSV)	Cold sores on genitals or mouth	Recurrent outbreaks	20 percent exposed	Uncurable; symptoms treatable by antivirals	Condoms, use of antivirals
Chlamydia	Bacterium	Chlamydia trachomatis	Burning during urination; often asymptomatic, especially in women	Infertility	3 million new cases per year	Curable with antibiotics	Condoms; routine screening
AIDS	Retrovirus	Human immuno-deficiency virus (HIV)	Flulike fever	Opportunistic infections	40,000 new cases per year	Antiretroviral combination therapy	Condoms
HPV	Virus	Human papilloma virus (HPV)	Genital warts; often asymptomatic	Cervical cancer	80 percent of women by age 50	Uncurable; warts and precancerous lesions removable by surgery	Condoms; routine screening; Merck is preparing for market a vaccine against HPV, the first STD vaccine ever.
Trichomoniasis	Protozoan	Trichomonas vaginalis	In women, yellow-green vaginal discharge; often asymptomatic in men	Secondary infections; premature birth	7 million new cases per year	Curable with metronidazole	Condoms; curing male sex partners of infected women
Gonorrhea	Bacterium	Neisseria gonorrhoeae	In men, burning during urination; discharge, in women often asymptomatic	Infertility, blindness in newborns	700,000 new infections per year	Curable with antibiotics	Condoms
Molluscum contagiosum	Virus	Molluscum contagiosum virus (MCV)	Lesions on thighs, buttocks, lower abdomen; can last for years	Can progress in immuno-compromised patients	Rare; <3 percent of STDs	May resolve; lesions can be removed surgically or chemically	Little protection from condoms; avoid skin-to-lesion contact
Chancroid	Bacterium	Haemophilus ducreyi	Soft, pus-filled sore, often painful in men; often undetect-able in women	Lymph-gland swelling, secondary infection	Very rare, but rising	Curable with antibiotics	Condoms; avoid touching sores
Human t-Cell lymphotropic virus	retrovirus	HTLV-I and HTLV-II	Generally none	Sometimes causes leukemia	Rare	Early detection to improve chances of curing leukemia	Condoms
Ectoparasites	lice	Pthirus pubis	Itching in genital region	No serious complications	3 million new cases per year	Curable with medicinal creams and shampoos	Avoid contact with infested people, bedding, clothing
Syphilis	bacterium	Treponema pallidum	Hard, rubbery chancre on genitals or skin	Blindness, mental illness, heart disease	7,000 new cases per year	Curable with antibiotics	Condoms

It is therefore exceedingly difficult for doctors and public-health officials to get forthright and honest answers from patients.

Since not all cases of STDs are even recognized by those affected, let alone referred to doctors and reported to public-health officials, it can be hard to get a full picture of the health burden they impose. The Centers for Disease Control (CDC) recently attempted to come up with a comprehensive estimate, pegged to the year 1998. It found that in that year, STDs accounted for some 20 million "adverse health consequences" in the U.S., including both the obvious (doctors' time, hospitalization) and the harder-to-gauge (lost work hours, infertility treatments). Almost half the burden fell on women, a figure that rises to nearly 90 percent if one removes AIDS from the equation. STDs also resulted in nearly 30,000 fatalities that year—enough to make them one of the country's top 10 causes of death.

Eliminating STDs, or at least reducing their prevalence, is therefore a worthy goal of the public-health community. To do so, scientists must first understand why these diseases behave the way they do. For although many STDs (particularly HIV) can be lethal to their hosts, they are most dangerous to society at large when in hosts who are still alive.

THE EVOLUTIONARY BIOLOGY OF STDs

There are many vehicles that allow disease-causing germs, or pathogens, to travel from one person to the next. Some of the most common are droplets in the air (measles, for instance); feces in untreated water (cholera); the transfer of mothers' blood to their newborn infants, in what is known as congenital transmission (syphilis, HIV); and intermediary hosts, called vectors—which can be anything from mosquitoes (malaria) to hypodermic needles (hepatitis, HIV). STDs use some of these pathways, but their main vehicle, of course, is sexual contact.

For any pathogen, each route poses trade-offs of its own. For instance, droplet-borne bugs have a short range but can spread very quickly. Mosquito-borne bugs tend to spread more slowly, but they do not require that their hosts be crowded together. Yet no matter how a pathogen spreads, the details of its environment make a big difference. If those details change, some strains can benefit and proliferate like mad, while others, less suited to the new conditions, can quickly fade away. In recent years, biologists who attempt to understand infectious diseases in these terms have created the discipline of evolutionary medicine. As an applied science, it aims to prevent the spread of pathogens by exploiting their weaknesses.

Consider the case of a pathogen that passes directly from one person to another via airborne droplets. A strain that reproduces quickly inside a host will have the advantage of inducing violent sneezes and coughs, making the sick person highly infectious. Such a strain will spread fast so long as its host comes into contact with fresh hosts, as in the crowded environment of a refugee camp, troop ship or hospital ward. But if patients are ill enough to retreat to their beds, far from other people, then the strains that make people this sick will die out, and forms that make their hosts less sick will win the race for survival. This is why most pathogens that spread on airborne droplets, such as the common cold, cause only minor sniffles and sneezes.

Now consider the malaria parasite, a one-celled animal that reaches a new host by hitching a ride on a mosquito. This parasite can still succeed if it multiplies as furiously as it can, because it can be transmitted even if the host is so feverish that he has to stay in bed. The parasite will still be able to jump to new hosts for a simple reason: mosquitoes pay house calls. Illnesses that spread through such vectors thus tend to make their patients very sick indeed.

The virulence of STDs similarly depends on the underlying details. In a world in which people very rarely come into sexual contact with new partners, no STD pathogen can propagate quickly. For a given strain to survive, therefore, its hosts must not only survive for a long time, but they must both feel well enough to want sex and remain attractive enough to be able to find partners.

Imagine, now, that the world changes, and people begin to come into contact with a great many sexual partners. Populations that had been isolated sexually come to form links, perhaps in the form of traveling salesmen, tourists and others who often cross social borders. With such changes, faster-reproducing STD strain—seven though they may kill their hosts very quickly—will predominate, because they can still spread rapidly to new hosts. These diseases will tend to make people sicker, faster. It is therefore no accident that some very serious STDs, notably HIV, arose when such economic and cultural changes took place worldwide.

Educating the public to avoid the more dangerous sexual practices—having vaginal or anal sex without a condom, or with multiple partners whose health status is unknown—therefore pays off in two ways. First, the policy is good medicine, because it lowers the chance of any given person getting the disease. Second, it is good public-health practice, because it should push STDs to evolve into milder forms.

Because their current hosts must remain alive for a while to allow disease propagation, most STDs have evolved into forms that do not reproduce too quickly. One might think that this constraint would keep the concentration of pathogens in the blood down, making hosts somewhat non-infectious. Yet people with STDs tend to be quite infectious. There are several reasons why this is so. Unlike most pathogens, those that cause STDs tend to live out their lives in a permanent envelope of bodily fluids that protects them from drying and other mishaps. Also, STD pathogens are injected into parts of the body that are only lightly policed by the body's immune system. What is more, as noted above, STDs sometimes spread by means other than sexual contact.

One other general advantage that STDs have over most other infectious diseases is their ability to re-infect people again and again. For reasons that are still not entirely understood, people do not naturally become immune to these microorganisms. Scientists have therefore had great difficulty in devising effective vaccines against them.

The different aspects of the public-health problems posed by STDs are best illustrated by specific examples. Let us look, first, at syphilis, the disease that sparked the development of the medical and public-health methodology that shapes policy options to this day, before moving on to examine some of the other most common STDs: chlamydia, herpes and HIV.

SYPHILIS

The mother of all STDs, syphilis was first noted by physicians in Europe in the late 15th century and hence blamed, by many, on Columbus's returning sailors, who were thought to have brought it back from the New World. Some scientists now argue, instead, that the agent of syphilis, a spiral-shaped bacterium called *Treponema pallidum,* had long caused related diseases that were transmitted by skin-to-skin contact (common when people huddled for warmth under shared blankets). If they are right, it may be speculated that something in the environment changed, favoring the success of those strains of *T. pallidum* that were particularly suited to spreading through sexual contact. (The origin of syphilis remains a controversial topic among anthropologists.)

When it first became an epidemic disease, syphilis progressed so quickly—sometimes killing a patient within months—that it was easy for physicians to deduce its sexual mode of transmission. Later, syphilis became milder, in the sense that the disease would persist in a chronic state over many years.

There is about a 30 percent chance of contracting syphilis from a single exposure during unprotected sex, although the risk varies depending on the stage of the infection, the nature of the sexual practices and the strength of the immune system. The first symptom comes several weeks after exposure, when a hard or rubbery sore, called a chancre, develops at the bacterium's point of entry into the body. Untreated, the chancre will heal within a few weeks; then the disease will enter a new phase, called secondary syphilis. Here the symptoms vary enormously but often include fatigue, rash and a sore throat. Because these symptoms often mimic those of other illnesses (hence the disease's nickname, "the Great Imitator"), diagnosis can be difficult.

Then the symptoms die away, and the patient has no clue that he is still sick, although he remains infectious. This so-called latent period can last just a few months, but in more than half of all cases, it goes on for the rest of the life of the patient. Those who do progress to the tertiary phase of the disease develop painful tumors, bone degeneration, blindness, cardiovascular problems and neurological disorders, which can lead to personality changes, dementia and death.

Syphilis was originally treated crudely, with mercury and other substances that proved nearly as toxic to the patient as to the pathogen. Early in the 20th century, German microbiologist Paul Ehrlich began a quest for what he termed the "magic bullet"—a drug that would kill the syphilis pathogen but spare the host. A year later, after screening hundreds of compounds, Ehrlich's colleague Sahachiro Hata discovered one containing arsenic that at least approached their ideal. The drug, called arsphenamine and trademarked Salvarsan, may be said to have spawned the modern, research-oriented pharmaceutical industry.

A later refinement of the arsphenamine regimen, incorporating other medicines, only had variable effects and took at least 60 weeks for patients to follow its full course. It was only in the 1940s, with the introduction of penicillin, that a certain cure was demonstrated. Even then, early penicillin regimens required repeated injections. Patients often failed to finish the course of treatment, which is terrible from the public-health viewpoint for two reasons: the patient can go on to infect others, and the surviving pathogen is given the opportunity to evolve resistance to the drug. Nowadays, versions of penicillin and other antibiotics are available in slow-release doses that need be administered only once to eliminate the infection.

In the end, then, with penicillin, the medical problem of syphilis could be considered solved. But in order to eliminate the disease utterly, it was also necessary to find and cure all its carriers. This goal required that public-health officials trace the sexual contacts of every known patient, a particularly intrusive practice. People tend to be secretive about their sexual relations, and are often reluctant to let sexual partners know that they have given them a notorious disease. Laws, therefore, were passed in the first half of the 20th century requiring physicians to report syphilitic patients to public-health authorities, who would then conduct contact-tracing assiduously and urge those who had been exposed to undergo treatment.

These methods have succeeded in nearly eliminating syphilis from entire countries for years at a time. In the U.S., the incidence rose in the 1980s and early 1990s, then fell considerably, dropping to 32,000 new cases in 2002, according to the CDC. (Worldwide, however, the disease remains a scourge: the World Health Organization estimates there are about 12 million new cases worldwide annually.) The number of syphilis infections serves as an indirect measure of the effectiveness of a public-health system, as well as of the prevalence of safe sexual practices. When syphilis gets out of hand, it is often a sign that other, perhaps harder-to-track STDs may also be spreading.

CHLAMYDIA

The public-health dimension of STDs is well illustrated by chlamydia, a rather common family of bacteria whose sexually transmitted form—*Chlamydia trachomatis*—often produces no symptoms, particularly in women. It thus constitutes a "silent" disease that is particularly hard to control. Chlamydia is the most common STD in the U.S., with some 830,000 reported cases a year, and an estimated three million cases in total. Worldwide, there are more than 90 million new cases a year.

In men the bacteria generally infect the urethra, sometimes causing a burning feeling during urination and producing a discharge, sometimes not. Long-term complications in male patients are rare. Women notice such symptoms in only about 15 percent of cases, but they are much more likely to suffer from complications. If the infection spreads to the fallopian tubes—which carry eggs from the ovaries to the uterus—it may provoke the body's immune system to mount a vigorous defense, in the form of widespread inflammation. It is this inflammation and the scarring it leaves behind, rather than the bacteria themselves, that block the tubes, causing infertility. Often a woman never knows she has had a chlamydia infection until she tries to get pregnant and fails. Both men and women who engage in anal sex can also get infections in the rectum, leading to pain and bleeding or some other discharge.

Treatment is easy; a single dose of the antibiotic drug azithromycin will clear up the infection. The hard part is diagnosing the ailment in time to head off complications and warn sexual partners to seek treatment. Sexually active people, particularly women, can be screened via samples of their urine or genital secretions. Yet even such screenings are foolproof only if conducted quite frequently.

Methods to prevent chlamydia infection in the first place are under study now. Some researchers are investigating gels and foam that could be applied to shield target tissues from infection. Others are working on vaccines. Still others are trying to find drugs that would slow the bacteria's propagation within the body and ease the task of the body's immune system. For now, though, the only proven method of fighting the disease is by sex education, in which half the job is simply motivating women to undergo screening.

HERPES

Genital herpes, though it rarely involves complications and can be effectively treated, if not cured, was nonetheless among the most dreaded STDs in the country at one time. In 1982 it was featured on the cover of *Time* magazine, which predicted that it would cause sweeping social changes, notably a return to chastity. Although the disease is in fact quite common—the CDC estimates that about 45 million Americans, or one in five adolescents and adults, have been exposed to it—neither the dread nor the predictions have proved justified.

The STD is caused by a strain of the herpes simplex virus, the cause of cold sores around the mouth. Indeed, the genital form is only a slight variation on the oral one, from which it appears to have evolved, perhaps under the influence of changing sexual practices. As with oral herpes, the first symptoms come some three to 10 days after exposure, when tiny sores break out near the virus's site of entry. The sores break, form a scab and heal without leaving a scar, all in a few weeks. Then the virus holes up in nerve fibers, becoming dormant for months or even years before staging new outbreaks. These "secondary" outbreaks are generally milder and shorter than the initial one, and patients can often feel them coming a day or two in advance. Such premonitions make it easier to minimize the symptoms with antiviral medications, such as acyclovir.

It is easier for a man to pass the virus to a woman than the reverse. One reason may be that the female genitalia are more likely to suffer tiny wounds during sex that facilitate the virus's entry. Condoms thus help to prevent the spread of herpes. The best preventative is to abstain from sex

during outbreaks and to quell symptoms (and infectiousness) with antiviral medications. But *Time's* prediction notwithstanding, genital herpes
poses a serious problem only to newborns—in whom it can spread to the
brain and internal organs—and to adults whose immune systems have
been compromised.

HIV

When, in 1984, AIDS was traced to a single causative agent, HIV, no one expected that the quest for a solution would be so arduous. Margaret Heckler, then secretary of the U.S. Department of Health and Human Services,
famously predicted that a protective vaccine would come within a matter
of years. In fact, immense progress has been made in understanding the
evolution of the virus and in devising chemotherapies against it, and as
a result, those who have access to modern medical treatment are now living much longer with the infection than was the case at first.

HIV reaches the bloodstream from its point of entry (the genitalia, rectum or elsewhere) and, in its initial stage, produces flulike symptoms. It
takes refuge in the lymph nodes, a critical part of the immune system,
where it attacks what are known as $CD4_+$ T cells (sometimes called helper
T cells), which serve as quarterbacks to coordinate the actions of other
immune cells. Normally $CD4_+$ T cells are present at a level of about 1,000
cells per cubic milliliter of blood; when this concentration drops below
about 200 per cubic milliliter, the body begins to lose its ability to combat
a range of normally harmless viruses and microbes. These agents then
stage opportunistic infections.

Some early observers understandably mistook these opportunists for
the ultimate cause of AIDS, especially since it is these infections, rather
than HIV itself, that normally sicken and kill the patient (although HIV
alone can do so, by infecting the brain, for instance). Much of the early
effort in chemotherapy centered on fighting these secondary infections,
especially Kaposi's sarcoma (a tumor on the skin); fungal infections of the
mouth; and cytomegalovirus infections of the eye, gastrointestinal tract
and other organs.

Untreated, HIV infection will lead to such AIDS-related symptoms in
10 to 12 years, on average, although some patients progress much more
quickly or slowly. A small minority, of great interest to researchers,
appears to be able to live with the infection for an indefinite period with
no ill effects at all. But untreated patients who develop full-blown AIDS
generally survive only one or two more years.

The critical scientific problem is the enormous mutability of the virus. HIV is a retrovirus, which means it codes its genetic information in RNA, a nucleic acid that transfers genetic codes from DNA to proteins. The way HIV copies that information to form new viruses in the host is highly prone to error, which makes it liable to natural mutations. A patient infected with a single line of the virus may quickly develop swarm of loosely related variants that respond differently to any treatment the patient undergoes. This variability is why HIV so readily develops resistance to drugs.

To fight this mutability, researchers have won a measure of success with an idea called combination therapy, pioneered in the 1940s for the treatment of tuberculosis and then used in the 1970s against cancer. The idea is that a chemical "cocktail" will have an effect greater than the sum of its parts, because a mutation that saves the virus from one chemical will tend to make it vulnerable to one of the others. The cocktail's elements—called antiretrovirals—interfere with the virus's ability to reproduce inside cells. HIV patients undergoing combination therapy must take many pills a day, at set times and either before or after meals, though this regimen is being eased by the introduction of more flexible drug formulations.

Combination therapy has greatly extended the life expectancy of people with HIV, particularly (but not exclusively) when begun early in the infection. Further improvements in the regimen are in the works, most of them involving the development of drugs to attack new molecular targets in HIV. But scientists must always be on the lookout for still newer targets, lest the virus keep evolving its way to drug resistance.

There is one drawback to these improved therapies, however: because HIV is now a disease that some people can live with, many of those at risk feel that they need not change their lives greatly to avoid it. Safe-sex practices, such as the regular use of condoms, have declined, particularly in the gay community. Meanwhile, public-health authorities worry that the epidemic will continue to edge into new populations.

The CDC estimates that 40,000 new HIV infections occur in the U.S. every year, 70 percent among men and 30 percent among women. It estimates that 40 percent of these infections stem from male-to-male sex, 30 percent from male-to-female or female-to-male sex and the rest from the sharing of hypodermic needles. In the world as a whole, there are some 40 million people with HIV, a number that grows by about five million a year. Three million die of AIDS every year.

These are truly appalling numbers, both in their size and their rate of growth. Most horrifying are the statistics from Africa, which accounts for three million of the five million new cases every year, and for 560,000 of the 640,000 annual new infections of children (through congenital transmission). What is more, HIV, by weakening the immune system, is also causing flare-ups of dormant cases of tuberculosis, which then is able to spread even to people who do not have HIV. This process appears to have increased the death rate from TB in Africa by as much as 20 percent.

The economic burden is even higher than these bare numbers suggest, because HIV, like other STDs, strikes people in the prime of life. Most of the investment in their rearing, education and job training has already been made; most of the expected return of that investment has not been realized. When they die, they hollow out the demographic structure of their countries, leaving them with a higher-than-usual proportion of dependents, both young and old. In Botswana, one of the hardest-hit countries of all, about 40 percent of all adults carry the HIV virus; the percentage of working-age adults who carry it is surely higher still.

Most people in the poor nations of the world have little access to antiretroviral therapy. What is more, most have had little education in the ways of safe sex, and those who have are often in no position to apply the knowledge. (Women in Africa, for instance, are often forced to endure unprotected sex with husbands who carry the virus.)

Discouraging as these data are, there is much that can be done. Antiretroviral therapy is increasingly being made available in poorer countries, especially for mothers giving birth, to prevent congenital transmission of HIV. Far greater gains can be expected from policies designed to spread awareness of the disease and techniques, such as condom use, that prevent it. A combination of treatment-based and prevention-based strategies appears likely to slow the infection and mortality rates most.

A complete solution to the problem of HIV, however, requires a vaccine. Even an imperfect immunization, one that lowered the chance of contracting the illness by, say, 50 percent, would be of immense benefit in slowing transmission. Unfortunately, no such vaccine is yet within reach.

WHAT LIES AHEAD

AIDS represents a daunting challenge to global public health, and a great deal of research, energy and resources are rightly being spent to combat and eventually eradicate the disease. But there is a larger issue to consider. It would be shortsighted to regard the fight against HIV and other STDs as a mere mop-up operation, a purely political question of mustering the will to use the tools we have to eliminate these diseases in all parts of the world.

The fact is that STDs will never disappear entirely, because new ones will always be popping up. Their mode of transmission offers an ecological niche that is ideal for microbes, and the increasing interconnection of the world's populations means that our entire planet will become more and more of an incubation chamber. We must regard ourselves as surrounded by infectious agents, any one of which could emerge as the next great STD scourge. Understanding how these diseases propagate, and heeding the lessons learned from the past battles against diseases like syphilis, as well as the present fight against AIDS, will be vital in future fights against as-yet-unknown STDs.

Can Chlamydia Be Stopped?

DAVID M. OJCIUS, TONI DARVILLE AND PATRIK M. BAVOIL

ORIGINALLY PUBLISHED IN MAY 2005

Ask the average American about chlamydia, and you will probably evoke an uneasy cringe. Most people think immediately of one of the world's most common sexually transmitted diseases (STDs). But the term actually refers to an entire genus of tiny bacteria that can ignite a variety of serious illnesses.

Ask a poor mother in Africa about chlamydia, and she may tell you that flies transmitting this infection gave her two young children the painful eye condition known as conjunctivitis. This illness—caused by a strain of *Chlamydia trachomatis* (the species that also causes STDs)—can lead to trachoma, a potentially blinding disease. In industrial countries, an airborne species, *C. pneumoniae,* causes colds, bronchitis and about 10 percent of pneumonias acquired outside of hospitals. Researchers have even drawn tentative links between *C. pneumoniae* and atherosclerosis, the artery-narrowing condition that leads to heart attacks and strokes.

Because chlamydiae are bacteria, antibiotics can thwart the infections they produce. Unfortunately, the illnesses often go undetected and untreated, for various reasons. The genital infections rarely produce symptoms early on. And in developing countries where trachoma is a concern, people often lack access to adequate treatment and hygiene. As a result, many of the estimated 600 million people infected with one or more *Chlamydia* strains will go without medical care until the consequences have become irreversible.

It is unrealistic to expect that doctors will ever identify all individuals who have the STD or that improved hygiene will soon wipe out the trachoma-causing bacteria in developing countries. For these reasons, the best hope for curtailing the spread of these ailments is to develop an effective vaccine or other preventive treatments. To discover agents able to block infections before they start, scientists need to know more about how chlamydiae replicate, incite disease and function at a molecular level. But that information has been hard to come by. These bugs are wily. Not only do they have varied strategies for evading the body's immune system, they also are notoriously difficult to study in the laboratory. In the past five years, however, new research—including the complete sequencing of the genomes of several *Chlamydia* strains—has helped scientists begin to address these obstacles. The resulting discoveries are renewing hope for developing new prevention strategies.

SILENT INJURY

One major impediment to the production of a vaccine is chlamydia's surreptitious way of wreaking havoc on the body. The microbes that cause tetanus or cholera swamp tissues with toxins that damage or kill vulnerable cells. Chlamydiae, in contrast, do not damage tissues directly. Rather they elicit an enthusiastic immune response that attempts to rein in the infection through inflammation for as long as the bacteria remain in the body—even at low levels. Ironically, this way of fighting the infection actually brings on the long-term damage. Vaccines prevent illness by priming the immune system to react strongly to specific disease-causing agents, but in this case, the inflammatory component of such a response could do more harm than good.

Whether, in the genital tract, eyelids or elsewhere, inflammation begins when certain cells of the host immune system secrete factors called cytokines—small signaling proteins that attract additional defensive cells to the site of infection. The attracted cells and the cytokines try to wall off the area to prevent the bacteria's spread. In the skin, this process gives rise to familiar outward manifestations of inflammation: redness, swelling and heat. At the same time, the inflammatory cytokines help to trigger the tissue repair response called fibrosis, which can lead to scarring.

In the genital tract, the early inflammation is not obvious. Of the 3.5 million Americans infected with sexually transmitted chlamydia every year, 85 to 90 percent show no symptoms. Men, whose inflammation occurs in the penis, may experience slight pain during urination; women may feel nothing as the bacteria move up the genital tract into the fallopian tubes. Unaware of the problem, these individuals inadvertently pass the bugs along. Indeed, a woman may not learn of her infection until she tries to become pregnant and realizes she is infertile. In other cases, persistent inflammation and scarring of the fallopian tubes causes chronic pelvic pain or increases the chances of ectopic, or tubal, pregnancy—the leading cause of first-trimester pregnancy-related deaths in the U.S.

Inflammation of the eyelids is more immediately obvious. Such infections afflict an estimated 150 million people living in developing countries with hot climates; there treatments may be scarce, and flies and gnats can readily transmit the bacteria between people's infrequently washed hands and faces. (Trachoma does not occur in the U.S. or western Europe because of better public health systems.) When infections scar the inside of the upper eyelid repeatedly over many years, the eyelid may begin to turn under, pointing the eyelashes inward where they can scratch the cornea. Unchecked, the corneal damage can cause blindness decades after the initial infection.

Given that inflammation accounts for most of chlamydia's ill effects, those who are striving to develop a vaccine must find a way to control the bacteria without inducing a strong inflammatory reaction. Ideally, any intervention would fine-tune the inflammatory response—evoking it just enough to help the body's other immune defenses eliminate the bacteria.

Much research on infections caused by chlamydia and other pathogens is focusing on factors that either initiate secretion of the inflammatory cytokines or dampen the inflammatory response once the infection has been cleared. Over the past few years, investigators have discovered small molecules that normally stimulate or inhibit these responses in the body. The next step will be to develop compounds that are able to regulate the activities of these molecules. These agents might be delivered to shut down inflammation artificially after an antibiotic has been administered to control the bacteria.

HANGING AROUND

Beyond inducing inflammation, chlamydiae have other properties that impede development of an effective vaccine. For instance, once you get mumps or measles—or the vaccines against them—you are immune for life. Not so with chlamydia. The body has a hard time eliminating the bacteria completely, and natural immunity after a bout with the microbes lasts only about six months. Hence, an infection that has apparently disappeared may flare up again months or years later, and little protection remains against new outbreaks. If the body's natural response to infection cannot confer long-term protection, it seems likely that a vaccine that merely mimicked this response would fail as well. To be successful, a vaccine would have to elicit defenses that were more powerful than those occurring naturally without triggering excessive inflammation.

One way that vaccines or natural immune responses to an initial infection protect against future colonization by certain microorganisms is by inducing the body to produce so-called memory B lymphocytes targeted to those specific invaders. These immune cells patrol the body throughout its lifetime, ready to secrete antibody molecules that can in turn latch onto any new bugs and mark them for destruction before they invade healthy cells. The antibody system works well against a number of disease-causing agents or pathogens—especially against the many bacteria that live outside a host's cells. In theory, antibodies could attack the microbes before they entered cells or when newly minted copies traveled from one cell to another. But the B lymphocyte system is not terribly effective at these tasks when it comes to chlamydiae, which live inside the cells, where circulating antibodies cannot reach them.

To prevent chlamydiae from lying dormant in cells and then proliferating a new, a vaccine would probably need to pump up the so-called cellular arm of the immune system in addition to evoking an antibody attack. This arm, critical to eradicating viruses (which also live inside cells), relies on killer and helper T cells as well as on scavenger cells known as macrophages to eliminate invaders. Unfortunately, even this trio of immune cells does an incomplete job of eliminating chlamydiae, too often allowing infected cells to survive and become bacteria-producing factories.

Developing a vaccine able to evoke a better cellular response than the body could mount on its own is a tall order. Most existing vaccines elicit a targeted antibody response, but safely activating cellular immunity against many infectious diseases remains a challenging task. The job is particularly difficult in the case of chlamydiae because these bacteria have special ways of protecting themselves from attack by the cellular branch of the immune system.

HIDDEN HIJACKERS

Like certain other bacterial pathogens, chlamydiae induce epithelial cells—in this case, those lining genital tracts, eyelids or lungs—to absorb them within a membrane-bound sac, or vacuole. Healthy cells typically attempt to kill internalized pathogens by having the entry vacuoles fuse with lysosomes, cellular structures containing enzymes that chop up proteins, lipids and DNA. All cells display the chopped-up pieces on proteins called major histocompatibility complex (MHC) molecules at the cell surface. Killer and helper T cells, which travel around the body continuously, will then glom on to MHC molecules that display bits of foreign proteins.

If the T cells also receive other indications of trouble, they will deduce that the cells are infected and will orchestrate an attack on them.

But chlamydiae somehow compel their entry vacuoles to avoid lysosomes, enabling the bacteria to proliferate freely while separated physically from the rest of the infected cell. If the lysosomes cannot provide bits of the bacteria for display on the cell surface, patrolling T cells will not recognize that a cell harbors invaders. Understanding how the bacteria grow and avoid lysosomes might suggest new ways to forestall or halt the infection. Recent findings, including the newly sequenced *Chlamydia* genomes, are aiding in that effort.

The sequence of genetic building blocks in an organism's DNA specifies the proteins that cells make; the proteins, in turn, carry out most cellular activities. Thus, the sequence of codes in a gene says a good deal about how an organism functions. Researchers, including Ru-ching Hsia and one of us (Bavoil) of the University of Maryland, discovered a particularly important element of chlamydiae by noting similarities between their genes and those of larger bacteria, such as *Salmonella typhimurium*, infamous for causing food poisoning. Scientists now generally agree that chlamydiae have everything they need to form a versatile, needlelike projection called a type III secretion apparatus. This apparatus, which spans the membrane of the entry vacuole, serves as a conduit between the bacteria and the cytoplasm of the host cell.

Such a connection implies that chlamydiae can inject proteins into the cytoplasm of the host cell. The apparatus may thus help chlamydiae resist interaction with lysosomes, because it can secrete proteins that remodel the vacuole membrane in ways that bar lysosome function. In addition, investigators have watched the chlamydiae-bearing vacuole divert artificially fluorescing lipids from certain compartments of the host cell, including the Golgi apparatus, to the vacuole membrane. Normally, the membrane of an entry vacuole bears molecules made by the pathogen inside. In this case, a membrane enclosing a bacterium would look foreign to the host cell, which would target the bacterium for immediate destruction by lysosomes. But the lipids that chlamydiae use to rebuild the membrane of their entry vacuole come from the host cell: the vacuoles are therefore indistinguishable from the host cell's organelles and invisible to lysosomes.

If scientists identify the proteins the bacteria secrete to camouflage vacuoles, they might be able to devise two kinds of infection-preventing treatments. One potential drug could interfere with the proteins' activity in a way that would force the entry vacuole to fuse with lysosomes, triggering an immune attack right after the chlamydiae invade the cell. Another drug might incapacitate the mechanisms the bacteria use to divert lipids from the host cell to the chlamydial vacuole, halting the trespassers' ability to hide. Hypothetically, such drugs could be incorporated into a topical microbicide that would thwart sexually transmitted chlamydiae.

Some of the proteins mentioned above—and any others that are unique to the bacteria and not made by human cells—might also be useful ingredients in vaccines. Newly sequenced genomes should be helpful in identifying good candidates.

SUICIDAL TENDENCIES

Recent findings about the role of T cells may open other doors. Biologists have long known that killer T cells normally destroy infected cells by inducing a type of cell death known as apoptosis or "cell suicide," during which cells use their own enzymes to lyse their proteins and DNA. Also known is that immune cells—including T cells and macrophages—stimulate the production of cytokines that help to cripple bacteria and to trigger an inflammatory response that stops their spread. One cytokine known to have this dual purpose is tumor necrosis factor-alpha (TNF-alpha). Laboratory investigations have shown, however, that some infected cells survive despite treatment with TNF-alpha and other apoptosis-inducing cytokines, leading to persistent infections. The problem is that the body does not give up easily. Cytokines continue to trigger chronic inflammation in an effort to contain the infection even if they cannot eliminate it outright.

But even persistently infected cells cannot live forever. Indeed, it appears that chlamydiae have developed their own way to elicit the death of a host cell, which they must do to ensure their own longevity. (The host cell must fall apart before the bacteria can infect other cells.) And as Jean-Luc Perfettini discovered while working as a graduate student with one of us (Ojcius) at the Pasteur Institute in Paris, chlamydiae can kill and exit the infected cells in a way that minimizes the host immune system's ability to sense any danger, thereby allowing the infection to spread essentially undetected in the body.

Addressing this final stage of the bacterial life cycle will require further investigation into the proteins involved in inducing apoptosis and in protecting persistently infected cells from suicidal signals. From what biologists know so far, the latter avenue may prove more fruitful in developing a vaccine. By rendering persistently infected cells more sensitive to apoptosis, it might:be possible to eliminate the bacteria that remain dormant in the system for long periods as well as decrease the lasting consequences of chronic infection.

MULTIPLE AVENUES OF ATTACK

Regardless of the discoveries that lie ahead, the ideal chlamydia vaccine will not be a simple one. It will have to activate both the antibody and cellular arms of the immune system more effectively than the body's natural response does yet somehow limit inflammation as well. For those concerned with preventing chlamydia-related STDs, an additional challenge is ensuring that memory lymphocytes remain in the genital tract poised to combat infection at all times. This tract does not contain the type of tissue that produces memory cells; such cells tend to vacate the area, leaving the person susceptible to infection after a brief period of immunity.

Recall that females bear the lasting effects of genital infection. One feasible goal of a vaccine might be to protect women from the disease rather than from infection per se. This aim might be achieved by vaccinating both men and women. In this scenario, the vaccine would have to generate only enough antibodies to reduce, rather than eliminate, the amount of bacteria men carry. Then, if a woman were exposed to a man's infection through intercourse, memory cells induced by her immunization would travel to the genital tract in numbers adequate for killing the relatively small number of organisms before they spread to her fallopian tubes.

Until researchers manage to develop such a vaccine, contraceptives that include antichlamydial drugs could pay off. These agents might take the form of compounds that either block the proteins chlamydiae use to bind to genital tract cells or target the proteins the microbes secrete to promote intracellular survival. For eye infections, the only vaccine likely to be useful is one that completely prevents infection.

While awaiting effective preventive strategies against chlamydia, it is worth remembering that current antibiotic treatment is highly successful when it is accessible. New details from genomic discoveries indicate that this efficacy will continue. Compared with free-living bacterial pathogens, which can share genes easily, the genomes of *Chlamydia* species have remained essentially the same for millions of years. This genetic stability implies that chlamydiae cannot easily acquire genes—including those for antibiotic resistance—from other bacteria.

It is also worth noting that antibiotics cannot undo the tissue damage caused by inflammation, and to be most useful, they must be given early.

Therefore, more widespread screening of high-risk individuals is needed. Researchers have already proved the feasibility of employing noninvasive urine screening of sexually active young men and women, particularly in settings such as high schools, military intake centers and juvenile detention facilities. Public health officials need to pursue such strategies in parallel with the ongoing search for effective vaccines.

FURTHER READING

CHLAMYDIAE PNEUMONIAE—AN INFECTIOUS RISK FACTOR FOR ATHEROSCLEROSIS? Lee Ann Campbell and Cho-cho Kuo in *Nature Reviews Microbiology*, Vol. 2, No. 1, pages 23–32; January 2004.
CHLAMYDIA AND APOPTOSIS: LIFE AND DEATH DECISIONS OF AN INTRACELLULAR PATHOGEN. Gerald I. Byrne and David M. Ojcius in *Nature Reviews Microbiology*, Vol. 2, No. 10, pages 802–808; October 2004.
Basic information on the infections, genomes, basic biology and immunology of chlamydia can be found at http://chlamydia-www. berkeley.edu:4231/ and www.chlamydiae .com/chlamydiae/

Tackling Malaria

CLAIRE PANOSIAN DUNAVAN

ORIGINALLY PUBLISHED IN DECEMBER 2005

Long ago in the Gambia, West Africa, a two-year-old boy named Ebrahim almost died of malaria. Decades later Dr. Ebrahim Samba is still reminded of the fact when he looks in a mirror. That is because his mother—who had already buried several children by the time he got sick—scored his face in a last-ditch effort to save his life. The boy not only survived but eventually became one of the most well-known leaders in Africa: Regional Director of the World Health Organization.

Needless to say, scarification is not what rescued Ebrahim Samba. The question is, What did? Was it the particular strain of parasite in his blood that day, his individual genetic or immunological makeup, his nutritional state? After centuries of fighting malaria—and conquering it in much of the world—it is amazing what we still do not know about the ancient scourge, including what determines life and death in severely ill children in its clutches. Despite such lingering questions, however, today we stand on the threshold of hope. Investigators are studying malaria survivors and tracking many other leads in efforts to develop vaccines. Most important, proven weapons—principally, insecticide-treated bed nets, other antimosquito strategies, and new combination drugs featuring an ancient Chinese herb—are moving to the front lines.

In the coming years the world will need all the malaria weapons it can muster. After all, malaria not only kills, it holds back human and economic development. Tackling it is now an international imperative.

A VILLAIN IN AFRICA

Four principal species of the genus *Plasmodium*, the parasite that causes malaria, can infect humans, and at least one of them still plagues every continent save Antarctica to a lesser or greater degree. Today, however, sub-Saharan Africa is not only the largest remaining sanctuary of *P. falciparum*—the most lethal species infecting humans—but the home of

Anopheles gambiae, the most aggressive of the more than 60 mosquito species that transmit malaria to people. Every year 500 million falciparum infections befall Africans, leaving one million to two million dead—mainly
children. Moreover, within heavily hit areas, malaria and its complications may account for 30 to 50 percent of inpatient admissions and up to
50 percent of outpatient visits.

The clinical picture of falciparum malaria, whether in children or
adults, is not pretty. In the worst-case scenario, the disease's trademark
fever and chills are followed by dizzying anemia, seizures and coma,
heart and lung failure—and death. Those who survive can suffer mental
or physical handicaps or chronic debilitation. Then there are people like
Ebrahim Samba, who come through their acute illness with no residual
effects. In 2002, at a major malaria conference in Tanzania where I met
the surgeon-turned–public health leader, this paradox was still puzzling
researchers more than half a century after Samba's personal clash with
the disease.

That is not to say we have learned nothing in the interim regarding in-
born and acquired defenses against malaria. We now know, for example,
that inherited hemoglobin disorders such as sickle cell anemia can limit
bloodstream infection. Furthermore, experts believe that antibodies and
immune cells that build up over time eventually protect many Africans
from malaria's overt wrath. Ebrahim Samba is a real-life example of this
transformed state following repeated infection; after his early brush with
death, he had no further malaria crises and to this day uses no preventive
measures to stave off new attacks. (As a tropical medicine doctor, all I can
say is: Don't try this on safari, folks, unless you, too, grew up immunized
by hundreds of malarial mosquitoes every year.)

Samba's story also has another lesson in it. It affirms the hope that
certain vaccines might one day mimic the protection that arises naturally
in people like him, thereby lessening malaria-related deaths and complications in endemic regions. A different malaria vaccine might work by
blocking infection altogether (for a short time, at least) in visitors such
as travelers, aid workers or military peacekeepers, whose need for protection is less prolonged.

On the other hand, the promise of vaccines should not be overstated.
Because malaria parasites are far more complex than disease-causing viruses and bacteria for which vaccines now exist, malaria vaccines may
never carry the same clout as, say, measles or polio shots, which protect
more than 90 percent of recipients who complete all recommended doses.
And in the absence of a vaccine, Africa's malaria woes could continue

to grow like a multiheaded Hydra. Leading the list of current problems are drug-resistant strains of *P. falciparum* (which first developed in South America and Asia and then spread to the African continent), followed by insecticide resistance among mosquitoes, crumbling public health infrastructures, and profound poverty that hobbles efforts to prevent infections in the first place. Finally, the exploding HIV/AIDS pandemic in Africa competes for precious health dollars and discourages the use of blood transfusions for severe malarial anemia.

Where does this leave us? With challenges, to be sure. But challenges should not lead to despair that Africa will always be shackled to malaria. Economic history, for one, teaches us it simply isn't so.

LESSONS OF HISTORY

When I lecture about malaria to medical students and other doctors, I like to show a map of its former geography. Most audiences are amazed to learn that malaria was not always confined to the tropics—until the 20th century, it also plagued such unlikely locales as Scandinavia and the American Midwest. The events surrounding malaria's exit from temperate zones and, more recently, from large swaths of Asia and South America reveal as much about its perennial ties to poverty as about its biology.

Take, for example, malaria's flight from its last U.S. stronghold—the poor, rural South. The showdown began in the wake of the Great Depression when the U.S. Army, the Rockefeller Foundation and the Tennessee Valley Authority (TVA) started draining and oiling thousands of mosquito breeding sites and distributing quinine (a plant-based antimalarial first discovered in South America) to purge humans of parasites that might otherwise sustain transmission. But the efforts did not stop there. The TVA engineers who brought hydroelectric power to the South also regulated dam flow to maroon mosquito larvae and installed acres of screen in windows and doors. As malaria receded, the local economies grew.

Then came the golden days of DDT (dichlorodiphenyltrichloroethane). After military forces used the wettable powder to aerially bomb mosquitoes in the malaria-ridden Pacific theater during World War II, public health authorities took the lead. Five years later selective spraying within houses became the centerpiece of global malaria eradication. By 1970 DDT spraying, elimination of mosquito breeding sites and the expanded use of anti-malarial drugs freed more than 500 million people, or roughly one third of those previously living under malaria's cloud.

Sub-Saharan Africa, however, was always a special case: with the exception of a handful of pilot programs, no sustained eradication efforts were ever mounted there. Instead the availability of chloroquine—a cheap, man-made relative of quinine introduced after World War II—enabled countries with scant resources to replace large, technical spraying operations with solitary health workers. Dispensing tablets to almost anyone with a fever, the village foot soldiers saved millions of lives in the 1960s and 1970s. Then chloroquine slowly began to fail against falciparum malaria. With little remaining infrastructure and expertise to counter Africa's daunting mosquito vectors, a rebound in deaths was virtually ordained.

Along the way, economists learned their lesson once again. Today in many African households, malaria not only limits income and robs funds for basic necessities such as food and youngsters' school fees, it fuels fertility because victims' families assume they will always lose children to the disease. On the regional level, it bleeds countries of foreign investment, tourism and trade. Continentwide, it costs up to $12 billion a year, or 4 percent of Africa's gross domestic product. In short, in many places malaria remains entrenched because of poverty and, at the same time, creates and perpetuates poverty.

BATTLING THE MOSQUITO

Years ago I thought everyone knew how malaria infected humans: nighttime bites of parasite-laden *Anopheles* mosquitoes. Today I know better. Some highly intelligent residents of malaria-plagued communities still believe that an evil spirit or certain foods cause the illness, a fact that underscores yet another pressing need: better malaria education. Nevertheless, long before Ronald Ross and Giovanni Batista Grassi learned in the late 19th century that mosquitoes transmit malaria, savvy humans were devising ways to elude mosquito bites. Writing almost five centuries before the common era, Herodotus described in *The Histories* how Egyptians living in marshy lowlands protected themselves with fishing nets: "Every man has a net which he uses in the daytime for fishing, but at night he finds another use for it: he drapes it over the bed. . . . Mosquitoes can bite through any cover or linen blanket . . . but they do not even try to bite through the net at all." Based on this passage, some bed-net advocates view nets steeped in fish oil as the world's earliest repellent-impregnated cloth.

It was not until World War II, however, when American forces in the South Pacific dipped nets and hammocks in 5 percent DDT, that insecticides and textiles were formally partnered. After public opinion swung against DDT, treating bed nets with a biodegradable class of insecticides—

the pyrethroids—was the logical next step. It proved a breakthrough. The first major use of pyrethroid-treated nets paired with antimalarial drugs, reported in 1991, halved mortality in children younger than five in the Gambia, and later trials, without the drugs, in Ghana, Kenya and Burkina Faso confirmed a similar lifesaving trend, plus substantial health gains in pregnant women. Moreover, with wide enough use, whole families and communities benefited from the nets—even people who did not sleep under them.

But insecticide-treated bed nets also have drawbacks. They work only if malaria mosquitoes bite indoors during sleeping hours—a behavior that is not universal. Nets make sleepers hot, discouraging use. Until recently, when PermaNet and Olyset—two long-lasting pyrethroid-impregnated nets—became available, nets had to be redipped every six to 12 months to remain effective. Finally, at $2 to $6 each, nets with or without insecticide are simply unaffordable for many people. A recent study in Kenya found that only 21 percent of households had even one bed net, of which 6 percent were insecticide-treated. A summary of 34 surveys conducted between 1999 and 2004 reached an even more depressing conclusion: a mere 3 percent of African youngsters were protected by insecticidal nets, although reports on the ground now suggest that use is quickly rising.

Insecticide resistance could also undermine nets as a long-term solution: mosquitoes genetically capable of inactivating pyrethroids have now surfaced in several locales, including Kenya and southern Africa, and some anophelines are taking longer to succumb to pyrethroids, a worrisome adaptive behavior known as knockdown resistance. Because precious few new insecticides intended for public health use are in sight (largely because of paltry economic incentives to develop them), one solution is rotating other agricultural insecticides on nets. Decoding the olfactory clues that attract mosquitoes to humans in the first place is another avenue of research that could yield dividends in new repellents. (Ironically, a change in body odor when *P. falciparum* parasites are present in the blood may also attract mosquito bites; according to a recent report, Kenyan schoolchildren harboring gametocytes—the malaria stage taken up by mosquitoes—drew twice as many bites as their uninfected counterparts.)

How about harnessing the winged creatures themselves to kill malaria parasites? In theory, genetic engineering could quell parasite multiplication before the protozoa ever left the insects' salivary glands. If such insects succeeded in displacing their natural kin in the wild, they could halt the spread of malaria parasites to people. Recently native genes hindering malaria multiplication within *Anopheles* mosquitoes have been identified, and genetically reengineered strains of several important species are now on the drawing board. Once they are reared in the laboratory, however, releasing these Trojan insects into the real world poses a whole new set of challenges, including ethical ones.

Bottom line: for the time being, old-fashioned, indoor residual spraying with DDT remains a valuable public health tool in many settings in Africa and elsewhere. Applied to surfaces, DDT is retained for six months or more. It reduces human-mosquito contact by two key mechanisms—repelling some mosquitoes before they ever enter a dwelling and killing others that perch on treated walls after feeding. A stunning example of its effectiveness surfaced in KwaZulu-Natal in 1999 and 2000. Pyrethroid-resistant *A. funestus* plus failing drugs had led to the largest number of falciparum cases there since the South African province launched its malaria-control program years ago. Re-introduction of residual spraying of DDT along with new, effective drugs yielded a 91 percent drop in cases within two years.

TREATING THE SICK

Antimosquito measures alone cannot win the war against malaria—better drugs and health services are also needed for the millions of youngsters and adults who, every year, still walk the malaria tightrope far from medical care. Some are entrusted to village herbalists and itinerant quacks. Others take pills of unknown manufacture, quality or efficacy (including counterfeits) bought by family members or neighbors from unregulated sources. In Africa, 70 percent of antimalarials come from the informal private sector—in other words, small, roadside vendors as opposed to licensed clinics or pharmacies.

Despite plummeting efficacy, chloroquine, at pennies per course, remains the top-selling antimalarial pharmaceutical downed by Africans. The next most affordable drug in Africa is sulfadoxine-pyrimethamine, an antibiotic that interferes with folic acid synthesis by the parasite. Unfortunately, *P. falciparum* strains in Africa and elsewhere are also sidestepping this compound as they acquire sequential mutations that will ultimately render the drug useless.

Given the looming specter of drug resistance, can lessons from other infectious diseases guide future strategies to beef up malaria drug therapy? In recent decades resistant strains of the agents responsible for tuberculosis, leprosy and HIV/AIDS triggered a switch to two- and three-drug regimens, which then helped to forestall further emergence of "superbugs." Now most experts believe that multidrug treatments can also combat drug resistance in falciparum malaria, especially if they include a form of *Artemisia annua,* a medicinal herb once used as a generic fever remedy in ancient China. *Artemisia*-derived drugs (collectively termed "artemisinins") beat back malaria parasites more quickly than any other treatment does and also block transmission from humans to mosquitoes. Because of these unequaled advantages, combining them with other effective antimalarial drugs in an effort to prevent or delay artemisinin resistance makes sense, not just for Africa's but for the entire world's sake. After all, there is no guarantee malaria will not return someday to its former haunts. We know it can victimize global travelers. In recent years *P. falciparum*-infected mosquitoes have even stowed away on international flights, infecting innocent bystanders within a few miles of airports, far from malaria's natural milieu.

Yet there is a hitch to the new combination remedies: their costs—currently 10 to 20 times higher than Africa's more familiar but increasingly impotent malaria drugs—are hugely daunting to most malaria victims and to heavily affected countries. Even if the new cocktails were more modest in price, the global supply of artemisinins is well below needed levels and requires donor dollars to jump-start the 18-month production cycle to grow, harvest and process the plants. Novartis, the first producer formally sanctioned by the WHO to manufacture a co-formulated artemisinin combination treatment (artemether plus lumefantrine), may not have enough funding and raw material to ship even a portion of the 120 million treatments it once hoped to deliver in 2006.

The good news? Cheaper, synthetic drugs that retain the distinctive chemistry of plant-based artemisinins (a peroxide bond embedded in a chemical ring) are on the horizon, possibly within five to 10 years. One prototype originating from research done in the late 1990s entered human trials in 2004. Another promising tactic that could bypass botanical extraction or chemical synthesis altogether is splicing *A. annua's* genes and yeast genes into *Escherichia coli,* then coaxing pharmaceuticals out of the bacterial brew. The approach was pioneered by researchers at the University of California, Berkeley.

Preventing, as opposed to treating, malaria in highly vulnerable hosts—primarily African children and pregnant women—is also gaining adherents. In the 1960s low-dose antimalarial prophylaxis given to pregnant Nigerians was found, for the first time, to increase their newborns' birthweight. Currently this approach has been superseded by a full course of sulfadoxine-pyrimethamine taken several times during pregnancy, infancy and, increasingly, childhood immunization visits. Right now the recipe works well in reducing infections and anemia, but once resistance truly blankets Africa, the question is, What preventive treatment will replace sulfadoxine-pyrimethamine? Although single-dose artemisinins might seem the logical answer at first blush, these agents are not suitable for prevention, because their levels in blood diminish so quickly. And repeated dosing of artemisinins in asymptomatic women and children—an untested practice so far—could also yield unsuspected side effects. In an ideal world, prevention equals vaccine.

WHERE WE STAND ON VACCINES

There is no doubt that creating malaria vaccines that deliver long-lasting protection has proved more difficult than scientists first imagined, although progress has occurred over several decades. At the root of the dilemma is malaria's intricate life cycle, which encompasses several stages in mosquitoes and humans; a vaccine effective in killing one stage may not inhibit the growth of another. A second challenge is malaria's complex genetic makeup: of the 5,300 proteins encoded by P. falciparum's genome, fewer than 10 percent trigger protective responses in naturally exposed individuals—the question is, Which ones? On top of that, several arms of the human immune system—antibodies, lymphocytes and the spleen, for starters—must work together to achieve an ideal response to malaria vaccination. Even in healthy people, much less populations already beset with malaria and other diseases, such responses do not always develop.

So far most experimental P. falciparum vaccines have targeted only one of malaria's three biological stages—sporozoite, merozoite or gametocyte [see illustration], although multistage vaccines, which could well prove more effective in the end, are also planned. Some of the earliest insights on attacking sporozoites (the parasite stage usually inoculated into humans through the mosquito's proboscis) came in the 1970s, when investigators at the University of Maryland found that x-ray-weakened falciparum sporozoites protected human volunteers, albeit only briefly. Presumably, the vaccine worked by inducing the immune system to neutralize naturally entering parasites before they escaped an hour later to their next way station, the liver.

The demonstration that antibodies artificially elicited against sporozoites could help fend off malaria prompted further work. Three decades later, in 2004, efforts bore fruit when a sporozoite vaccine more than

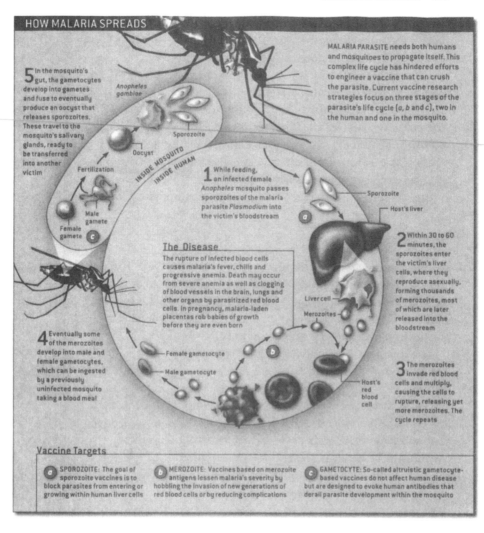

HOW MALARIA SPREADS

MALARIA PARASITE needs both humans and mosquitoes to propagate itself. This complex life cycle has hindered efforts to engineer a vaccine that can crush the parasite. Current vaccine research strategies focus on three stages of the parasite's life cycle (*a, b* and *c*), two in the human and one in the mosquito.

Anopheles gambiae

5 In the mosquito's gut, the gametocytes develop into gametes and fuse to eventually produce an oocyst that releases sporozoites. These travel to the mosquito's salivary glands, ready to be transferred into another victim

Sporozoite

Oocyst

Fertilization

INSIDE MOSQUITO
INSIDE HUMAN

Male gamete

Female gamete **c**

1 While feeding, an infected female *Anopheles* mosquito passes sporozoites of the malaria parasite *Plasmodium* into the victim's bloodstream **a**

Sporozoite

Host's liver

2 Within 30 to 60 minutes, the sporozoites enter the victim's liver cells, where they reproduce asexually, forming thousands of merozoites, most of which are later released into the bloodstream

The Disease

The rupture of infected blood cells causes malaria's fever, chills and progressive anemia. Death may occur from severe anemia as well as clogging of blood vessels in the brain, lungs and other organs by parasitized red blood cells. In pregnancy, malaria-laden placentas rob babies of growth before they are even born

Liver cell

Merozoites

Host's red blood cell

3 The merozoites invade red blood cells and multiply, causing the cells to rupture, releasing yet more merozoites. The cycle repeats

4 Eventually some of the merozoites develop into male and female gametocytes, which can be ingested by a previously uninfected mosquito taking a blood meal

Female gametocyte

Male gametocyte

Vaccine Targets

a SPOROZOITE: The goal of sporozoite vaccines is to block parasites from entering or growing within human liver cells

b MEROZOITE: Vaccines based on merozoite antigens lessen malaria's severity by hobbling the invasion of new generations of red blood cells or by reducing complications

c GAMETOCYTE: So-called altruistic gametocyte-based vaccines do not affect human disease but are designed to evoke human antibodies that derail parasite development within the mosquito

halved serious episodes of malaria in 2,000 rural Mozambican children between the ages of one and four, the years when African children are most susceptible to dying from the disease. The formula used in this clinical trial (the most promising to date) included multiple copies of a *P. falciparum* sporozoite protein fragment fused to a hepatitis B viral protein added for extra potency. Even so, subjects required three separate immunizations, and the period of protection was short (only six months). Realistically, the earliest that an improved version of the vaccine known as RTS,S (or any of its roughly three dozen vaccine brethren currently in clinical development) might come to market is in 10 years, at a final price tag that leaves even Big Pharma gasping for air. Because of the anticipated costs, public-private partnerships such as the Seattle-based Malaria Vaccine Initiative are now helping to fund ongoing trials.

There is just one more thing to keep in mind about malaria vaccines. Even when they do become available—with any luck, sooner rather than later—effective treatments and antimosquito strategies will still be needed. Why? First of all, because rates of protection will never reach anywhere near 100 percent in those who actually receive the vaccines. Other malaria-prone individuals, especially the rural African poor, may not have access to the shots at all. Therefore, at least for the foreseeable future, all preventive and salvage measures must remain in the arsenal.

INVESTING IN MALARIA

Once again the world is coming to terms with the truth about malaria: the ancient enemy still claims at least one million lives every year while, at the same time, imposing tremendous physical, mental and economic hardships. Given our current tools and even more promising weapons on the horizon, the time has come to fight back.

The past decade has already witnessed significant milestones. In 1998 the WHO and the World Bank established the Roll Back Malaria partnership. In 2000 the G8 named malaria as one of three pandemics they hoped to curb, if not vanquish. The United Nations subsequently created the Global Fund to Fight AIDS, Tuberculosis and Malaria and pledged to halt and reverse the rising tide of malaria within 15 years. In 2005 the World Bank declared a renewed assault on malaria, and President George W. Bush announced a $1.2-billion package to fight malaria in Africa over five years, using insecticide-treated nets, indoor spraying of insecticides and combination drug treatments. More recently, the World Bank has begun looking for ways to subsidize artemisinin combination treatments. As this issue of *Scientific American* went to press, the Bill and Melinda Gates Foundation announced three grants totaling $258.3 million to support advanced development of a malaria vaccine, new drugs and improved mosquito-control methods.

Despite these positive steps, the dollars at hand are simply not equal to the task. Simultaneously with the announcement from the Gates Foundation, a major new analysis of global malaria research and development funding noted that only $323 million was spent in 2004. This amount falls far short of the projected $3.2 billion a year needed to cut malaria deaths in half by 2010. Perhaps it is time to mobilize not only experts and field-workers but ordinary folk. At roughly $5, the price of a lunch in the U.S. could go a long way toward purchasing an insecticide-treated bed net or a three-day course of artemisinin combination treatment for an African child.

In considering their potential return on investment, readers might also recall a small boy with scars on his cheeks who made it through malaria's minefield, then devoted his adult life to battling disease. Decades from now, how many other children thus spared might accomplish, equally wondrous feats?

FURTHER READING

WHAT THE WORLD NEEDS NOW IS DDT. Tina Rosenberg in *New York Times Magazine*, pages 38–43; April 11, 2004.
Medicines for Malaria Venture:www.mmv.org/
World Health Organization, Roll Back Malaria Department: www.who.int/malaria

Attacking Anthrax

JOHN A. T. YOUNG AND R. JOHN COLLIER

ORIGINALLY PUBLISHED IN MARCH 2002

The need for new anthrax therapies became all too clear last fall When five people died of inhalation anthrax, victims of the first purposeful release of anthrax spores in the U.S. Within days of showing initially unalarming symptoms, the patients were gone, despite intensive treatment with antibiotics. Six others became seriously ill as well before pulling through.

Fortunately, our laboratories and others began studying the causative bacterium, *Bacillus anthracis,* and seeking antidotes long before fall 2001. Recent findings are now pointing the way to novel medicines and improved vaccines. Indeed, in the past year alone, the two of us and our collaborators have reported on three promising drug prototypes.

AN ELUSIVE KILLER

The new ideas for fighting anthrax have emerged from ongoing research into how *B. anthracis* causes disease and death. Anthrax does not spread from individual to individual. A person (or animal) gets sick only after incredibly hardy spores enter the body through a cut in the skin, through contaminated food or through spore-laden air. Inside the body the spores molt into "vegetative," or actively dividing, cells.

Anthrax bacteria that colonize the skin or digestive tract initially do damage locally and may cause self-limited ailments: black sores and swelling in the first instance; possibly vomiting and abdominal pain and bleeding in the second. If bacterial growth persists unchecked in the skin or gastrointestinal tract, however, the microbes may eventually invade the bloodstream and thereby cause systemic disease.

Inhaled spores that reach deep into the lungs tend to waste little time where they land. They typically convert to the vegetative form and travel quickly to lymph nodes in the middle of the chest, where many of the cells find ready access to the blood. (Meanwhile bacteria that remain in the chest set the stage for a breath-robbing buildup of fluid around the lungs.)

Extensive replication in the blood is generally what kills patients who succumb to anthrax. *B. anthracis's* ability to expand so successfully derives from its secretion of two substances, known as virulence factors, that can profoundly derail the immune defenses meant to keep bacterial growth in check. One of these factors encases the vegetative cells in a polymer capsule that inhibits ingestion by the immune system's macrophages and neutrophils—the scavenger cells that normally degrade disease-causing bacteria. The capsule's partner in crime is an extraordinary toxin that works its way into those scavenger cells, or phagocytes, and interferes with their usual bacteria-killing actions.

The anthrax toxin, which also enters other cells, is thought to contribute to mortal illness not only by dampening immune responses but also by playing a direct role. Evidence for this view includes the observation that the toxin alone, in the absence of bacteria, can kill animals. Conversely, inducing the immune system to neutralize the toxin prevents *B. anthracis* from causing disease.

TERRIBLE TOXIN

Harry Smith and his co-workers at the Microbiological Research Establishment in Wiltshire, England, discovered the toxin in the 1950s. Aware of its central part in anthrax's lethality, many researchers have since focused on learning how the substance "intoxicates" cells—gets into them and disrupts their activities. Such details offer essential clues to blocking its effects. Stephen H. Leppla and Arthur M. Friedlander, while at the U.S. Army Medical Research Institute of Infectious Diseases, initiated that effort with their colleagues in the 1980s; the two of us and others took up the task somewhat later.

The toxin turns out to consist of three proteins: protective antigen, edema factor and lethal factor. These proteins cooperate but are not always joined together physically. They are harmless individually until they attach to and enter cells, which they accomplish in a highly orchestrated fashion.

First, protective antigen binds to the surface of a cell, where an enzyme trims off its outermost tip. Next, seven of those trimmed molecules combine to form a ring-shaped structure, or heptamer, that captures the two factors and is transported to an internal membrane-bound compartment called an endosome. Mild acidity in this compartment causes the heptamer to change shape in a way that leads to the transport of edema factor and lethal factor across the endosomal membrane into the cytosol (the internal matrix of cells), where they do their mischief. In essence, the heptamer is like a syringe loaded with edema factor and lethal factor, and the slight acidity of the endosome causes the syringe to pierce the membrane of the endosome and inject the toxic factors into the cytosol.

Edema factor and lethal factor catalyze different molecular reactions in cells. Edema factor upsets the controls on ion and water flow across cell membranes and thereby promotes the swelling of tissues. In phagocytes, it also saps energy that would otherwise be used to engulf bacteria.

The precise behavior of lethal factor, which could be more important in causing patient deaths, is less clear. Scientists do know that it is a protease (a protein-cutting enzyme) and that it cleaves enzymes in a family known as MAPKKs. Now they are trying to tease out the molecular events that follow such cleavage and to uncover the factor's specific contributions to disease and death.

THERAPEUTIC TACTICS

Certainly drugs able to neutralize the anthrax toxin would help the immune system fight bacterial multiplication and would probably reduce a patient's risk of dying. At the moment, antibiotics given to victims of inhalation anthrax may control microbial expansion but leave the toxin free to wreak havoc.

In principle, toxin activity could be halted by interfering with any of the steps in the intoxication process. An attractive approach would stop the sequence almost before it starts, by preventing protective antigen from attaching to cells. Scientists realized almost 10 years ago that this protein initiated toxin entry by binding to some specific protein on the surface of cells; when cells were treated with enzymes that removed all their surface proteins, protective antigen found no footing. Until very recently, though, no one knew which of the countless proteins on cells served as the crucial receptor.

The two of us, with our colleagues Kenneth Bradley, Jeremy Mogridge and Michael Mourez, found the receptor last summer. Detailed analysis of this molecule (now named ATR, for anthrax toxin receptor) then revealed that it spans the cell membrane and protrudes from it. The protruding part contains an area resembling a region that serves in other receptors as an attachment site for particular proteins. This discovery suggested that the area was the place where protective antigen latched onto ATR, and indeed it is.

We have not yet learned the normal function of the receptor, which surely did not evolve specifically to allow the anthrax toxin into cells. Nevertheless, knowledge of the molecule's makeup is enabling us to begin testing inhibitors of its activity. We have had success, for instance, with a compound called sATR, which is a soluble form of the receptor domain that binds to protective antigen. When sATR molecules are mixed into the medium surrounding cells, they serve as effective decoys, tricking protective antigen into binding to them instead of to its true receptor on cells.

We are now trying to produce sATR in the amounts needed for evaluating its ability to combat anthrax in rodents and nonhuman primates—experiments that must be done before any new drug can be considered for fighting anthrax in people. Other groups are examining whether carefully engineered antibodies (highly specific molecules of the immune system) might bind tightly to protective antigen in ways that will keep it from coupling with its receptor.

MORE TARGETS

Scientists are also seeking ways to forestall later steps in the intoxication pathway. For example, a team from Harvard has constructed a drug able to clog the regions of the heptamer that grasp edema and lethal factors. The group—from the laboratories of one of us (Collier) and George M. Whitesides—reasoned that a plugged heptamer would be unable to draw the factors into cells.

We began by screening randomly constructed peptides (short chains of amino acids) to see if any of them bound to the heptamer. One did, so we examined its ability to block toxin activity. It worked, but weakly. Assuming that fitting many plugs into the heptamer's binding domains for edema and lethal factor would be more effective, we took advantage of chemical procedures devised by Whiteside's group and linked an average of 22 copies of the peptide to a flexible polymer. That construction showed itself to be a strong inhibitor of toxin action—more than 7,000 times better than the free peptide—both in cell cultures and in rats.

Another exciting agent, and the one probably closest to human testing, would alter the heptamer itself. This compound was discovered after Bret R. Sellman in Collier's group noted that when certain mutant forms of protective antigen were mixed with normal forms, the heptamers formed on cells as usual but were unable to inject edema and lethal factors into the cytosol. Remarkably, some of these mutants were so disruptive that a single copy in a heptamer completely prevented injection.

In a study reported last April, these mutants—known as dominant negative inhibitors, or DNIs—proved to be potent blockers of the anthrax toxin in cell cultures and in rats. Relatively small amounts of selected DNIs neutralized an amount of protective antigen and lethal factor that would otherwise kill a rat in 90 minutes. These findings suggest that each mutant copy of protective antigen is capable of inactivating six normal copies in the bloodstream and that it would probably reduce toxin activity in patients dramatically.

Of course, as more and more questions about the toxin are answered, scientists should discover further treatment ideas. Now that die receptor for protective antigen has been identified, researchers can use it as a target in screening tests aimed at finding drugs able to bar the receptor from binding to protective antigen. And understanding of the receptor's three-dimensional structure would reveal the precise contact points between protective antigen and the receptor, enabling drugmakers to custom-design receptor blocking agents.

Scientists would also like to uncover the molecular interactions that enable protective antigen heptamers to move from the cell surface into endosomes inside the cell. Impeding that migration should be very useful. And what happens after lethal factor cleaves MAPKK enzymes? How do those subsequent events affect cells? Although the latter question remains a vexing challenge, recent study of lethal factor has brightened the prospects for finding drugs able to inactivate it. Last November, Robert C. Liddington of the Burnham Institute in La Jolla, Calif., and his colleagues in several laboratories published the three-dimensional structure of the part of lethal factor that acts on MAPKK molecules. That site can now become a target for drug screening or design.

New leads for drugs should also emerge from the recent sequencing of the code letters composing the B. anthracis genome. By finding genes that resemble those of known functions in other organisms, biologists are likely to discover additional information about how the anthrax bacterium causes disease and how to stop it.

The continuing research should yield several antitoxins. To be most effective, such drugs will probably be used with antibiotics, much as cocktails of antiviral drugs are recommended for treating HIV infection.

PROMISING PREVENTIVES

As plans to improve therapies proceed, so does work on better vaccines. Vaccines against toxin-producing bacteria often prime the immune system to neutralize the toxin of concern as soon as it appears in the body, thus preventing disease. Livestock in parts of the U. S. receive preparations consisting of B. *anthracis* cells that lack the protective capsule and thus replicate poorly. A similar vaccine for humans has been used in the former Soviet Union. But preparations that contain whole microbes often cause side effects, and they raise the specter that renegade cells might at times give rise to the very diseases they were meant to prevent.

The only anthrax vaccine approved for human use in the U. S. takes a different form. It consists primarily of toxin molecules that have been chemically treated to prevent them from making people ill. It is produced by growing the weakened strain of B. *anthracis* in culture, filtering the bacterial cells from the culture medium, adsorbing the toxin proteins in the remaining filtrate onto an adjuvant (a substance that enhances immune responses) and treating the mixture with formaldehyde to inactivate the proteins. Injection of this preparation, known as AVA (for anthrax vaccine adsorbed), stimulates the immune system to produce antibodies that specifically bind to and inactivate the toxin's components. Most of the antibodies act on protective antigen, however, which explains the protein's name: it is the component that best elicits protective immunity.

AVA is given to soldiers and certain civilians but is problematic as a tool for shielding the general public against biological warfare. Supplies are limited. And even if AVA were available in abundance, it would be cumbersome to deliver on a large scale; the standard protocol calls for six shots delivered over 18 months followed by annual boosters. The vaccine has not been licensed for use in people already exposed to anthrax spores. But late last year officials, worried that spores might sometimes survive in the lungs for a long time, began offering an abbreviated, three-course dose on an experimental basis to postal workers and others who had already taken 60 days of precautionary antibiotics. People who accepted the offer were obliged to take antibiotics for an additional 40 days, after which the immunity stimulated by the vaccine would presumably be strong enough to provide adequate protection on its own.

In hopes of producing a more powerful, less cumbersome and faster-acting vaccine, many investigators are focusing on developing inoculants composed primarily of protective antigen produced by recombinant DNA technology. By coupling the recombinant protein with a potent new-generation adjuvant, scientists may be able to evoke good protective immunity relatively quickly with only one or two injections. The dominant negative inhibitors discussed earlier as possible treatments could be useful forms of protective antigen to choose. Those molecules retain their ability to elicit immune responses. Hence, they could do double duty: disarming the anthrax toxin in the short run while building up immunity that will persist later on.

We have no doubt that the expanding research on the biology of B. *anthracis* and on possible therapies and vaccines will one day provide a range of effective anthrax treatments. We fervently hope that these efforts will mean that nobody will have to die from anthrax acquired either naturally or as a result of biological terrorism.

FURTHER READING

ANTHRAX AS A BIOLOGICAL WEAPON: MEDICAL AND PUBLIC HEALTH MANAGEMENT. Thomas V. Inglesby et al. in *Journal of the American Medical Association*, Vol. 281, No. 18, pages 1735–1745; May 12,1999.

DOMINANT-NEGATIVE MUTANTS OF A TOXIN SUBUNIT: AN APPROACH TO THERAPY OF ANTHRAX. Bret R. Sellman, Michael Mourez and R. John Collier in *Science*, Vol. 292, pages 695–697; April 27, 2001.

DESIGNING A POLYVALENT INHIBITOR OF ANTHRAX TOXIN. Michael Mourez et al. in *Nature Biotechnology*, Vol. 19, pages 958–961; October 2001.

IDENTIFICATION OF THE CELLULAR RECEPTOR FOR ANTHRAX TOXIN. Kenneth A. Bradley, Jeremy Mogridge, Michael Mourez, R. John Collier and John A. T. Young in *Nature*, Vol. 414, pages 225–229; November 8, 2001.

The U.S. Centers for Disease Control and Prevention maintain a Web site devoted to anthrax at www.cdc.gov/ncidod/dbmd/diseaseinfo/anthrax_g.htm

The Prion Diseases

STANLEY B. PRUSINER

ORIGINALLY PUBLISHED IN JANUARY 1995

Fifteen years ago I evoked a good deal of skepticism when I proposed that the infectious agents causing certain degenerative disorders of the central nervous system in animals and, more rarely, in humans might consist of protein and nothing else. At the time, the notion was heretical. Dogma held that the conveyers of transmissible diseases required genetic material, composed of nucleic acid (DNA or RNA), in order to establish an infection in a host. Even viruses, among the simplest microbes, rely on such material to direct synthesis of the proteins needed for survival and replication.

Later, many scientists were similarly dubious when my colleagues and I suggested that these "proteinaceous infectious particles"—or "prions," as I called the disease-causing agents—could underlie inherited, as well as communicable, diseases. Such dual behavior was then unknown to medical science. And we met resistance again when we concluded that prions (pronounced "preeons") multiply in an incredible way; they convert normal protein molecules into dangerous ones simply by inducing the benign molecules to change their shape.

Today, however, a wealth of experimental and clinical data has made a convincing case that we are correct on all three counts. Prions are indeed responsible for transmissible and inherited disorders of protein conformation. They can also cause sporadic disease, in which neither transmission between individuals nor inheritance is evident. Moreover, there are hints that the prions causing the diseases explored thus far may not be the only ones. Prions made of rather different proteins may contribute to other neurodegenerative diseases that are quite prevalent in humans. They might even participate in illnesses that attack muscles.

The known prion diseases, all fatal, are sometimes referred to as spongiform encephalopathies. They are so named because they frequently cause the brain to become riddled with holes. These ills, which can brew for years (or even for decades in humans) are widespread in animals.

The most common form is scrapie, found in sheep and goats. Afflicted animals lose coordination and eventually become so incapacitated that they cannot stand. They also become irritable and, in some cases, develop an intense itch that leads them to scrape off their wool or hair (hence the name "scrapie"). The other prion diseases of animals go by such names as transmissible mink encephalopathy, chronic wasting disease of mule deer and elk, feline spongiform encephalopathy and bovine spongiform encephalopathy. The last, often called mad cow disease, is the most worrisome.

Gerald A. H. Wells and John W. Wile smith of the Central Veterinary Laboratory in Weybridge, England, identified the condition in 1986, after it began striking cows in Great Britain, causing them to became uncoordinated and unusually apprehensive. The source of the emerging epidemic was soon traced to a food supplement that included meat and bone meal from dead sheep. The boro of the NIH Rocky Mountain Laboratories made his own probes and established that mouse cells harbor the gene as well. That work made it possible to isolate the gene and to establish that it resides not in prions but in the chromosomes of hamsters, mice, humans and all other mammals that have been examined. What is more, most of the time, these animals make PrP without getting sick.

One interpretation of such findings was that we had made a terrible mistake: PrP had nothing to do with prion diseases. Another possibility was that PrP could be produced in two forms, one that generated disease and one that did not. We soon showed the latter interpretation to be correct.

The critical clue was the fact that the PrP found in infected brains resisted breakdown by cellular enzymes called proteases. Most proteins in cells are degraded fairly easily. I therefore suspected that if a normal, non-threatening form of PrP existed, it too would be susceptible to degradation. Ronald A. Barry in my laboratory then identified this hypothetical protease-sensitive form. It thus became dear that scrapie-causing PrP is a variant of a normal protein.We therefore called the normal protein "cellular PrP" and the infectious (protease–resistant) form "scrapie PrP." The latter term is now used to refer to the protein molecules that constitute the prions causing all scrapielike diseases of animals and humans.

PRION DISEASES CAN BE INHERITED

Early on we had hoped to use the PrP gene to generate pure copies of PrP. Next, we would inject the protein molecules into animals, secure in the knowledge that no elusive virus was clinging to them. If the injections caused scrapie in the animals, we would have shown that protein molecules could, as we had proposed, transmit disease. By 1986, however, we knew the plan would not work. For one thing, it proved very difficult to induce the gene to make the high levels of PrP needed for conducting studies. For another thing, the protein that was produced was the normal, cellular form. Fortunately, work on a different problem led us to an alternative approach for demonstrating that prions could transmit scrapie without the help of any accompanying nucleic acid.

In many cases, the scrapielike illnesses of humans seemed to occur without having been spread from one host to another, and in some families they appeared to be inherited. (Today researchers know that about 10 percent of human prion diseases are familial, felling half of the members of the affected families.) It was this last pattern that drew our attention. Could it be that prions were more unusual than we originally thought? Were they responsible for the appearance of both hereditary and transmissible illnesses?

In 1988 Karen Hsiao in my laboratory and I uncovered some of the earliest data showing that human prion diseases can certainly be inherited. We acquired clones of a PrP gene obtained from a man who had Gerstmann-Sträussler-Scheinker disease in his family and was dying of it himself. Then we compared his gene with PrP genes obtained from a healthy population and found a tiny abnormality known as a point mutation.

To grasp the nature of this mutation, it helps to know something about the organization of genes. Genes consist of two strands of the DNA building blocks called nucleotides, which differ from one another in the bases they carry. The bases on one strand combine with the bases on the other strand to form base pairs: the "rungs" on the familiar DNA "ladder." In addition to holding the DNA ladder together, these pairs spell out the sequence of amino adds that must be strung together to make a particular protein. Three base pairs together—a unit called a codon—specify a single amino acid. In our dying patient, just one base pair (out of more than 750) had been exchanged for a different pair. The change, in turn, had altered the information carried by codon 102, causing the amino add leucine to be substituted for the amino acid proline in the man's PrP protein.

With the help of Tim J. Crow of Northwick Park Hospital in London and Jurg Ott of Columbia University and their colleagues, we discovered the same mutation in genes from a large number of patients with Gerstmann-Sträussler-Scheinker disease, and we showed that the high incidence in the affected families was statistically significant. In other words, we established genetic linkage between the mutation and the disease—a finding that strongly implies the mutation is the cause. Over the past six years work by many investigators has uncovered 18 mutations in families with inherited prion diseases; for five of these mutations, enough cases have now been collected to demonstrate genetic linkage.

The discovery of mutations gave us a way to eliminate the possibility that a nucleic acid was traveling with prion proteins and directing their multiplication. We could now create genetically altered mice carrying a mutated PrP gene. If the presence of the altered gene in these "transgenic" animals led by itself methods for processing sheep carcasses had been changed in the late 1970s. Where once they would have eliminated the scrapie agent in the supplement, now they apparently did not. The British government banned the use of animal-derived feed supplements in 1988, and the epidemic has probably peaked. Nevertheless, many people continue to worry that they will eventually fall ill as a result of having consumed tainted meat.

The human prion diseases are more obscure. Kuru has been seen only among the Fore highlanders of Papua New Guinea. They call it the "laughing death." Vincent Zigas of the Australian Public Health Service and D. Carleton Gajdusek of the U.S. National Institutes of Health described it in 1957, noting that many highlanders became afflicted with a strange, fatal disease marked by loss of coordination (ataxia) and often later by dementia. The affected individuals probably acquired kuru through ritual cannibalism: the Fore tribe reportedly honored the dead by eating their brains. The practice has since stopped, and kuru has virtually disappeared.

Creutzfeldt-Jakob disease, in contrast, occurs worldwide and usually becomes evident as dementia. Most of the time it appears sporadically, striking one person in a million, typically around age 60. About 10 to 15 percent of cases are inherited, and a small number are, sadly, iatrogenic— spread inadvertently by the attempt to treat some other medical problem. Iatrogenic Creutzfeldt-Jakob disease has apparently been transmitted by corneal transplantation, implantation of dura mater or electrodes in the brain, use of contaminated surgical instruments, and injection of growth hormone derived from human pituitaries (before recombinant growth hormone became available).

The two remaining human disorders are Gerstmann-Sträussler-Scheinker disease (which is manifest as ataxia and other signs of damage to the cerebellum) and fatal familial insomnia (in which dementia follows difficulty sleeping). Both these conditions are usually inherited and typically appear in mid-life. Fatal familial insomnia was discovered

DISEASE	TYPICAL SYMPTOMS	ROUTE OF ACQUISITION	DISTRIBUTION	SPAN OF OVERT ILLNESS
Kuru	Loss of coordination, often followed by dementia	Infection (probably through cannibalism, which stopped by 1958)	Known only in highlands of Papua New Guinea; some 2,600 cases have been identified since 1957	Three months to one year
Creutzfeldt-Jakob disease	Dementia, followed by loss of coordination, although sometimes the sequence is reversed	Usually unknown (in "sporadic" disease) Sometimes (in 10 to 15 percent of cases) inheritance of a mutation in the gene coding for the prion protein (PrP) Rarely, infection (as an inadvertent consequence of a medical procedure)	*Sporadic form:* 1 person per million worldwide *Inherited form:* some 100 extended families have been identified *Infectious form:* about 80 cases have been identified	Typically about one year; range is one month to more than 10 years
Gerstmann-Sträussler-Scheinker disease	Loss of coordination, often followed by dementia	Inheritance of a mutation in the PrP gene	Some 50 extended families have been identified	Typically two to six years
Fatal familial insomnia	Trouble sleeping and disturbance of autonomic nervous system, followed by insomnia and dementia	Inheritance of a mutation in the PrP gene	Nine extended families have been identified	Typically about one year

Prion diseases of humans, which may incubate for 30 years or more, can all cause progressive decline in cognition and motor function; hence, the distinctions among them are sometimes blurry. As the genetic mutations underlying familial forms of the diseases are found, those disorders are likely to be identified by their associated mutations alone.

only recently, by Elio Lugaresi and Rossella Medori of the University of Bologna and Pierluigi Gambetti of Case Western Reserve University.

IN SEARCH OF THE CAUSE

I first became intrigued by the prion diseases in 1972, when as a resident in neurology at the University of California School of Medicine at San Francisco, I lost a patient to Creutzfeldt-Jakob disease. As I reviewed the scientific literature on that and related conditions, I learned that scrapie, Creutzfeldt-Jakob disease and kuru had all been shown to be transmissible by injecting extracts of diseased brains into the brains of healthy animals. The infections were thought to be caused by a slow-acting virus, yet no one had managed to isolate the culprit.

In the course of reading, I came across an astonishing report in which Tikvah Alper and her colleagues at the Hammersmith Hospital in London suggested that the scrapie agent might lack nucleic acid, which usually can be degraded by ultraviolet or ionizing radiation. When the nucleic acid in extracts of scrapie-infected brains was presumably destroyed by those treatments, the extracts retained their ability to transmit scrapie. If the organism did lack DNA and RNA, the finding would mean that it was not a virus or any other known type of infectious agent, all of which contain genetic material. What, then, was it? Investigators had many ideas— including, jokingly, linoleum and kryptonite—but no hard answers.

I immediately began trying to solve this mystery when I set up a laboratory at U.C.S.F. in 1974. The first step had to be a mechanical one—purifying the infectious material in scrapie-infected brains so that its composition could be analyzed. The task was daunting; many investigators had tried and failed in the past. But with the optimism of youth, I forged ahead [see "Prions," by Stanley B. Prusiner; Scientific American, October 1984]. By 1982 my colleagues and I had made good progress, producing extracts of hamster brains consisting almost exclusively of infectious material. We had, furthermore, subjected the extracts to a range of tests designed to reveal the composition of the disease-causing component.

AMAZING DISCOVERY

All our results pointed toward one startling conclusion: the infectious agent in scrapie (and presumably in the related diseases) did indeed lack nucleic acid and consisted mainly, if not exclusively, of protein. We deduced that DNA and RNA were absent because, like Alper, we saw that procedures known to damage nucleic acid did not reduced infectivity. And we knew protein was an essential component because procedures that denature (unfold) or degrade protein reduced infectivity. I thus introduced the term "prion" to distinguish this class of disease conveyer from viruses, bacteria, fungi and other known pathogens. Not long afterward, we determined that scrapie prions contained a single protein that we called PrP, for "prion protein."

Now the major question became, Where did the instructions specifying the sequence of amino acids in PrP reside? Were they carried by an undetected piece of DNA that traveled with PrP, or were they, perhaps, contained in a gene housed in the chromosomes of cells? The key to this riddle was the identification in 1984 of some 15 amino acids at one end of the PrP protein. My group identified this short amino acid sequence in collaboration with Leroy E. Hood and his co-workers at the California Institute of Technology.

Knowledge of the sequence allowed us and others to construct molecular probes, or detectors, able to indicate whether mammalian cells carried the PrP gene. With probes produced by Hood's team, Bruno Oesch, working in the laboratory of Charles Weissmann at the University of Zurich, showed that hamster cells do contain a gene for PrP. At about the same time, Bruce

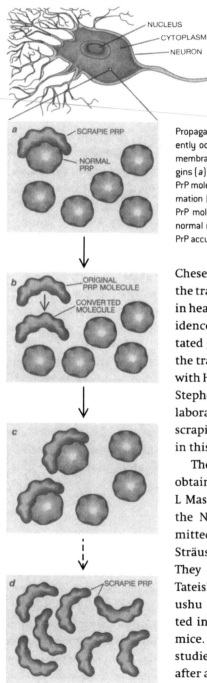

NUCLEUS
CYTOPLASM
NEURON

a SCRAPIE PRP
NORMAL PRP

b ORIGINAL PRP MOLECULE
CONVERTED MOLECULE

c

d SCRAPIE PRP

Propagation of scrapie PrP in neurons of the brain apparently occurs by a kind of domino effect on some internal membrane. A favored hypothesis holds that the process begins (*a*) when one molecule of scrapie PrP contacts a normal PrP molecule and induces it to refold into the scrapie conformation (*b*). Then the scrapie particles attack other normal PrP molecules (*c*). Those molecules, in turn, attack other normal molecules, and so on (*broken arrow*), until scrapie PrP accumulates to dangerous levels (*d*).

Chese to scrapie, and if the brain tissue of the transgenic animals then caused scrapie in healthy animals, we would have solid evidence that the protein encoded by the mutated gene had been solely responsible for the transfer of disease. Studies I conducted with Hsiao, Darlene Groth in my group and Stephen J. De-Armond, head of a separate laboratory at U.C.S.F., have now shown that scrapie can be generated and transmitted in this way.

These results in animals resemble those obtained in 1981, when Gajdusek, Colin L Masters and Clarence J. Gibbs, Jr., all at the National Institutes of Health, transmitted apparently inherited Gerstmann-Sträussier-Scheinker disease to monkeys. They also resemble the findings of Jun Tateishi and Tetsuyuki Kitamoto of Kyushu University in Japan, who transmitted inherited Creutzfeldt-Jakob disease to mice. Together the collected transmission studies persuasively argue that prions do, after all, represent an unprecedented class of infectious agents, composed only of a modified mammalian protein. And the conclusion is strengthened by the fact that assiduous searching for a scrapie-specific nucleic acid (especially by Detlev H. Riesner of Heinrich Heine University in Düsseldorf) has produced no evidence that such genetic material is attached to prions.

Scientists who continue to favor the virus theory might say that we still have not proved our case. If the PrP gene coded for a protein that, when mutated, facilitated infection by a ubiquitous virus, the mutation would lead to viral infection of the brain. Then injection of brain extracts from the mutant animal would spread the infection to another host. Yet in the absence of any evidence of a virus, this hypothesis looks to be untenable.

In addition to showing that a protein can multiply and cause disease without help from nucleic acids, we have gained insight into how scrapie PrP propagates in cells. Many details remain to be worked out, but one aspect appears quite clear: the main difference between normal PrP and scrapie PrP is conformational. Evidently, the scrapie protein propagates itself by contacting normal PrP molecules and somehow causing them to unfold and flip from their usual conformation to the scrapie shape. This change initiates a cascade in which newly converted molecules change the shape of other normal PrP molecules, and so on. These events apparently occur on a membrane in the cell interior.

We started to think that the differences between cellular and scrapie forms of PrP must be conformational after other possibilities began to seem unlikely. For instance, it has long been known that the infectious form often has the same amino acid sequence as the normal type. Of course, molecules that start off being identical can later be chemically modified in ways that alter their activity. But intensive investigations by Neil Stahl and Michael A. Baldwin in my laboratory have turned up no differences of this kind.

ONE PROTEIN, TWO SHAPES

How, exactly, do the structures of normal and scrapie forms of PrP differ? Studies by Keh-Ming Pan in our group indicate that the normal protein consists primarily of alpha helices, regions in which the protein backbone twists into a specific kind of spiral; the scrapie form, however, contains beta strands, regions in which the backbone is fully extended. Collections of these strands form beta sheets. Fred E. Cohen, who directs another laboratory at U.C.S.F., has used molecular modeling to try to predict the structure of the normal protein based on its amino acid sequence. His calculations imply that the protein probably folds into a compact structure having four helices in its core. Less is known about the structure, or structures, adopted by scrapie PrP.

The evidence supporting the proposition that scrapie PrP can induce an alpha-helical PrP molecule to switch to a beta-sheet form comes primarily from two important studies by investigators in my group. Maria Gasset learned that synthetic peptides (short strings of amino acids) corresponding to three of the four putative alpha-helical regions of PrP can fold into beta sheets. And Jack Nguyen has shown that in their beta-sheet conformation, such peptides can impose a beta-sheet structure on helical PrP peptides. More recently Byron W. Caughey of the Rocky Mountain Laboratories and Peter T. Lansbury of the Massachusetts Institute of Technology have reported that cellular PrP can be converted into scrapie PrP in a test tube by mixing the two proteins together.

PrP molecules arising from mutated genes probably do not adopt the scrapie conformation as soon as they are synthesized. Otherwise, people carrying mutant genes would become sick in early childhood. We suspect that mutations in the PrP gene render the resulting proteins susceptible to flipping from an alpha-helical to a beta-sheet shape. Presumably, it takes time until one of the molecules spontaneously flips over and still more time for scrapie PrP to accumulate and damage the brain enough to cause symptoms.

Fred Cohen and I think we might be able to explain why the various mutations that have been noted in PrP genes could facilitate folding into the beta-sheet form. Many of the human mutations give rise to the substitution of one amino acid for another within the four putative helices or at their borders. Insertion of incorrect amino acids at those positions might destabilize a helix, thus increasing the likelihood that the affected helix and its neighbors will refold into a beta-sheet conformation. Conversely, Hermann Schätzel in my laboratory finds that the harmless differences distinguishing the PrP gene of humans from those of apes and monkeys affect amino acids lying outside of the proposed helical domains—where the divergent amino acids probably would not profoundly influence the stability of the helical regions.

TREATMENT IDEAS EMERGE

No one knows exactly how propagation of scrapie PrP damages cells. In cell cultures, the conversion of normal PrP to the scrapie form occurs inside neurons, after which scrapie PrP accumulates in intracellular vesicles known as lysosomes. In the brain, filled lysosomes could conceivably burst and damage cells. As the diseased cells died, creating holes in the brain, their prions would be released to attack other cells.

We do know with certainty that cleavage of scrapie PrP is what produces PrP fragments that accumulate as plaques in the brains of some patients. Those aggregates resemble plaques seen in Alzheimer's disease, although the Alzheimer's clumps consist of a different protein. The PrP plaques are a useful sign of prion infection, but they seem not to be a major cause of impairment. In many people and animals with prion disease, the plaques do not arise at all.

Even though we do not yet know much about how PrP scrapie harms brain tissue, we can foresee that an understanding of the three-dimensional structure of the PrP protein will lead to therapies. If, for example, the four-helix-bundle model of PrP is correct, drug developers might be able to design a compound that would bind to a central pocket that could be formed by the four helices. So bound, the drug would stabilize these helices and prevent their conversion into beta sheets.

Another idea for therapy is inspired by research in which Weissmann and his colleagues applied gene-targeting technology to create mice that lacked the PrP gene and so could not make PrP. By knocking out a gene and noting the consequences of its loss, one can often deduce the usual functions of the gene's protein product. In this case, however, the animals missing PrP displayed no detectable abnormalities. If it turns out that PrP is truly inessential, then physicians might one day consider delivering so-called antisense or antigene therapies to the brains of patients with prion diseases. Such therapies aim to block genes from giving rise to unwanted proteins and could potentially shut down production of cellular PrP [see "The New Genetic Medicines," by Jack S. Cohen and Michael E. Hogan; *Scientific American*, December 1994]. They would thereby block PrP from propagating itself.

It is worth noting that the knockout mice provided a welcomed opportunity to challenge the prion hypothesis. If the animals became ill after inoculation with prions, their sickness would have indicated that prions could multiply even in the absence of a preexisting pool of PrP molecules. As I expected, inoculation with prions did not produce scrapie, and no evidence of prion replication could be detected.

The enigma of how scrapie PrP multiplies and causes disease is not the only puzzle starting to be solved. Another long-standing question—the mystery of how prions consisting of a single kind of protein can vary markedly in their effects—is beginning to be answered as well. Iain H. Pattison of the Agriculture Research Council in Compton, England, initially called attention to this phenomenon. Years ago he obtained prions from two separate sets of goats. One isolate made inoculated animals drowsy, whereas the second made them hyperactive. Similarly, it is now evident that some prions cause disease quickly, whereas others do so slowly.

THE MYSTERY OF "STRAINS"

Alan G. Dickinson, Hugh Fraser and Moira E. Bruce of the Institute for Animal Health in Edinburgh, who have examined the differential effects of varied isolates in mice, are among those who note that only pathogens containing nucleic acids are known to occur in multiple strains. Hence, they and others assert, the existence of prion "strains" indicates the prion hypothesis must be incorrect; viruses must be at the root of scrapie and its relatives. Yet because efforts to find viral nucleic acids have been unrewarding, the explanation for the differences must lie elsewhere.

One possibility is that prions can adopt multiple conformations. Folded in one way, a prion might convert normal PrP to the scrapie form highly efficiently, giving rise to short incubation times. Folded another way, it might work less efficiently. Similarly, one "conformer" might be attracted to neuronal populations in one part of the brain, whereas another might be attracted to neurons elsewhere, thus producing different symptoms. Considering that PrP can fold in at least two ways, it would not be surprising to find it can collapse into other structures as well.

Since the mid-1980s we have also sought insight into a phenomenon known as the species barrier. This concept refers to the fact that something makes it difficult for prions made by one species to cause disease in animals of another species. The cause of this difficulty is of considerable interest today because of the epidemic of mad cow disease in Britain. We and others have been trying to find out whether the species barrier is strong enough to prevent the spread of prion disease from cows to humans.

BREAKING THE BARRIER

The barrier was discovered by Pattison, who in the 1960s found it hard to transmit scrapie between sheep and rodents. To determine the cause of the trouble my colleague Michael R. Scott and I later generated transgenic mice expressing the PrP gene of the Syrian hamster—that is, making the hamster PrP protein. The mouse gene differs from that of the hamster gene at 16 codons out of 254. Normal mice inoculated with hamster prions rarely acquire scrapie, but the transgenic mice became ill within about two months.

We thus concluded that we had broken the species barrier by inserting the hamster genes into the mice. Moreover, on the basis of this and other experiments, we realized that the barrier resides in the amino acid sequence of PrP: the more the sequence of a scrapie PrP molecule resembles the PrP sequence of its host, the more likely it is that the host will acquire prion disease. In one of those other experiments, for example, we examined transgenic mice carrying the Syrian hamster PrP gene in addition to their own mouse gene. Those mice make normal forms of both hamster and mouse PrP. When we inoculated the animals with mouse prions, they made more mouse prions. When we inoculated them with hamster prions, they made hamster prions. From this behavior, we learned that prions preferentially interact with cellular PrP of homologous, or like, composition.

The attraction of scrapie PrP for cellular PrP having the same sequence probably explains why scrapie managed to spread to cows in England from food consisting of sheep tissue: sheep and bovine PrP differ only at seven positions. In contrast, the sequence difference between human and bovine PrP is large: the molecules diverge at more than 30 positions. Because the variance is great, the likelihood of transmission from cows to people would seem to be low. Consistent with this assessment are epidemiological studies by W. Bryan Matthews, a professor emeritus at the University of Oxford. Matthews found no link between scrapie in sheep and the occurrence of Creutzfeldt-Jakob disease in sheep-farming countries.

On the other hand, two farmers who had "mad cows" in their herds have recently died of Creutzfeldt-Jakob disease. Their deaths may have nothing to do with the bovine epidemic, but the situation bears watching. It may turn out that certain parts of the PrP molecule are more important than others for breaking the species barrier. If that is the case, and if cow PrP closely resembles human PrP in the critical regions, then the likelihood of danger might turn out to be higher than a simple comparison of the complete amino acid sequences would suggest.

We began to consider the possibility that some parts of the PrP molecule might be particularly important to the species barrier after a study related to this blockade took an odd turn. My colleague Glenn C. Telling had created transgenic mice carrying a hybrid PrP gene that consisted of human codes flanked on either side by mouse codes; this gene gave rise to a hybrid protein. Then he introduced brain tissue from patients who had died of Creutzfeldt-Jakob disease or Gerstmann-Sträussler-Scheinker disease into the transgenic animals. Oddly enough, the animals became ill much more frequently and faster than did mice carrying a full human PrP gene, which diverges from mouse PrP at 28 positions. This outcome implied that similarity in the central region of the PrP molecule may be more critical than it is in the other segments.

The result also lent support to earlier indications—uncovered by Shu-Lian Yang in DeArmond's laboratory and Albert Taraboulos in my group—that molecules made by the host can influence the behavior of scrapie PrP. We speculate that in the hybrid-gene study, a mouse protein, possibly a "chaperone" normally involved in folding nascent protein chains, recognized one of the two mouse-derived regions of the hybrid PrP protein. This chaperone bound to that region and helped to refold the hybrid molecule into the scrapie conformation. The chaperone did not provide similar help in mice making a totally human PrP protein, presumably because the human protein lacked a binding site for the mouse factor.

THE LIST MAY GROW

An unforeseen story has recently emerged from studies of transgenic mice making unusually high amounts of normal PrP proteins. DeArmond, David Westaway in our group and George A. Carlson of the McLaughlin Laboratory in Great Falls, Mont., became perplexed when they noted that some older transgenic mice developed an illness characterized by rigidity and diminished grooming. When we pursued the cause, we found that making excessive amounts of PrP can eventually lead to neurodegeneration and, surprisingly, to destruction of both muscles and peripheral nerves. These discoveries widen the spectrum of prion diseases and are prompting a search for human prion diseases that affect the peripheral nervous system and muscles.

Investigations of animals that overproduce PrP have yielded another benefit as well. They offer a clue as to how the sporadic form of Creutzfeldt-Jakob disease might arise. For a time I suspected that sporadic disease might begin when the wear and tear of living led to a mutation of the PrP gene in at least one cell in the body. Eventually, the mutated protein might switch to the scrapie form and gradually propagate itself, until the buildup of scrapie PrP crossed the threshold to overt disease. The mouse studies suggest that at some point in the lives of the one in a million individuals who acquire sporadic Creutzfeldt-Jakob disease, cellular PrP may spontaneously convert to the scrapie form. The experiments also raise the possibility that people who become afflicted with sporadic Creutzfeldt-Jakob disease overproduce PrP, but we do not yet know if, in fact, they do.

All the known prion diseases in humans have now been modeled in mice. With our most recent work we have inadvertently developed an animal model for sporadic prion disease. Mice inoculated with brain extracts from scrapie-infected animals and from humans afflicted with Creutzfeldt-Jakob disease have long provided a model for the infectious forms of prion disorders. And the inherited prion diseases have been modeled in transgenic mice carrying mutant PrP genes. These murine representations of the human prion afflictions should not only extend understanding of how prions cause brain degeneration, they should also create opportunities to evaluate therapies for these devastating maladies.

STRIKING SIMILARITIES

Ongoing research may also help determine whether prions consisting of other proteins play a part in more common neurodegenerative conditions, including Alzheimer's disease, Parkinson's disease and amyotrophic lateral sclerosis. There are some marked similarities in all these disorders. As is true of the known prion diseases, the more widespread ills mostly occur sporadically but sometimes "run" in families. All are also usually diseases of middle to later life and are marked by similar pathology: neurons degenerate, protein deposits, can accumulate as plaques, and glial cells (which support and nourish nerve cells) grow larger in reaction to damage to neurons. Strikingly, in none of these disorders do white blood cells—those ever present warriors of the immune system—infiltrate the brain. If a virus were involved in these illnesses, white cells would be expected to appear.

Recent findings in yeast encourage speculation that prions unrelated in amino acid sequence to the PrP protein could exist. Reed B. Wicknor of the NIH reports that a protein called Ure2p might sometimes change its conformation, thereby affecting its activity in the cell. In one shape, the protein is active; in the other, it is silent.

The collected studies described here argue persuasively that the prion is an entirely new class of infectious pathogen and that prion diseases result from aberrations of protein conformation. Whether changes in protein shape are responsible for common neurodegenerative diseases, such as Alzheimer's, remains unknown, but it is a possibility that should not be ignored.

FURTHER READING

SCRAPIE DISEASE IN SHEEP. Herbert B. Parry. Edited by D. R. Oppenheimer. Academic Press, 1983.

MOLECULAR BIOLOGY OF PRION DISEASES. S. B. Prusiner in *Science*, Vol. 252, pages 1515–1522; June 14, 1991.

PRION DISEASES OF HUMANS AND ANIMALS. Edited by S. B. Prusiner, J. Collinge, J. Powell and B. Anderton. Ellis Horwood, 1992.

FATAL FAMILIAL INSOMNIA: INHERITED PRION DISEASES, SLEEP, AND THE THALAMUS. Edited by C. Guilleminault et al. Raven Press, 1994.

MOLECULAR BIOLOGY OF PRION DISEASES. Special issue of *Philosophical Transactions of the Royal Society of London, Series B*, Vol. 343, No. 1306; March 29, 1994.

STRUCTURAL CLUES TO PRION REPLICATION. F. E. Cohen, K.-M. Pan, Z. Huang, M. Baldwin, R. J. Fletterick and S. B. Prusiner in *Science*, Vol. 264, pages 530–531; April 22, 1994.

Detecting Mad Cow Disease

STANLEY B. PRUSINER

ORIGINALLY PUBLISHED IN JULY 2004

Last December mad cow disease made its U.S debut when federal officials announced that a holstein from Mabton, Wash., had been stricken with what is formally known as bovine spongiform encephalopathy (BSE). The news kept scientists, government officials, the cattle industry and the media scrambling for information well past New Year's. Yet the discovery of the sick animal came as no surprise to many of us who study mad cow disease and related fatal disorders that devastate the brain. The strange nature of the prion—the pathogen at the root of these conditions—made us realize long ago that controlling these illnesses and ensuring the safety of the food supply would be difficult.

As researchers learn more about the challenges posed by prions—which can incubate without symptoms for years, even decades—they uncover strategies that could better forestall epidemics. Key among these tools are highly sensitive tests, some available and some under development, that can detect prions even in asymptomatic individuals; currently BSE is diagnosed only after an animal has died naturally or been slaughtered. Researchers have also made some headway in treating a human prion disorder called Creutzfeldt-Jakob disease (CJD), which today is uniformly fatal.

IDENTIFYING THE CAUSE

Although mass concern over mad cow disease is new in the U.S., scientific efforts to understand and combat it and related disorders began heating up some time ago. In the early 1980s I proposed that the infectious pathogen causing scrapie (the sheep analogue of BSE) and CJD consists only of a protein, which I termed the prion. The prion theory was greeted with great skepticism in most quarters and with outright disdain in others, as it ran counter to the conventional wisdom that pathogens capable of reproducing must contain DNA or RNA [see "Prions," by Stanley B. Prusiner; *Scientific American*, October 1984]. The doubt I encountered was healthy and important, because most dramatically novel ideas are eventually shown to be incorrect. Nevertheless, the prion concept prevailed.

In the years since my proposal, investigators have made substantial progress in deciphering this fascinating protein. We know that in addition to causing scrapie and CJD, prions cause other spongiform encephalopathies, including BSE and chronic wasting disease in deer and elk [see "The Prion Diseases," by Stanley B. Prusiner; pp. 155–169 in this volume]. But perhaps the most startling finding has been that the prion protein, or PrP, is not always bad. In fact, all animals studied so far have a gene that codes for PrP. The normal form of the protein, now called PrP^c (C for cellular), appears predominantly in nerve cells and may help maintain neuronal functioning. But the protein can twist into an abnormal, disease-causing shape, denoted PrP^{Sc} (Sc for scrapie, the prion disease that until recently was the most studied).

Unlike the normal version, PrP^{Sc} tends to form difficult-to-dissolve clumps that resist heat, radiation and chemicals that would kill other pathogens. A few minutes of boiling wipes out bacteria, viruses and molds, but not PrP^{Sc}. This molecule makes more of itself by converting normal prion proteins into abnormal forms: PrP^{Sc} can induce PrP^c to refold and become PrP^{Sc}. Cells have the ability to break down and eliminate misfolded proteins, but if they have difficulty clearing PrP^{Sc} faster than it forms, PrP^{Sc} builds up, ruptures cells and creates the characteristic pathology of these diseases—namely, masses of protein and microscopic holes in the brain, which begins to resemble a sponge. Disease symptoms appear as a result.

Prion diseases can afflict people and animals in various ways. Most often the diseases are "sporadic"—that is, they happen spontaneously for no apparent reason. Sporadic CJD is the most common prion disease among humans, striking approximately one in a million, mostly older, people. Prion diseases may also result from a mutation in the gene that codes for the prion protein; many families are known to pass on CJD and two other disorders, Gerstmann-Sträussler-Scheinker disease and fatal insomnia. To date, researchers have uncovered more than 30 different mutations in the PrP gene that lead to the hereditary forms of the sickness—all of which are rare, occurring about once in every 10 million people. Finally, prion disease may result from an infection, through, for instance, the consumption of bovine prions.

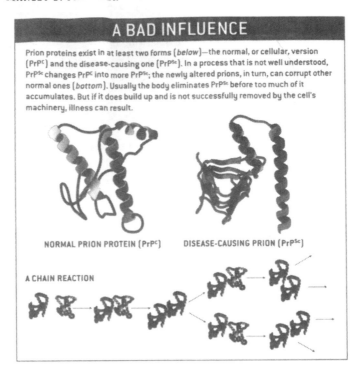

A BAD INFLUENCE

Prion proteins exist in at least two forms (below)—the normal, or cellular, version (PrP^C) and the disease-causing one (PrP^{Sc}). In a process that is not well understood, PrP^{Sc} changes PrP^C into more PrP^{Sc}; the newly altered prions, in turn, can corrupt other normal ones (bottom). Usually the body eliminates PrP^{Sc} before too much of it accumulates. But if it does build up and is not successfully removed by the cell's machinery, illness can result.

NORMAL PRION PROTEIN (PrP^C)   DISEASE-CAUSING PRION (PrP^{Sc})

A CHAIN REACTION

TRACING THE MAD COW EPIDEMIC

The world awoke to the dangers of prion disease in cows after the BSE outbreak that began ravaging the British beef industry in the mid-1980s. The truly novel concepts emerging from prion science forced researchers and society to think in unusual ways and made coping with the epidemic difficult. Investigators eventually learned that prions were being transmitred to cattle through meat-and-bone meal, a dietary supplement prepared from the parts of sheep, cattle, pigs and chickens that are processed, or rendered, for industrial use. High heat eliminated conventional pathogens, but PrP^{Sc} survived and went on to infect cattle.

As infected cattle became food for other cattle, BSE began appearing throughout the U.K. cattle population, reaching a high of 37,280 identified cases in 1992. The British authorities instituted some feed bans beginning in 1989, but it was not until 1996 that a strict ban on cannibalistic feeding finally brought BSE under control in the U.K.; the country saw 612 cases last year. Overall the U.K. has identified about 180,000 mad cows, and epidemiological models suggest that another 1.9 million were infected but undetected.

For many people, the regulations came too late. Despite the British government's early assurances to the contrary, mad cow disease proved transmissible to humans. In March 1996 Robert Will, James Ironside and Jeanne Bell, who were working in the National CJD Surveillance Unit in Edinburgh, reported that 11 British teenagers and young adults had died of a variant of Creutzfeldt-Jakob disease (vCJD). In these young patients the patterns of PrP^{Sc} deposition in the brain differed markedly from that found in typical CJD patients.

Many scientists, including myself, were initially dubious of the presumed link between BSE and vCJD. I eventually changed my mind, under the weight of many studies. One of these was conducted by my colleagues at the University of California at San Francisco, Michael Scott and Stephen DeArmond, who collected data in mice genetically engineered to resemble cattle, at least from a prion protein point of view (the PrP gene from cattle was inserted into the mouse genome). These mice became ill approximately nine months after receiving injections of prions from cattle with BSE or people with vCJD, and the resulting disease looked the same whether the prions originated from cows or vCJD patients.

As of February 2004, 146 people have been diagnosed with vCJD in the U.K. and another 10 elsewhere. No one knows exactly how many other people are incubating prions that cause vCJD. Epidemiological models suggest that only a few dozen more individuals will develop vCJD, but these models are based on assumptions that may prove wrong. One assumption, for example, is that vCJD affects only those with a particular genetic makeup. Because prions incubate for so long, it will take some time before we know the ultimate number of vCJD cases and whether they share similar genetics.

In vCJD, PrP^{Sc} builds up, not just in the brain but also in the lymphoid system, such as the tonsils and appendix, suggesting that PrP^{Sc} enters the bloodstream at some point. Animal studies have shown that prions can be transmitted to healthy animals through blood transfusions from infected animals. In response to this information, many nations have enacted stricter blood donation rules. In the U.K., people born after 1996, when the tough feed ban came into force, can receive blood only from overseas (those born before are considered already exposed). In the U.S., those who spent three months or more in the U.K. between 1980 and 1996 cannot give blood.

Although such restrictions have contributed to periodic blood shortages, the measures appear justified. Last December the U.K. announced the vCJD death of one of 15 individuals who received transfusions from donors who later developed vCJD. The victim received the transfusion seven and a half years before his death. It is possible that he became infected through prion-tainted food, but his age argues against that: at 69, he was much older than the 29 years of typical vCJD patients. Thus, it seems fairly likely that vCJD is not limited to those who have eaten prion-infected beef.

Since the detection of mad cow disease in the U.K., two dozen other nations have uncovered cases. Canada and the U.S. are the latest entrants. On May 20, 2003, Canadian officials reported BSE in an eight-and-a-half-year-old cow that had spent its life in Alberta and Saskatchewan. (The country's only previous mad cow had arrived as a U.K. import 10 years earlier.) Although the animal had been slaughtered in January 2003, slow processing meant that officials did not test the cow remains until April. By then, the carcass had been turned into pet food exported to the U.S.

Seven months later, on December 23, the U.S. Department of Agriculture announced the country's first case of BSE, in Washington State. The six-and-a-half-year-old dairy cow had entered the U.S. at the age of four. The discovery means that U.S. officials can no longer labor under the misconception that the nation is free of BSE. like Canada, U.S. agricultural interests want the BSE problem to disappear. Financial woes stem primarily from reduced beef exports: 58 other countries are keeping their borders shut, and a $3-billion export market has largely evaporated.

A WORLDVIEW

Many countries have reported cattle afflicted with BSE, but cases of human infection thought to stem from eating meat from sick cows—variant Creutzfeldt-Jakob disease (vCJD)—remain low, for now at least.

COUNTRY	BSE CASES	vCJD DEATHS (current cases)
Austria	1	0
Belgium	125	0
Canada	2	1
Czech Republic	9	0
Denmark	13	0
Falkland Islands	1	0
Finland	1	0
France	891	6
Germany	312	0
Greece	1	0
Hong Kong	0	1*
Ireland	1,353	1
Israel	1	0
Italy	117	1
Japan	11	0
Lichtenstein	2	0
Luxembourg	2	0
Netherlands	75	0
Oman	2	0
Poland	14	0
Portugal	875	0
Slovakia	15	0
Slovenia	4	0
Spain	412	0
Switzerland	453	0
U.S.	1	0 (1)†
U.K.	183,803	141 (5)
Worldwide		151 (6)

*Awaiting confirmation †British subject

DESIGNING DIAGNOSTICS

The most straightforward way to provide this assurance—both for foreign nations and at home—appears to be simple: just test the animals being slaughtered for food and then stop the infected ones from entering the food supply, where they could transmit pathogenic prions to humans. But testing is not easy. The USDA uses immunohistochemistry, an old technique that is cumbersome and extremely time-consuming (taking a few days to complete) and so is impractical for universal application.

Accordingly, others and I have been working on alternatives. In the mid-1980s researchers at my lab and elsewhere produced new kinds of antibodies that can help identify dangerous prions in the brain more efficiently. These antibodies, similar to those used in the standard test, recognize any prion—normal or otherwise. To detect PrP^{Sc}, we need to first remove any trace of PrP^{c}, which is done by applying a protease (protein-degrading enzyme) to a brain sample. Because PrP^{Sc} is generally resistant to the actions of proteases, much of it remains intact. Antibodies then added to the sample will reveal the amount of PrP^{Sc} present. Using a similar approach, a handful of companies, including Prionics in Switzerland and Bio-Rad in France, have developed their own antibodies and commercial kits. The results can be obtained in a few hours, which is why such kits are proving useful in the mass screenings under way in Europe and Japan. (Japan discovered its first case of BSE in 2001 and by this past April had reported 11 infected cows in total.)

These rapid tests, however, have limitations. They depend on PrP^{Sc} accumulating to detectable amounts—quite often, relatively high levels—in an animal's brain. Yet because BSE often takes three to five years to develop, most slaughterage cattle, which tend to be younger than two years, usually do not test positive, even if they are infected. Therefore, these tests are generally most reliable for older bovines, regardless of whether they look healthy or are "downers." At the moment, downer cattle, which cannot stand on their own, are the group most likely to be tested.

Until new regulations came out in January, the U.S. annually sent roughly 200,000 downers to slaughter for human consumption. Of these, only a fraction were tested. Over the next year, however, the USDA will examine at least 200,000 cows for BSE. (Whether milk can be affected remains open; my laboratory is testing milk from BSE-infected cattle.)

Because of the limitations of the existing tests, developing one that is able to detect prions in the bulk of the beef supply—that is, in asymptomatic young animals destined for slaughter—continues to be one of the most important weapons in confronting prion disease outbreaks.

Scientists have pursued several strategies. One tries to boost the amount of PrPS^{Sc} in a sample so the prions are less likely to escape detection. If such an amplification system can be created, it might prove useful in developing a blood test for use in place of ones that require an animal's

TESTING FOR MAD COW

Four kinds of tests are now used to detect dangerous prions (PrP^{Sc}) in brain tissue from dead cattle. By identifying infected animals, public health officials and farmers can remove them from the food supply. Some of these tests, however, are time-consuming and expensive, so researchers are working to develop the ideal diagnostic: one that could quickly detect even tiny amounts of PrP^{Sc} in blood and urine and thus could work for live animals and people. The hope is to forestall outbreaks by catching infection as early as possible and, eventually, to treat infection before it progresses to disease.

BIOASSAY

The bioassay can take up to 36 months to provide results and can be very expensive. Its advantage is that it can reveal particular strains of prions as well as how infectious the sample is based on the time it takes for the test animal to become sick.

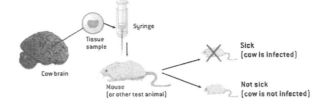

Cow brain · Tissue sample · Syringe · Mouse (or other test animal) · Sick (cow is infected) · Not sick (cow is not infected)

IMMUNOHISTOCHEMISTRY

The first test used to specifically detect prions, immunohistochemistry is considered the gold standard that other tests must meet. But because technicians must examine each slide, the process is very time-consuming, often taking as many as seven days, and is not useful for mass screenings.

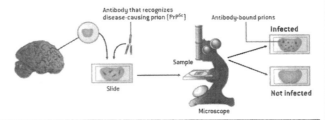

Antibody that recognizes disease-causing prion (PrP^{Sc}) · Slide · Sample · Microscope · Antibody-bound prions · Infected · Not infected

IMMUNOASSAY

Many companies produce immunoassays, and these rapid tests are now in widespread use in Europe; they have just been introduced to the U.S. Results can be had within eight hours, and hundreds of samples can be run simultaneously. These tests work well only with high levels of PrP^{Sc}.

Infection is determined by adding marked antibodies that recognize prions or by sorting proteins by weight on a gel, where darks bands indicate the presence of PrP^{Sc}.

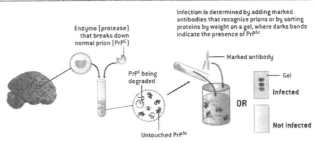

Enzyme (protease) that breaks down normal prion (PrP^{C}) · PrP^{C} being degraded · Untouched PrP^{Sc} · Marked antibody · Gel · Infected · OR · Not infected

CONFORMATION-DEPENDENT IMMUNOASSAY (CDI)

This automated test can detect very low levels of PrP^{Sc} and reveal how much of the dangerous prion is present in the sample without first having to degrade PrP^{C}. It informs experts about the animal's level of infection within five hours or so. It has been approved for testing in Europe. CDI is being tested on tissue from live animals and might one day serve as a blood test.

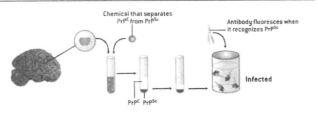

Chemical that separates PrP^{C} from PrP^{Sc} · Antibody fluoresces when it recognizes PrP^{Sc} · PrP^{C} PrP^{Sc} · Infected

sacrifice (not enough PrP^{Sc} circulates in the blood to be detectable by current methods). Claudio Soto of Serono Pharmaceuticals and his colleagues have attempted to carry this out. They mixed brain preparations from normal and scrapie-infected hamsters and then subjected the mixture to sound pulses to break apart clumps of PrP^{Sc} so it could convert the normal form of the prion protein into the rogue version. The experiment resulted in a 10-fold increase in protease-resistant prion protein. Surachai Supattapone of Dartmouth Medical School has obtained similar results.

Another strategy focuses on the intricacies of the protein shape instead of trying to bolster amounts. A test I developed with my U.C.S.F. colleague Jiri Safar, for example, is based on the ability of some antibodies to react with either PrPc or PrP^{Sc}, but not both. Specifically, the antibody targets a portion of the prion protein that is accessible in one conformation but that is tucked away in the alternative conformation, much like a fitted corner of a sheet is hidden when folded. This specificity means that test samples do not have to be subjected to proteases. Removing the protease step is important because we now know that a form of PrP^{Sc} is sensitive to the action of proteases, which means that tests that eliminate PrPc probably also eliminate most or all of the protease-sensitive PrP^{Sc}—and so these tests could well underestimate the amount of PrP^{Sc} present by as much as 90 percent.

Our test—the conformation-dependent immunoassay (CDI)—gained approval for use in Europe in 2003 and might be sensitive enough to detect PrP^{Sc} in blood. The CDI has already shown promise in screening young cows. In the fall of 2003 Japan reported two cases of BSE in cows 21 and 23 months of age. Neither animal showed outward signs of neurological dysfunction. In the case of the 23-month-old cow, two commercially available tests for PrP^{Sc} returned inconclusive, borderline-positive results, but the CDI showed that the brain stem harbored malevolent prions.

Neither of these cases would have been discovered in Europe, where only cattle older than 30 months (24 months in Germany) must be tested if they are destined for human consumption. Initially the Japanese government proposed adopting the European Union's testing protocol. But consumer advocates forced the government to change its policy and test every slaughtered animal. Given that seemingly healthy animals can potentially carry pathogenic prions, I believe that testing all slaughtered animals is the only rational policy. Until now, the tests have been inadequately sensitive. But the advent of rapid, sensitive tests means universal screening can become the norm. (I understand that this statement could seem self-serving because I have a financial interest in the company making the CDI test. But I see no other option for adequately protecting the human food supply.)

SOME NEW INSIGHTS

During our work on the CDI, we discovered a surprising fact about the development of prion disease. As alluded to above, we found that the prion is actually a collection of proteins having different degrees of resistance to protease digestion. We also learned that protease-sensitive forms of PrP^{Sc} appear long before the protease-resistant forms appear. Whether protease-sensitive PrP^{Sc} is an intermediate in the formation of protease-resistant PrP^{Sc} remains to be determined. Regardless, a test that could identify protease-sensitive forms should be able to detect infection before symptoms appear, so the food supply can have maximum protection and infected patients can be assisted as early as possible. Fortunately, by using the CDI, my colleagues and I have been able to detect low levels of the protease-sensitive forms of PrP^{Sc} in the blood of rodents and humans.

Hunting for prions in blood led us to another surprise as well. Patrick Bosque, now at the University of Colorado's Health Sciences Center, and I found prions in the hind limb muscles of mice at a level 100,000 times as high as that found in blood; other muscle groups had them, too, but at much lower levels. Michael Beekes and his colleagues at the Robert Koch Institute in Berlin discovered PrP^{Sc} in virtually all muscles after they fed prions to their hamsters, although they report high levels of prions in all muscles, not just the hind limbs. (We do not know why our results differ or why the hind limbs might be more prone to supporting prions than other muscles are.) These findings were not observed in rodents exclusively but in human patients as well. U.C.S.F. scientists Safar and DeArmond found PrP^{Sc} in the muscles of some CJD patients, and Adriano Aguzzi and his colleagues at the University of Zurich identified PrP^{Sc} in the muscles of 25 percent of the CJD patients they examined.

Of course, the ideal way to test for prions would be a noninvasive method, such as a urine test. Unfortunately, so far the only promising lead—discovery of protease-resistant PrP in urine—could not be confirmed in later studies.

NOVEL THERAPIES

Although new diagnostics will improve the safety of the food and blood supply, they will undoubtedly distress people who learn that they have a fatal disease. Therefore, many investigators are looking at ways to block prion formation or to boost a cell's ability to clear existing prions. So far researchers have identified more than 20 compounds that can either inhibit prion formation or enhance prion clearance in cultured cells. Several compounds have been shown to extend the lives of mice or hamsters when administered around the time they were inoculated with prions, but none have been shown to alter the course of disease when administered well after the initial infection occurred. Furthermore, many of these agents require high doses to exert their effects, suggesting that they would be toxic in animals.

Beyond the problem of potential toxicity that high doses might entail lies the challenge of finding drugs that can cross the blood-brain barrier and travel from the bloodstream into brain tissue. Carsten Korth, now at Heinrich Heine University in Dusseldorf, and I—and, independently, Katsumi Doh-ura of Kyushu University in Japan and Byron Caughey of the National Institute of Allergy and Infectious Diseases—have found that certain drugs known to act in the brain, such as thorazine (used in the treatment of schizophrenia), inhibit prion formation in cultured cells. Another compound, quinacrine, an antimalarial drug with a structure that resembles thorazine, is approximately 10 times as powerful.

Quinacrine has shown some efficacy in animals. My co-workers and I administered quinacrine to mice, starting 60 days after we injected prions into their brain, and found that the incubation time (from the moment of infection to the manifestation of disease) was prolonged nearly 20 percent compared with untreated animals. Such an extension might be quite significant for humans with prion diseases if they could be made to tolerate the high levels of quinacrine needed or if more potent relatives of the drug could be made. My U.C.S.F. colleagues Barnaby May and Fred E. Cohen are pursuing the potency problem. In cell cultures, they have boosted the effectiveness of quinacrine 10-fold by joining two of its molecules together.

Another therapeutic approach involves the use of antibodies that inhibit PrP^{Sc} formation cultured cells. Several teams have had some success using this strategy. In mice inoculated with prions in the gut and then given antibodies directed against prion proteins, the incubation period was prolonged. So far, however, only a few patients have received anti-prion drugs. Quinacrine has been administered orally to patients with vCJD and to individuals who have the sporadic or genetic forms of prion disease. It has not cured them, but it may have slowed the progression of disease; we await further evidence.

Physicians have also administered pentosan polysulfate to vCJD patients. Generally prescribed to treat a bladder condition, the molecule is highly charged and is unlikely to cross the blood-brain barrier, so it has been injected directly into a ventricle of the brain. The drug has apparently slowed the progression of vCJD in one young man, but it seems unlikely that it will diffuse throughout the brain because similarly charged drugs—administered in the same way—have not.

A controlled clinical trial is needed before any assessment of efficacy can be made for quinacrine and other antiprion drugs. Even an initial clinical trial may prove to be insufficient because we have no information about how delivery of die drug should be scheduled. For example, many cancer drugs must be given episodically, where the patient alternates periods on and off the drug, to minimize toxicity.

Although the road to a successful treatment seems long, we have promising candidates and strategies that have brought us much further along than we were just five years ago. Investigators are also hopeful that when a successful therapy for prion disease is developed, it will suggest effective therapies for more common neurodegenerative diseases, including Alzheimer's, Parkinson's and amyotrophic lateral sclerosis (ALS). Aberrant, aggregated proteins feature in all these diseases, and so lessons learned from prions may be applicable to, them as well.

FURTHER READING

PRIONS. Stanley B. Prusiner. 1997 Nobel Prize lecture. Available from the Nobel Foundation site at: www.nobel.se/medicine/laureates/1997/index.html

PRION BIOLOGY AND DISEASES. Second edition. Edited by Stanley B. Prusiner. Cold Spring Harbor Laboratory Press, 2004.

ADVANCING PRION SCIENCE: GUIDANCE FOR THE NATIONAL PRION RESEARCH PROGRAM. Edited by Rick Erdtmann and Laura B. Sivitz. National Academy Press, 2004.

THE IMMUNE SYSTEM

Infectious Diseases and the Immune System

WILLIAM E. PAUL

ORIGINALLY PUBLISHED IN SEPTEMBER 1993

Throughout the world infectious diseases have always been the leading cause of human death. Malaria, tuberculosis, infectious diarrhea and many other illnesses still exact an awful toll in suffering and mortality, particularly in the developing world. For a time, it was widely assumed that infectious diseases had been brought under control in at least the industrialized nations. Yet the appearance of AIDS and the recent resurgence of tuberculosis, including the evolution of strains resistant to many drugs, vividly illustrate that the monster was not slain but merely asleep.

Notwithstanding those grim truths, the power of the immune system to deal with infection is remarkable—particularly when enhanced by modern vaccine technology. Because of a concerted global vaccination effort, smallpox has been completely eradicated: the last case from a natural infection occurred in Somalia in 1977. There is hope for a similar success in the control of polio. The World Health Organization has set a goal of eradicating polio by the year 2000; the disease may have already been eliminated from the Western Hemisphere. Those triumphs underscore the need to use the vaccines we now possess to their full effectiveness and to develop vaccines against those diseases that still remain great public health problems.

Perhaps the key to further success lies in a keener appreciation of how the immune system responds to infectious agents. The highly sophisticated immunologic defenses seen in humans and other higher organisms were shaped through evolution by the perpetual struggle between diverse, extremely mutable microorganisms and their hosts. The struggle is reenacted within each individual: a person's immune armament meets innumerable challenges in a life time, foiling countless opponents, and a fatal infection often represents the only unqualified loss in a generally victorious campaign.

An individual's immune response can be tailored to the challenges that person encounters because its mechanisms, like its foes, are diverse and specific. Lymphocytes can detect invading organisms because they are equipped with surface receptor molecules; the genes for those receptors can be shuffled and varied to produce structures that match virtually any foreign substance. In addition, the varied cells that make up the immune system specialize in their functions. These specializations endow vertebrates with the capacity to recognize and eliminate (or at least control) microorganisms that establish themselves in different microenvironments within the body. The complexity of the immune response—the very feature that gives it extraordinary flexibility and clout—poses daunting challenges to those who would decipher it.

With that in mind, it is instructive to describe the current understanding of the immune response in relation to various types of infection. In a broad sense, each of the immune system's components appears to be directed against agents that infect one niche in the human body. The immunologic proteins called antibodies, for example, are especially effective at destroying bacteria that live outside human cells—in, say, the blood or the fluid surrounding lung cells. The white blood cells known as CD4 T lymphocytes are of central importance in defeating the bacteria and other parasites that live within cells, especially those found in the organelles that are their pathway of entry into the cells. Another class of white blood cells, the CD8 T lymphocytes, routs pathogens such as viruses that associate even more intimately with the cellular machinery of the host.

Of course, this view of the immune system is a vast oversimplification; the most protective immune responses involve all the system's components acting in concert. In practice, each part may be involved, directly or indirectly, in repelling almost any type of infection. The CD4 T cells, for example, are often loosely called helper T cells because they secrete substances that amplify and control virtually all aspects of immunity. Nevertheless, this simplified scheme does afford insights into the principal responses in the battle against pathogens.

A look at the infection that causes pneumococcal pneumonia highlights the protective importance of antibodies. When the bacterium *Streptococcus pneumoniae* (also referred to as the pneumococcus) enters the lungs, it colonizes the space in the alveoli, the microscopic sacs where oxygen is transferred into the blood and carbon dioxide is removed. The pneumococcus multiplies there, causing tissue damage and inflammation that can impair breathing. If the bacterial infection goes unchecked, complications can develop, and in a substantial fraction of cases the patient will die.

Because they live outside of cells and very near the bloodstream, pneumococci would seem to be easy prey for the macrophages, neutrophils and other phagocytic cells of the immune system that scavenge the body for bacteria and debris. Instead the pneumococcal bacteria escape detection because they are surrounded by capsules of complex sugar molecules, or polysaccharides. Phagocytes cannot bind to these polysaccharides and consequently cannot ingest S. pneumoniae.

The capsular polysaccharides offer a much more attractive target, however, for the antibody-producing white blood cells called B lymphocytes. The surface membranes of B cells bristle with receptors for certain determinants of foreign substances, or antigens. Each B cell carries receptors for only one kind of antigenic determinant. But because the body produces billions of B cells, most with receptors of distinct structure and specificity, the chances are good that at least some B cells will bear receptors that can bind to the antigens presented by a microbial invader.

Invariably, then, some B cells have receptors for antigenic determinants of the polysaccharide capsule of S. pneumoniae. Because the molecular structures of the polysaccharides are repetitive, the same capsule antigens appear frequently. Consequently, many receptors on a single B cell will latch onto the capsule simultaneously, which brings them into proximity on the cell surface. Such receptor aggregation is essential for the activation of the B cells in response to polysaccharide antigens, and it elicits a potent B cell response.

Once a B cell has been activated in this way, rapid biochemical events occur within it. Enzymes called tyrosine kinases catalyze changes in intracellular signal molecules, such as some proteins associated with the receptors. Ultimately, this cascade of reactions prompts the B cells to divide and to secrete antibodies against the capsule polysaccharides. (Certain chemical signals, or cytokines, supplied by helper T cells also seem to be essential to the full mobilization of the B cell defense.)

The antibodies released by an activated B cell bind to the capsule around the pneumococci and improve the ability of phagocytes to ingest them. The precise means by which this result is achieved depends on the structural and chemical characteristics of the antibody molecules. Antibodies belonging to the immunoglobulin G (IgG) class of proteins have a region designated Fc-gamma. Macrophages and other phagocytes have receptor molecules that bind specifically to the Fc-gamma region and that, when engaged, signal the macrophage to ingest the attached particle. These antibodies thus give the phagocytes the "handholds" they need to attack the pneumococci and end the infection.

Polysaccharide-bound antibodies can also make the pneumococci more vulnerable by activating a cascade of circulating enzymes referred to as the complement system. Fragments of one of the enzymes, known as C3, can bind firmly to the bacterial surface. Receptors on the phagocytes can then recognize the C3 fragments—much as the Fc-gamma receptors bind to IgG antibodies—and augment their action against the pneumococci. Because the binding of even one antibody to the pneumococcal capsule can set off complement reactions, antibodies act as a powerful amplifying signal for inducing phagocytosis. Furthermore, researchers have recently shown that the presence of complement fragments on an antigen also markedly increases the efficiency with which B cells will be stimulated by that antigen, which means that the complement system also increases the production of antibodies. Because of the positive feedback inherent in this arrangement, the line of defense offered by antibodies is extraordinarily rapid and effective.

The antibody response is in many respects the simplest in the immune system's repertoire. It is a straightforward race between antibody production and pathogen replication, a case of "the quick or the dead." The response is well suited to fighting pneumococcal pneumonia and other infections caused by extracellular bacteria; B cells and antibodies specialize in nabbing conspicuous invaders. The resistance to infection that develops in people who have been vaccinated or previously exposed to an infectious agent also depends primarily on the antibody response.

Many microbes, however, establish infections inside cells, where antibodies and complement cannot reach them. Antibodies may have a chance to control such an invader while it is en route to its intracellular destination, but that response may not be prompt or vigorous enough to prevent the pathogen from gaining entry to cells—especially if the host has never before been exposed to the pathogen. The situation calls for different protective strategies, ones that can recognize a covert attack.

Such intracellular infections can be thought of as taking two forms. In one, the infectious microorganisms are found within the membrane-bound organelles (endosomes and lysosomes) through which they entered the cell. This behavior is typical of the bacteria that cause tuberculosis and leprosy. In the other form of infection, the microorganism gains access to the fluid part of the cell (the cytosol) and the cell nucleus. Viruses are the most common of these intracellular pathogens. The T cells provide the main defense against both types of infections, although the means by which they control or eradicate each one are quite different.

Perhaps the simplest and clearest illustration of how T cells fight intracellular infections can be drawn from studies involving the parasitic protozoa *Leishmania*. The principal diseases caused by *Leishmania*, visceral and cutaneous leishmaniasis, may be less well known to Western readers than is tuberculosis, but they are common ailments in much of the developing world. Patients who have visceral leishmaniasis, or kala-azar, as it is often called, suffer from enlarged spleens and livers and low white blood cell counts; they lose their appetites and begin to waste away. Left untreated, the disease can be fatal. Cutaneous leishmaniasis is marked by ulcerated skin lesions that generally heal but that may take a year or more to do so. In some cases, the lesions may spread to the mucous membranes of the nose and throat, with disfiguring consequences.

The primary target of the *Leishmania* parasite is the macrophage. During their routine scavenging in the bloodstream, macrophages engulf *Leishmania* organisms and package them in vacuoles. Those vacuoles fuse with others that contain proteolytic (protein-splitting) enzymes that can kill and digest most microbes. *Leishmania*, however, differentiate into a new form that not only can endure this chemical assault but actually can thrive during it. The parasites multiply inside the vacuole until the infected macrophage host can no longer sustain them all and dies.

Fortunately, the body has a method for eliminating intracellular parasites sequestered in this way. Vertebrate organisms possess sets of molecules that bind to peptides produced within the cell and bring them to the cell surface, where they can be recognized by the immune system. These major histocompatibility complex (MHC) molecules have a groove in their structure that can bind to a range of antigenic peptides, or protein fragments. There are two classes of MHC molecules; in the case of *Leishmania* infections of macrophages, it is the class II MHC molecules that pick up peptides from the microbes.

Class II MHC molecules are imported into the vacuoles containing the *Leishmania* organisms and other extracellular antigens ingested by the macrophage. The MHC molecules become loaded with peptides shed by the parasites or cleaved from them by the proteolytic enzymes. Not all the peptides present will be able to associate with the class II MHC molecules, but from an antigen as complex as *Leishmania*, at least several will. These complexes of MHC molecules and peptides then move to the macrophage's outer membrane.

Once displayed on the surface, the complexes can alert passing CD4 T cells to the presence of the intracellular infection. These T cells have receptor molecules that can recognize one particular peptide-class II MHC combination. All the receptors on a T cell are identical, as are those on B cells, but the great diversity of receptors made by the T cell population ensures that a match can be found for virtually any peptide-MHC combination. Thus, with the help of MHC molecules, T cells can recognize antigens from pathogens that hide inside cells.

This recognition event develops into an immunologic response if the macrophage also provides an additional signal to the T cell. One surface molecule that can provide this "accessory function" is B7, which macrophages and similar cells express when they become infected. B7 is recognized by a separate protein, CD28, on the T cell's surface. Interactions both between the T cell receptor and the peptide-class II complex and between B7 and CD28 are necessary for the CD4 T cells to mobilize an optimal response.

Indeed, in the absence of the accessory B7 signal, a CD4 T cell may become anergized, or inactivated, by its exposure to the antigenic peptide. Thereafter, that cell may be unable to respond to the antigen. The induction of B7 expression on macrophages displaying foreign antigens seems to be very important for eliciting protective immune responses by CD4 T cells against intracellular pathogens.

When a CD4 T cell does receive the dual signal, it releases cytokines that increase the macrophage's ability to destroy the enemy within. The most critical of these cytokines, gamma-interferon, prompts the macrophage to produce other cytokines, such as tumor necrosis factor, and chemicals such as nitric oxide and toxic forms of oxygen, which lead to the microbe's destruction.

Studies have revealed, however, that the type of response by the CD4 T cells can vary, thereby significantly altering the efficacy of the protective reaction. Much of the detailed work on *Leishmania* infections has been conducted in mice, which are susceptible to the parasite *L. major*. In most strains of mice, experimental infections with *L. major* are transitory: the animals' immune system can purge them. Their CD4 T cells, when activated by leishmanial antigens, produce gamma-interferon. Yet some inbred mice, such as those belonging to the BALB/c strain, cannot control *Leishmania* infections: instead they develop progressive lesions and eventually succumb. The reason for that failure seems to be that, when stimulated, their CD4 T cells predominantly secrete the cytokines called interleukin-4 and interleukin-10, and not gamma-interferon. The combination of those two interleukins is an especially powerful blocker of the microbe-killing activities that gamma-interferon induces.

Clearly, which cytokine the T cells "choose" to make in response to intracellular infection is critical to the course of the disease. Subsequent research has revealed details about how CD4 T cells make that decision. In general, when T cells in culture are exposed to antigenic peptides displayed on macrophages or other presenting cells, they are stimulated to develop into cells that can secrete large amounts of gamma-interferon and interleukin-2, a cytokine that prompts T cell proliferation, but little or no interleukin-4. Yet if interleukin-4 is present in the growth medium when the T cells first recognize the displayed antigens, they produce more of that cytokine instead of gamma-interferon. The choice between making interleukin-4 or gamma-interferon seems to involve a commitment by the lymphocyte. Once a cultured T cell has responded to an antigen by producing one of these cytokines, it and its progeny will not produce the other.

Those observations may partially explain why the outcomes of some infectious diseases differ from one person to the next. *Leishmania* infections in humans take divergent courses. Most people are able to control the parasite without becoming sick, but in a few the infection develops into leishmaniasis. Evidence now being collected suggests that differences in the patterns of cytokines produced by a patient's T cells may contribute to the severity of the infections.

A similar divergence seems to be at work in leprosy. In tuberculoid leprosy, the milder of the two major forms of the disease, the skin lesions contain few if any bacilli. The T cells of people who have tuberculoid leprosy vigorously produce gamma-interferon. In contrast, those with the severe lepromatous form of the disease have lesions containing vast numbers of intracellular bacilli; their immune responses are dominated by the production of interleukin-4.

Detailed investigations of such responses in tuberculosis patients have not yet been completed. It is nonetheless well known that most individuals can stave off the infections because their highly effective immune response prevents the tuberculosis bacilli from spreading beyond small lesions ringed by white blood cells. Only in a minority of patients does the disease progress and become fatal if left untreated. It is tempting to speculate that the outcomes of those infections may be determined in part by whether the response of their CD4 T cells is dominated by the release of protective gamma-interferon or of macrophage-incapacitating interleukin-4 and interleukin-10.

Insights from this work could someday be a boon to vaccine researchers. By blocking the effects of interleukin-4 at the time of inoculation, experimenters might be able to coax the antigen-challenged T cells to produce the protective gamma-interferon. Experiments have shown that if susceptible mice receive an injection of monoclonal antibodies against interleukin-4 at the onset of a *Leishmania* infection, the animals can control the spread of the parasite. The monoclonal antibodies seem to neutralize the interleukin and allow the *T* cells to differentiate into gamma-interferon producers.

Conversely, when *Leishmania* resistant strains of mice have been injected simultaneously with the parasite and interleukin-4, the animals develop more severe infections. Good vaccine strategies should therefore be aimed at maximizing the production of gamma-interferon at the time of immunization and either blocking or eliminating the action of interleukin-4. Recently the newly described cytokine interleukin-12 has been shown to increase strikingly the capacity of CD4 *T* cells to develop into gamma-interferon producers. The use of interleukin-12 in vaccines therefore deserves study.

Like many bacteria and parasites such as *Leishmania*, viruses establish infections inside the cells of the body, beyond the reach of antibodies. Unlike *Leishmania*, however, viruses live in the fluid interior of the cell and not inside a vacuole. They interact freely with many cellular components. Viruses use the protein-synthesizing apparatus of human cells, for example, to manufacture their own proteins. Consequently, the viral proteins intermingle with the normal cellular proteins instead of staying within a neat vacuole bundle and so present a less easily isolated target for the molecules of the immune system.

Despite the intimacy of this arrangement, MHC molecules in all the body's cells can still find and display peptide fragments from viruses. The process is fundamentally similar to the one that reveals *Leishmania* infection, but it has some important differences. First, the MHC molecules that can bind to peptides from the cytosol are class I proteins, which differ in structure from the class II molecules. When viral and cellular proteins are fragmented in the cytosol, transporter molecules carry them into the organelle called the rough endoplasmic reticulum. There the peptides are loaded onto the class I MHC molecules. After further processing, the peptide-class I MHC complexes are shipped to the cell's surface in secretory vesicles. Once they are inserted into the outer membrane, the complexes can be examined by *T* cells. In this case, however, the lymphocytes are CD8 *T* cells, which bear receptors, specific for the class I complexes.

When CD8 *T* lymphocytes detect antigenic peptides, they often act, directly and indirectly, to kill the infected cells. These *T* cells can destroy their infected targets by secreting perform and other proteins that disrupt the integrity of the cellular membrane. Recent work indicates that the killer *T* cells may also act by producing molecules that elicit a form of cell death called apoptosis—in effect, these signals tell the infected cell that it should kill itself. In addition, activated CD8 cells release potent cytokines, including gamma-interferon and tumor necrosis factor. Those molecules limit viral replication inside a cell, while also attracting macrophages and other phagocytes that can destroy the cell.

The control of viral infections through the destruction of the body's own cells has some powerful advantages. If the recruitment of antigenic peptides by the class I MHC molecules and the subsequent T cell response are fast enough, the infected cells can be destroyed even before the viral particles inside them have been completely assembled. Any virus particles that may be released when the cells are killed will not be capable of infecting other cells, and so the infection will be terminated before it can propagate.

On the other hand, the immunologic response mediated by CD8 T cells carries the potential for extensive harm to the host. If a virus multiplies and radiates quickly, the immune system's attempts to contain it may do no more than leave a path of destruction in the wake of the virus, while never quite catching up to it. The tissue damage associated with the infection would therefore result from the effects of both the virus and the immune reaction. In general, the amount of tissue damage caused by such an infection will be largely determined by how fast the immune response occurs in relation to the rate of spread of the virus.

The antiviral immune response becomes even more problematic when the viral infection itself does little or no damage to the cells—and many viruses do indeed infect cells without seriously impairing cellular function. Such noncytopathic infections can still provoke forceful reactions by CD8 cells. If the harmless virus spreads relatively quickly, the T cells may end up attacking a very large number of the host's cells. In these cases, the disease stems not from the virus at all but rather from the immune response.

One experimental demonstration of the harm that such immune responses can do comes from work with the lymphocytic choriomeningitis virus (LCMV), which infects tissues in the nervous system but has relatively little intrinsic pathogenicity. If newborn mice are inoculated with LCMV, the infection disperses speedily through their tissues but causes no evident disease. The reason is that their immature immune systems learn to tolerate the viral antigens as harmless constituents of the body; consequently, their T lymphocytes ignore the LCMV-infected cells. If cytotoxic T cells that respond to LCMV antigens are injected into these mice, however, the immune response is drastic and often kills the animals.

A variety of noncytopathic human infections show a similar pattern of tissue injury. Chronic carriers of the hepatitis B virus, for example, typically suffer liver damage even though it is a fairly harmless pathogen. The destruction of the infected liver cells is almost certainly a consequence of actions waged by cytotoxic T cells, which can be found in both the blood and the liver of the patients.

There are indications that the immune system may sometimes subvert its own reaction to viral Infections if that response would hurt the host more than the pathogen. Investigators have shown that if mice are injected with overwhelming numbers of LCMV, the CD8 T cells that should mobilize against the infection become activated but then die. Indeed, it seems likely that such cell deaths are the common result of a highly exuberant T cell response to an antigenic stimulus.

The elimination of those antigen-specific T cells leaves a strategic hole in the immunologic defense. Against a non-cytopathic virus, this deficiency may be to the host's benefit because it allows the cells harboring the virus to survive. One might argue that the elimination of these T cells after an exposure to overwhelming numbers of a virus is an adaptation of the immune system to an infection that it cannot control without causing irreparable injury to the host. As long as the persistence of the virus does not lead to the death of its host cells or to malignant abnormalities, the lack of an answering immune response will protect against disease.

Such examples illustrate the fallibility of this elegant but imperfect defense system: the very mechanisms that provide protection against certain kinds of disease will sometimes abet the pathology of others. Perhaps the cruelest demonstration of this principle comes from viral infections that exploit the cells and interactions of the immune system to propagate themselves. In such infections the immune response actually assists the replication of the virus rather than limiting it.

That is exactly what happens when people become infected by the human immunodeficiency virus (HIV) that causes AIDS. The virus resides preferentially in CD4 T lymphocytes and other cells of the immune system. As it turns out, activated T cells are much more hospitable to growing viruses than are resting T cells; consequently, the more agitated the immune system becomes, the better the conditions for viral replication. In addition, cytokines such as tumor necrosis factor, which T cells produce when they detect viral antigens, can actually stimulate the replication of viruses in CD4 T cells. Thus, HIV uses the most sophisticated defenses of the immune system to further its own survival.

Diseases such as AIDS are a painful reminder of the challenges that pathogens continue to pose to human immunity. The tremendous diversity and mutability of infectious agents ensure that such challenges will not abate. Nevertheless, the understanding of the molecular basis of cellular responses is developing rapidly and promises to illuminate new ways to minimize tissue damage and to control infection. Harnessing our knowledge to boost immune responses will be critical to efforts to conquer the present and future microbial threats to humanity.

FURTHER READING

THE PATHOGENESIS OF INFECTIOUS DISEASE. Cedric A. Mims. Academic Press, 1987.

VACCINES. Gordon L. Ada in *Fundamental Immunology*. Edited by William E. Paul. Raven Press, 1989.

REGULATION OF IMMUNITY TO PARASITES BY T CELLS AND T CELL-DERIVED CYTOKINES. A. Sher and R. L. Coffman in *Annual Review of Immunology*, Vol. 10, pages 385–409; 1992.

ROLES OF $\alpha\beta$ AND $\gamma\delta$ T CELL SUBSETS IN VIRAL IMMUNITY. P. C. Doherty, W. Allan, M. Eichelberger and S. R. Carding in *Annual Review of Immunology*, Vol. 10, pages 123–151; 1992.

IMMUNITY TO INTRACELLULAR BACTERIA. Stefan H. E. Kaufmann in *Annual Review of Immunology*, Vol. 11, pages 129–163; 1993.

Immunity's Early-Warning System

LUKE A. J. O'NEILL

ORIGINALLY PUBLISHED IN JANUARY 2005

Awoman is riding an elevator when her fellow passengers start to sneeze. As she wonders what sort of sickness the other riders might be spreading, her immune system swings into action. If the bug being dispersed by the contagious sneezers is one the woman has met before, a battalion of trained immune cells—the foot soldiers of the so-called adaptive immune system—will remember the specific invader and clear it within hours. She might never realize she had been infected.

But if the virus or bacterium is one that our hapless rider has never wrestled, a different sort of immune response comes to the rescue. This "innate" immune system recognizes generic classes of molecules produced by a variety of disease-causing agents, or pathogens. When such foreign molecules are detected, the innate system triggers an inflammatory response, in which certain cells of the immune system attempt to wall off the invader and halt its spread. The activity of these cells—and of the chemicals they secrete—precipitates the redness and swelling at sites of injury and accounts for the fever, body aches and other flulike symptoms that accompany many infections.

The inflammatory assault, we now know, is initiated by Toll-like receptors (TLRs): an ancient family of proteins that mediate innate immunity in organisms from horseshoe crabs to humans. If TLRs fail, the entire immune system crashes, leaving the body wide open to infection. If they work too hard, however, they can induce disorders marked by chronic, harmful inflammation, such as arthritis, lupus and even cardiovascular disease.

Discovery of TLRs has generated an excitement among immunologists akin to that seen when Christopher Columbus returned from the New World. Scores of researchers are now setting sail to this new land, where they hope to find explanations for many still mysterious aspects of immunity, infection and disorders involving abnormal defensive activity. Study of these receptors, and of the molecular events that unfold after they encounter a pathogen, is already beginning to uncover targets for pharmaceuticals that may enhance the body's protective activity, bolster vaccines, and treat a range of devastating and potentially deadly disorders.

CINDERELLA IMMUNITY

Until about five years ago, when it came to the immune system, the adaptive division was the star of the show. Textbooks were filled with details about B cells making antibodies that latch onto specific proteins, or antigens, on the surface of an invading pathogen and about T cells that sport receptors able to recognize fragments of proteins from pathogens. The response is called adaptive because over the course of an infection, it adjusts to optimally handle the particular microorganism responsible for the disease.

Adaptive immunity also grabbed the spotlight because it endows the immune system with memory. Once an infection has been eliminated, the specially trained B and T cells stick around, priming the body to ward off subsequent attacks. This ability to remember past infections allows vaccines to protect us from diseases caused by viruses or bacteria. Vaccines expose the body to a disabled form of a pathogen (or harmless pieces of it), but the immune system reacts as it would to a true assault, generating protective memory cells in the process. Thanks to T and B cells, once an organism has encountered a microbe and survived, it becomes exempt from being overtaken by the same bug again.

The innate immune system seemed rather drab in comparison. Its components—including antibacterial enzymes in saliva and an interlocking set of proteins (known collectively as the complement) that kill bacteria in the bloodstream—were felt to be less sophisticated than targeted antibodies and killer T cells. What is more, the innate immune system does not tailor its response in the same way that the adaptive system does.

In dismissing the innate immune response as dull and uninteresting, however, immunologists were tiptoeing around a dirty little secret: the adaptive system does not work in the absence of the allegedly more crude innate response. The innate system produces certain signaling proteins called cytokines that not only induce inflammation but also activate the B and T cells that are needed for the adaptive response. The posh sister, it turns out, needs her less respected sibling to make her shine.

By the late 1990s immunologists knew a tremendous amount about how the adaptive immune system operates. But they had less of a handle on innate immunity. In particular, researchers did not understand how

THE DIVISIONS OF THE IMMUNE SYSTEM

The mammalian immune system has two overarching divisions. The innate part (*left side*) acts near entry points into the body and is always at the ready. If it fails to contain a pathogen, the adaptive division (*right side*) kicks in, mounting a later but highly targeted attack against the specific invader.

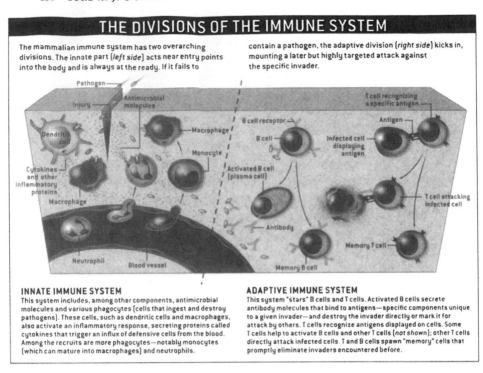

INNATE IMMUNE SYSTEM
This system includes, among other components, antimicrobial molecules and various phagocytes (cells that ingest and destroy pathogens). These cells, such as dendritic cells and macrophages, also activate an inflammatory response, secreting proteins called cytokines that trigger an influx of defensive cells from the blood. Among the recruits are more phagocytes—notably monocytes (which can mature into macrophages) and neutrophils.

ADAPTIVE IMMUNE SYSTEM
This system "stars" B cells and T cells. Activated B cells secrete antibody molecules that bind to antigens—specific components unique to a given invader—and destroy the invader directly or mark it for attack by others. T cells recognize antigens displayed on cells. Some T cells help to activate B cells and other T cells (*not shown*); other T cells directly attack infected cells. T and B cells spawn "memory" cells that promptly eliminate invaders encountered before.

microbes activate the innate response—or exactly how this stimulation helps to drive the adaptive response of T and B cells. Soon after, though, they would learn that much of the answer lay with the TLRs, which are produced by various immune system cells. But the path scientists traveled to get to these proteins was a circuitous one, winding through studies of fruit fly development, the search for drugs to treat arthritis, and the dawn of the genomic era.

WEIRD PROTEIN

The path actually had its beginnings in the early 1980s, when immunologists started to study the molecular activity of cytokines. These protein messengers are produced by various immune cells, including macrophages and dendritic cells. Macrophages patrol the body's tissues, searching for signs of infection. When they detect a foreign protein, they set off the inflammatory response. In particular, they engulf and destroy the invader bearing that protein and secrete a suite of cytokines, some of which raise an alarm that recruits other cells to the site of infection and puts the immune system in general on full alert. Dendritic cells ingest invading microbes and head off to the lymph nodes, where they present fragments of the pathogen's proteins to armies of T cells and release cytokines— activities that help to switch on the adaptive immune response.

To study the functions of various cytokines, researchers needed a way to induce the molecules' production. They found that the most effective way to get macrophages and dendritic cells to make cytokines in the laboratory was to expose them to bacteria—or more important, to selected components of bacteria. Notably, a molecule called lipopolysaccharide (LPS), made by a large class of bacteria, stimulates a powerful immune response. In humans, exposure to LPS causes fever and can lead to septic shocka deadly vascular shutdown triggered by an overwhelming, destructive action of immune cells. LPS, it turns out, evokes this inflammatory response by prompting macrophages and dendritic cells to release the cytokines tumor necrosis factor-alpha (TNF-alpha) and interleukin-1 (IL-1).

Indeed, these two cytokines were shown to rule the inflammatory response, prodding immune cells into action. If left unchecked, they can precipitate disorders such as rheumatoid arthritis, an autoimmune condition characterized by excessive inflammation that leads to destruction of the joints. Investigators therefore surmised that limiting the effects of TNF-alpha and IL-1 might slow the progress of the disease and alleviate the suffering of those with arthritis. To design such a therapy, though, they needed to know more about how these molecules work. And the first step was identifying the proteins with which they interact.

In 1988 John E. Sims and his colleagues at Immunex in Seattle discovered a receptor protein that recognizes IL-1. This receptor resides in the membranes of many different cells in the body, including macrophages and dendritic cells. The part of the receptor that juts out of the cell binds to IL-1, whereas the segment that lies inside the cell relays the message that IL-1 has been detected. Sims examined the inner part of the IL-1 receptor carefully, hoping it would yield some clue as to how the protein transmits its message—revealing, for example, which signaling molecules it activates within cells. But the inner domain of the human IL-1 receptor was unlike anything researchers had seen before, so he was stymied.

Then, in 1991, Nick J. Gay of the University of Cambridge—working on a completely unrelated problem—made a strange discovery. He was looking for proteins that were similar to a fruit-fly protein called Toll. Toll had been identified by Christiane Nusslein-Volhard in Tübingen, Germany, who gave the protein its name because flies that lack Toll look weird (*Toll* being the German word for "weird"). The protein helps the developing *Drosophila* embryo to differentiate its top from its bottom, and flies without Toll look jumbled, as if they have lost their sidedness.

Gay searched the database containing all the gene sequences then known. He was looking for genes whose sequences closely matched that of Toll and thus might encode Toll-like proteins. And he discovered that part of the Toll protein bears a striking resemblance to the inner part of the human IL-1 receptor, the segment that had mystified Sims.

At first the finding didn't make sense. Why would a protein involved in human inflammation look like a protein that tells fly embryos which end is up? The discovery remained puzzling until 1996, when Jules A. Hoffmann and his collaborators at CNRS in Strasbourg showed that flies use their Toll protein to defend themselves from fungal infection. In *Drosophila*, it seems, Toll multitasks and is involved in both embryonic development and adult immunity.

WORMS, WATER FLEAS AND YOU

The IL-1 receptor and the Toll protein are similar only in the segments that are tucked inside the cell; the bits that are exposed to the outside look quite different. This observation led researchers to search for human proteins that resemble Toll in its entirety. After all, evolution usually conserves designs that work well—and if Toll could mediate immunity in flies, perhaps similar proteins were doing the same in humans.

Acting on a tip from Hoffmann, in 1997 Ruslan Medzhitov and the late Charles A. Janeway, Jr., of Yale University discovered the first of these proteins, which they called human Toll. Within six months or so, Fernando Bazan and his colleagues at DNAX in Palo Alto, Calif., had identified five human Tolls, which they dubbed Toll-like receptors (TLRs). One, TLR4, was the same human Toll described by Medzhitov and Janeway.

At that point, researchers still did not know exactly how TLRs might contribute to human immunity. Janeway had found that stuffing the membranes of dendritic cells with TLR4 prompted the production of cytokines. But he could not say how TLR4 became activated during an infection.

The answer came in late 1998, when Bruce Beutler and his co-workers at the Scripps Institute in La Jolla, Calif., found that mutant mice unable to respond to LPS harbor a defective version of TLR4. Whereas normal mice die of sepsis within an hour of being injected with LPS, these mutant mice survive and behave as if they have not been exposed to the

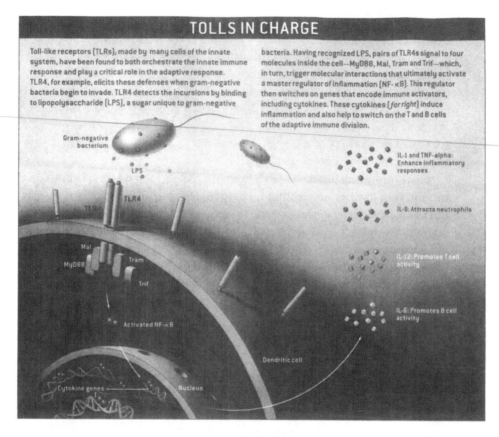

TOLLS IN CHARGE

Toll-like receptors (TLRs), made by many cells of the innate system, have been found to both orchestrate the innate immune response and play a critical role in the adaptive response. TLR4, for example, elicits these defenses when gram-negative bacteria begin to invade. TLR4 detects the incursions by binding to lipopolysaccharide (LPS), a sugar unique to gram-negative

bacteria. Having recognized LPS, pairs of TLR4s signal to four molecules inside the cell—MyD88, Mal, Tram and Trif—which, in turn, trigger molecular interactions that ultimately activate a master regulator of inflammation (NF-κB). This regulator then switches on genes that encode immune activators, including cytokines. These cytokines (far right) induce inflammation and also help to switch on the T and B cells of the adaptive immune division.

Gram-negative bacterium

LPS

TLR4

Mal

MyD88

Tram

Trif

Activated NF-κB

Cytokine genes

Nucleus

Dendritic cell

IL-1 and TNF-alpha: Enhance inflammatory responses

IL-8: Attracts neutrophils

IL-12: Promotes T cell activity

IL-6: Promotes B cell activity

molecule at all; that is, the mutation in the TLR4 gene renders these mice insensitive to LPS.

This discovery made it clear that TLR4 becomes activated when it interacts with LPS. Indeed, its job is to sense LPS. That realization was a major breakthrough in the field of sepsis, because it revealed the molecular mechanism that underlies inflammation and provided a possible new target for treatment of a disorder that sorely needed effective therapies. Within two years, researchers determined that most TLRs—of which 10 are now known in humans—recognize molecules important to the survival of bacteria, viruses, fungi and parasites. TLR2 binds to lipoteichoic acid, a component of the bacterial cell wall. TLR3 recognizes the genetic material of viruses; TLR5 recognizes flagellin, a protein that forms the whiplike tails used by bacteria to swim; and TLR9 recognizes a signature genetic sequence called CpG, which occurs in bacteria and viruses in longer stretches and in a form that is chemically distinct from the CpG sequences in mammalian DNA.

TLRs, it is evident, evolved to recognize and respond to molecules that are fundamental components of pathogens. Eliminating or chemically altering any one of these elements could cripple an infectious agent, which means that the organisms cannot dodge TLRs by mutating until these components are unrecognizable. And because so many of these elements are shared by a variety of microbes, even as few as 10 TLRs can protect us from virtually every known pathogen.

Innate immunity is not unique to humans. In fact, the system is quite ancient. Flies have an innate immune response, as do starfish, water fleas and almost every organism that has been examined thus far. And many use TLRs as a trigger. The nematode worm has one that allows it to sense and swim away from infectious bacteria. And plants are rife with TLRs. Tobacco has one called N protein that is required for fighting tobacco mosaic virus. The weed *Arabidopsis* has more than 200. The first Toll-like protein most likely arose in a single-celled organism that was a common ancestor of plants and animals. Perhaps these molecules even helped to

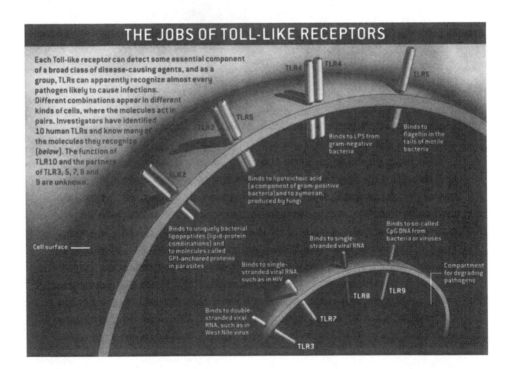

THE JOBS OF TOLL-LIKE RECEPTORS

Each Toll-like receptor can detect some essential component of a broad class of disease-causing agents, and as a group, TLRs can apparently recognize almost every pathogen likely to cause infections. Different combinations appear in different kinds of cells, where the molecules act in pairs. Investigators have identified 10 human TLRs and know many of the molecules they recognize (*below*). The function of TLR10 and the partners of TLR3, 5, 7, 8 and 9 are unknown.

TLR4   TLR4   TLR5

TLR6

TLR2   Binds to LPS from gram-negative bacteria   Binds to flagellin in the tails of motile bacteria

TLR2   Binds to lipoteichoic acid (a component of gram-positive bacteria) and to zymosan, produced by fungi

Binds to so-called CpG DNA from bacteria or viruses

Cell surface ———   Binds to uniquely bacterial lipopeptides (lipid-protein combinations) and to molecules called GPI-anchored proteins in parasites   Binds to single-stranded viral RNA   Compartment for degrading pathogens

Binds to single-stranded viral RNA, such as in HIV

TLR8   TLR9

Binds to double-stranded viral RNA, such as in West Nile virus   TLR7

TLR3

facilitate our evolution. Without an efficient means of defense against infection, multicellular organisms might never have survived.

STORMING THE CASTLE

The innate system was once thought to be no more elaborate than the wall of a castle. The real action, researchers believed, occurred once the wall had been breached and the troops inside—the T and B cells—became engaged. We now know that the castle wall is studded with sentries— TLRs—that identify the invader and sound the alarm to mobilize the troops and prepare the array of defenses needed to fully combat the attack. TLRs, in other words, unleash both the innate and adaptive systems.

The emerging picture looks something like this. When a pathogen first enters the body, one or more TLRs, such as those on the surface of patrolling macrophages and dendritic cells, latch onto the foreign molecules—for example, the LPS of gram-negative bacteria. Once engaged, the TLRs prompt the cells to unleash particular suites of cytokines. These protein messengers then recruit additional macrophages, dendritic cells and other immune cells to wall off and nonspecifically attack the marauding microbe. At the same time, cytokines released by all these busy cells can produce the classic symptoms of infection, including fever and flulike feelings.

Macrophages and dendritic cells that have chopped up a pathogen display pieces of it on their surface, along with other molecules indicating that a disease-causing agent is present. This display, combined with the cytokines released in response to TLRs, ultimately activates B and T cells that recognize those specific pieces, causing them—over the course of several days—to proliferate and launch a powerful, highly focused attack on the particular invader. Without the priming effect of TLRs, B and T cells would not become engaged and the body would not be able to mount a full immune response. Nor could the body retain any memory of previous infections.

Following the initial infection, enough memory T and B cells are left behind so that the body can deal more efficiently with the invader should it return. This army of memory cells can act so quickly that inflammation might not occur at all. Hence, the victim does not feel as ill and might not even notice the infection when it recurs.

Innate and adaptive immunity are thus part of the same system for recognizing and eliminating microbes. The interplay between these two systems is what makes our overall immune system so strong.

CHOOSE YOUR WEAPON

To fully understand how TLRs control immune activity, immunologists need to identify the molecules that relay signals from activated TLRs on the cell surface to the nucleus, switching on genes that encode cytokines and other immune activators. Many investigators are now pursuing this search intensively, but already we have made some fascinating discoveries.

We now know that TLRs, like many receptors that reside on the cell surface, enlist the help of a long line of signaling proteins that carry their message to the nucleus, much as a bucket brigade shuttles water to a fire. All the TLRs, with the exception of TLR3, hand off their signal to an adapter protein called MyD88. Which other proteins participate in the relay varies with the TLR: my laboratory studies Mal, a protein we discovered that helps to carry signals generated by TLR4 and TLR2. TLR4 also requires two other proteins—Tram and Trif—to relay the signal, whereas TLR3 relies on Trif alone. Shizuo Akira of Osaka University in Japan has shown that mice engineered so that they do not produce some of these intermediary signaling proteins do not respond to microbial products, suggesting that TLR-associated proteins could provide novel targets for new anti-inflammatory or antimicrobial agents.

Interaction with different sets of signaling proteins allows TLRs to activate different sets of genes that hone the cell's response to better match the type of pathogen being encountered. For example, TLR3 and TLR7 sense the presence of viruses. They then trigger a string of molecular interactions that induce the production and release of interferon, the major antiviral cytokine. TLR2, which is activated by bacteria, stimulates the release of a blend of cytokines that does not include interferon but is more suited to activating an effective antibacterial response by the body.

The realization that TLRs can detect different microbial products and help to tailor the immune response to thwart the enemy is now overturning long-held assumptions that innate immunity is a static, undiscriminating barrier. It is, in fact, a dynamic system that governs almost every aspect of inflammation and immunity.

FROM *LEGIONELLA* TO LUPUS

On recognizing the central role that TLRs play in initiating immune responses, investigators quickly began to suspect that hobbled or overactive versions of these receptors could contribute to many infectious and immune-related disorders. That hunch proved correct. Defects in innate immunity lead to greater susceptibility to viruses and bacteria. People with an underactive form of TLR4 are five times as likely to have severe bacterial infections over a five-year period than those with a normal TLR4. And people who die from Legionnaire's disease often harbor a mutation in TLR5 that disables the protein, compromising their innate immune response and rendering them unable to fight off the *Legionella* bacterium. On the other hand, an overzealous immune response can be equally destructive. In the U.S. and Europe alone, more than 400,000 people die annually from sepsis, which stems from an overactive immune response led by TLR4.

Other studies are pointing to roles for TLRs in autoimmune diseases such as systemic lupus erythematosus and rheumatoid arthritis. Here TLRs might respond to products from damaged cells, propagating an inappropriate inflammatory response and promoting a misguided reaction by the adaptive immune system. In lupus, for example, TLR9 has been found to react to the body's own DNA.

Innate immunity and the TLRs could also play a part in heart disease. People with a mutation in TLR4 appear to be less prone to developing cardiovascular disease. Shutting down TLR4 could protect the heart because inflammation appears to contribute to the formation of the plaques that clog coronary arteries. Manipulation of TLR4 might therefore be another approach to preventing or limiting this condition.

VOLUME CONTROL

Many of the big pharmaceutical companies have an interest in using TLRs and their associated signaling proteins as targets for drugs that could treat infections and immune-related disorders. With the spread of antibiotic resistance, the emergence of new and more virulent viruses, and the rising threat of bioterrorism, the need to come up with fresh ways to help our bodies fight infection is becoming more pressing.

Work on TLRs could, for example, guide the development of safer, more effective vaccines. Most vaccines depend on the inclusion of an adjuvant, a substance that kick-starts the inflammatory response, which in turn pumps up the ability of the adaptive system to generate the desired memory cells. The adjuvant used in most vaccines today does not provoke a full adaptive response; instead it favors B cells over T cells. To elicit a stronger response, several companies have set their sights on compounds that activate TLR9, a receptor that recognizes a broad range of bacteria and viruses and drives a robust immune response.

And TLRs are teaching us how to defend ourselves against biological weapons, such as poxviruses. A potential staple in the bioterrorist arsenal, these viruses can shut down TLRs and thereby avoid detection and elimination. In collaboration with Geoffrey L. Smith of Imperial College London, my lab found that by removing the viral protein that disables TLRs, we could generate a weakened virus that could serve as the basis of a vaccine unlikely to provoke an unintended fatal pox infection.

Armed with an understanding of TLRs and innate immunity, physicians might be able to predict which patients will fare poorly during infection and treat them more aggressively. If, for instance, patients came to a clinic with a bacterial infection and were found to have a mutant TLR4, the doctor might bombard them with antibiotics or with agents that could somehow bolster their immune response to prevent the infection from doing lasting damage.

Of course, a balance must be struck between stimulating an immune response that is sufficient to clear a microbe and precipitating an inflam-

TLRs AS DRUG TARGETS

Agents that activate TLRs and thus enhance immune responses could increase the effectiveness of vaccines or protect against infection. They might even prod the immune system to destroy tumors. In contrast, drugs that block TLR activity might prove useful for dampening inflammatory disorders. Drugs of both types are under study (*below*).

DRUG TYPE	EXAMPLES
TLR4 activator	MPL, an allergy treatment and vaccine adjuvant (immune system activator) from Corixa (Seattle), is in large-scale clinical trials
TLR7 activator	ANA245 (isatoribine), an antiviral agent from Anandys (San Diego), is in early human trials for hepatitis C
TLR7 and TLR8 activator	Imiquimod, a treatment for genital warts, basal cell skin cancer and actinic keratosis from 3M (St. Paul, Minn.), is on the market
TLR9 activator	ProMune, a vaccine adjuvant and treatment for melanoma skin cancer and non-Hodgkin's lymphoma from Coley (Wellesley, Mass.), is in large-scale clinical trials
TLR4 inhibitor	E5564, an antisepsis drug from Eisai (Teaneck, N.J.), is in early human trials
General TLR inhibitor	RDP58, a drug for ulcerative colitis and Crohn's disease from Genzyme (Cambridge, Mass.), is entering large-scale clinical trials
General TLR inhibitor	OPN201, a drug for autoimmune disorders from Opsona Therapeutics (Dublin, Ireland), is being tested in animal models of inflammation

matory response that will do more harm than good. Similarly, any medications that aim to relieve inflammation by quelling TLR activity and cytokine release must not, at the same time, undercut the body's defense against infection.

Anti-inflammatory drugs that interfere with TNF-alpha, one of the cytokines produced as a result of TLR4 activation, offer a cautionary tale. TNF-alpha produced during infection and inflammation can accumulate in the joints of patients with rheumatoid arthritis. The anti-inflammatory compounds alleviate the arthritis, but some people taking them wind up with tuberculosis. The infection is probably latent, but reining in the inflammatory response can also dampen the pathogen-specific responses and allow the bacterium to reemerge.

In short, TLRs are like the volume knob on a stereo, balancing adaptive immunity and inflammation. Researchers and pharmaceutical companies are now looking for ways to tweak these controls, so they can curtail inflammation without disabling immunity.

Given that TLRs were unheard of seven years ago, investigators have made enormous progress in understanding the central role these proteins play in the body's first line of defense. Innate immunity, long shrouded in oblivion, has suddenly become the belle of the ball.

FURTHER READING

INNATE IMMUNITY. Ruslan Medzhitov and Charles Janeway in *New England Journal of Medicine,* Vol. 343, No. 5, pages 338–344; August 3, 2000.

INFERENCES, QUESTIONS AND POSSIBILITIES IN TOLL-LIKE RECEPTOR SIGNALING. Bruce Beutler in *Nature,* Vol. 430, pages 257–263; July 8, 2004.

TOLL-LIKE RECEPTOR CONTROL OF THE ADAPTIVE IMMUNE RESPONSES. Akiko Iwasaki and Ruslan Medzhitov in *Nature Immunology,* Vol. 5, No. 10, pages 987–995; October 2004.

TLRS: PROFESSOR MECHNIKOV, SIT ON YOUR HAT. L. A. J. O'Neill in *Trends in Immunology,* Vol. 25, No. 12, pages 687–693; December 2004.

The Long Arm of the Immune System

JACQUES BANCHEREAU

ORIGINALLY PUBLISHED IN NOVEMBER 2002

They lie buried—their long, tentaclelike arms outstretched—in all the tissues of our bodies that interact with the environment. In the lining of our nose and lungs, lest we inhale the influenza virus in a crowded subway car. In our gastrointestinal tract, to alert our immune system if we swallow a dose of salmonella bacteria. And most important, in our skin, where they lie in wait as stealthy sentinels should microbes breach the leathery fortress of our epidermis.

They are dendritic cells, a class of white blood cells that encompasses some of the least understood but most fascinating actors in the immune system. Over the past several years, researchers have begun to unravel the mysteries of how dendritic cells educate the immune system about what belongs in the body and what is foreign and potentially dangerous. Intriguingly, they have found that dendritic cells initiate and control the overall immune response. For instance, the cells are crucial for establishing immunological "memory," which is the basis of all vaccines. Indeed, physicians, including those at a number of biotechnology companies, are taking advantage of the role that dendritic cells play in immunization by "vaccinating" cancer patients with dendritic cells loaded with bits of their own tumors to activate their immune system against their cancer. Dendritic cells are also responsible for the phenomenon of immune tolerance, the process through which the immune system learns not to attack other components of the body.

But dendritic cells can have a dark side. The human immunodeficiency virus (HIV) hitches a ride inside dendritic cells to travel to lymph nodes, where it infects and wipes out helper T cells, causing AIDS. And those cells that become active at the wrong time might give rise to autoimmune disorders such as lupus. In these cases, shutting down the activity of dendritic cells could lead to new therapies.

RARE AND PRECIOUS

Dendritic cells are relatively scarce: they constitute only 0.2 percent of white blood cells in the blood and are present in even smaller proportions in tissues such as the skin. In part because of their rarity, their true function eluded scientists for nearly a century after they were first identified in 1868 by German anatomist Paul Langerhans, who mistook them for nerve endings in the skin.

In 1973 Ralph M. Steinman of the Rockefeller University rediscovered the cells in mouse spleens and recognized that they are part of the immune system. The cells were unusually potent in stimulating immunity in experimental animals. He renamed the cells "dendritic" because of their spiky arms, or dendrites, although the subset of dendritic cells that occur in the epidermis layer of the skin are still commonly called Langerhans cells.

For almost 20 years after the cells' rediscovery, researchers had to go through a painstakingly slow process to isolate them from fresh tissue for study. But in 1992, when I was at the Schering-Plough Laboratory for Immunology Research in Dardilly, France, my co-workers and I devised methods for growing large amounts of human dendritic cells from bone marrow stem cells in culture dishes in the laboratory. At roughly the same time, Steinman—in collaboration with Kayo Inaba of Kyoto University in Japan and her colleagues—reported that he had invented a technique for culturing dendritic cells from mice.

In 1994 researchers led by Antonio Lanzavecchia, now at the Institute for Research in Biomedicine in Bellinzona, Switzerland, and Gerold Schuler, now at the University of Erlangen-Nuremberg in Germany, found a way to grow the cells from white blood cells called monocytes. Scientists now know that monocytes can be prompted to become either dendritic cells, which turn the immune system on and off, or macrophages, cells that crawl through the body scavenging dead cells and microbes.

The ability to culture dendritic cells offered scientists the opportunity to investigate them in depth for the first time. Some of the initial discoveries expanded the tenuous understanding of how dendritic cells function.

There are several subsets of dendritic cells, which arise from precursors that circulate in the blood and then take up residence in immature form in the skin, mucous membranes, and organs such as the lungs and spleen. Immature dendritic cells are endowed with a wealth of mechanisms for capturing invading microbes: they reel in invaders using suction cup-like receptors on their surfaces, they take microscopic sips of the fluid surrounding them, and they suck in viruses or bacteria by engulfing them in sacks known as vacuoles. Yong-Jun Liu, a former colleague of mine from Schering-Plough who is now at DNAX Research Institute in Palo Alto, Calif., has found that some immature dendritic cells can also zap viruses immediately by secreting a substance called interferon-alpha.

Once they devour foreign objects, the immature cells chop them into fragments (antigens) that can be recognized by the rest of the immune system. The cells use pitchfork-shaped molecules termed the major histocompatibility complex (MHC) to display the antigens on their surfaces. The antigens fit between the tines of the MHC, which comes in two types, class I and class II. The two types vary in shape and in how they acquire their antigen cargo while inside cells.

Dendritic cells are very efficient at capturing and presenting antigens: they can pick up antigens that occur in only minute concentrations. As they process antigens for presentation, they travel to the spleen through the blood or to lymph nodes through a clear fluid known as lymph. Once at their destinations, the cells complete their maturation and present their antigen-laden MHC molecules to naive helper T cells, those that have never encountered antigens before. Dendritic cells are the only cells that can educate naive helper T cells to recognize an antigen as foreign or dangerous. This unique ability appears to derive from co-stimulatory molecules on their surfaces that can bind to corresponding receptors on the T cells.

Once educated, the helper T cells go on to prompt so-called B cells to produce antibodies that bind to and inactivate the antigen. The dendritic cells and helper cells also activate killer T cells, which can destroy cells infected by microbes. Some of the cells that have been educated by dendritic cells become "memory" cells that remain in the body for years—perhaps decades—to combat the invader in case it ever returns.

Whether the body responds with antibodies or killer cells seems to be determined in part by which subset of dendritic cell conveys the message and which of two types of immune-stimulating substances, called cytokines, they prompt the helper T cells to make. In the case of parasites or some bacterial invaders, type 2 cytokines are best because they arm the immune system with antibodies; type 1 cytokines are better at mustering killer cells to attack cells infected by other kinds of bacteria or by viruses.

If a dendritic cell prompts the wrong type of cytokine, the body can mount the wrong offense. Generating the appropriate kind of immune response can be a matter of life or death: when exposed to the bacterium that causes leprosy, people who mount a type 1 response develop a mild, tuberculoid form of the disease, whereas those who have a type 2 response can end up with the potentially fatal lepromatous form.

CANCER KILLERS

Activating naive helper T cells is the basis of vaccines for everything from pneumonia to tetanus to influenza. Scientists are now turning the new knowledge of the role that dendritic cells play in immunity against microbes and their toxins into a strategy to fight cancer.

Cancer cells are abnormal and as such are thought to generate molecules that healthy cells don't. If researchers could devise drugs or vaccines that exclusively targeted those aberrant molecules, they could combat cancer more effectively while leaving normal cells and tissues alone—thereby eliminating some of the pernicious side effects of chemotherapy and radiation, such as hair loss, nausea and weakening of the immune system caused by destruction of the bone marrow.

Antigens that occur only on cancerous cells have been hard to find, but researchers have succeeded in isolating several of them, most notably from the skin cancer melanoma. In the early 1990s Thierry Boon of the Ludwig Cancer Institute in Brussels, Steven A. Rosenberg of the National Cancer Institute and their colleagues independently identified melanoma-specific antigens that are currently being targeted in a variety of clinical trials involving humans.

Such trials generally employ vaccines made of dendritic cell precursors that have been isolated from cancer patients and grown in the laboratory together with tumor antigens. During this process, the dendritic cells pick up the antigens, chop them up and present them on their surfaces. When injected back into the patients, the antigen-loaded dendritic cells are expected to ramp up patients' immune response against their own tumors.

Various researchers—including Frank O. Nestle of the University of Zurich and Ronald Levy and Edgar G. Engleman of Stanford University, as well as scientists at several biotechnology companies—are testing this approach against cancers as diverse as melanoma, B cell lymphoma, and tumors of the prostate and colon. There have been glimmers of success. In September 2001, for instance, my co-workers and I, in collaboration with Steinman's group, reported that 16 of 18 patients with advanced melanoma to whom we gave injections of dendritic cells loaded with

Dendritic Cell Cancer Vaccines under Development

COMPANY NAME	HEADQUARTERS	STOCK SYMBOL	CANCER TYPE	STATUS*
ML Laboratories	Warrington, England	LSE: MLB	Melanoma	Entering phase I tests
Dendreon	Seattle	Nasdaq: DNDN	Prostate, breast, ovary, colon, multiple myeloma	Phase III (prostate), phase II (prostate, multiple myeloma), phase I (breast, ovary, colon)
Genzyme	Framingham, Mass.	Nasdaq: GZMO	Kidney, melanoma	Phase I (kidney), phase I/II (melanoma)
Immuno-Designed Molecules	Paris	Privately held	Prostate, melanoma	Phase II tests
Merix Bioscience	Durham, N.C.	Privately held	Melanoma	Entering phase I
Oxford BioMedica	Oxford, England	LSE: OXB	Colorectal	Phase I/II
Zycos	Lexington, Mass.	Privately held	DNA-based vaccine against various cancers	Phases I and II

*Phase I tests evaluate safety in a small number of patients; phases II and III assess ability to stimulate the immune system and effectiveness in larger numbers of patients.

melanoma antigens showed signs in laboratory tests of an enhanced immune response to their cancer. What is more, tumor growth was slowed in the nine patients who mounted responses against more than two of the antigens.

Scientists are now working to refine the approach and test it on larger numbers of patients. So far cancer vaccines based on dendritic cells have been tested only in patients with advanced cancer. Although researchers believe that patients with earlier-stage cancers may respond better to the therapy—their immune systems have not yet tried and failed to eradicate their tumor—several potential problems must first be considered.

Some researchers fear that such vaccines might induce patients' immune systems to attack healthy tissue by mistake. For instance, vitiligo—white patches on the skin caused by the destruction of normal pigment-producing melanocytes—has been observed in melanoma patients who have received the earliest antimelanoma vaccines. Conversely, the tumors might mutate to "escape" the immune onslaught engendered by a dendritic cell vaccine. Tumor cells could accomplish this evasion by no longer making the antigens the vaccine was designed to stimulate the immune system against. This problem is not unique to dendritic cells, though: the same phenomenon can occur with traditional cancer therapies.

In addition, tailoring a dendritic cell vaccine to fight a particular patient's tumors might not be economically feasible. But many scientists are working to circumvent the costly and time-consuming steps of isolating cells from patients and manipulating them in the laboratory for reinjection.

One approach involves prompting dendritic cell precursors already present in a person's body to divide and start orchestrating an immune response against their tumors. David H. Lynch of Immunex in Seattle (recently acquired by Amgen in Thousand Oaks, Calif.) and his co-workers have discovered a cytokine that causes mice to make more dendritic cells, which eventually induce the animals to reject grafted tumors. Other scientists, including Drew M. Pardoll of Johns Hopkins University, have observed that tumor cells that have been genetically engineered to secrete large amounts of cytokines that activate dendritic cells have the most potential as cancer vaccines.

SHUTTING IMMUNITY DOWN

In the meantime, other scientists are looking at ways to turn off the activity of dendritic cells in instances where they exacerbate disease instead of fighting it. Usually, in a phenomenon known as central tolerance, an organ in the chest called the thymus gets rid of young T cells that happen to recognize the body's own components as foreign before they have a chance to circulate. Some inevitably slip through, however, so the body has a backup mechanism for restraining their activity.

But this mechanism, termed peripheral tolerance, appears to be broken in patients with autoimmune disorders such as rheumatoid arthritis, type 1 diabetes and systemic lupus erythematosus. Last year my colleagues and I reported that dendritic cells from the blood of people with lupus are unnaturally active. Cells from these patients release high amounts of interferon-alpha, an immune-stimulating protein that causes precursors to grow into mature dendritic cells while still in the bloodstream. The mature cells then ingest DNA, which is present in unusual amounts in the blood of people with lupus, and that in turn causes the individual's immune system to generate antibodies against his or her own DNA. These antibodies result in the life-threatening complications of lupus when they lodge in the kidneys or the walls of blood vessels. Accordingly, we propose that blocking interferon-alpha might lead to a therapy for lupus by preventing dendritic cell activation. A similar strategy might prevent organ transplant recipients from rejecting their new tissues.

A new treatment for AIDS might also rest on a better understanding of dendritic cells. In 2000 Carl G. Figdor and Yvette van Kooyk of the University Medical Center St. Radboud in Nijmegen, the Netherlands, identified a subset of dendritic cells that makes DC-SIGN, a molecule that can bind to the outer coat of HIV. These cells pick up HIV as they regularly prowl the mucous membranes and deep tissues. When they travel to the lymph nodes, they unwittingly deliver the virus to a large concentration of

T cells. Drugs that block the interaction between DC-SIGN and HIV might slow the progression of AIDS.

Other infectious diseases—including malaria, measles and cytomegalovirus—also manipulate dendritic cells for their own ends. Red blood cells that have been infected by malaria parasites, for instance, bind to dendritic cells and prevent them from maturing and alerting the immune system to the presence of the invaders. Several groups of researchers are now devising approaches to prevent such microbes from hijacking dendritic cells; some are even seeking to use supercharged dendritic cells to fight the infections.

As we learn more about the molecules that control dendritic cells, we will find ways to harness their therapeutic potential. The increasing number of scientists and corporations working on dendritic cells portends that we will soon be able to maximize the biological power of these cells to treat and prevent the diseases that plague humankind.

FURTHER READING

DENDRITIC CELLS AND THE CONTROL OF IMMUNITY. Jacques Banchereau and Ralph M. Steinman in *Nature*, Vol. 392, pages 245–252; March 19, 1998.

DENDRITIC CELLS AS VECTORS FOR THERAPY. Jacques Banchereau, Beatrice Schuler-Thurner, A. Karolina Palucka and Gerold Schuler in *Cell*, Vol. 106, No. 3, pages 271–274; August 10, 2001.

Background information on the immune system and on experimental cancer therapies such as those using dendritic cells can be found on the American Cancer Society's Web site: www.cancer.org

Intrigue at the Immune Synapse

DANIEL M. DAVIS

ORIGINALLY PUBLISHED IN FEBRUARY 2006

Comic-book fans know well that the most sought after editions are those in which a superhero appears for the first time. A comic book published in 1962 featuring the first appearance of Spider-Man, for example, recently sold at auction for $122,000. Sadly, publications representing the first appearance of an important scientific fact generally do not command similar prices, but to scientists these firsts are equally treasured.

Just such a moment occurred in 1995, when Abraham "Avi" Kupfer of the National Jewish Medical and Research Center in Denver stood before an unsuspecting group of a few hundred immunologists gathered for one of the prestigious Keystone symposia, named for a U. S. ski resort. Kupfer's presentation included the first three-dimensional images of immune cells interacting with one another. As the crowd watched in stunned silence, Kupfer showed them image after image of proteins organized into bull's-eye patterns at the area of contact between the cells.

To the group, the pictures were instantly understandable and unequivocal: like the synapses that form the critical junctures between neurons in neural communication networks, the contacts between the immune cells involved organized aggregates of proteins. Both outer rings of molecules keeping the cells adhered to one another and inner clusters of interacting proteins particular to the discussion between the cells were clearly visible.

The idea that immune cells—which must exchange and store information in the course of searching for and responding to disease—might share mechanisms with those consummate communicators, the cells of the nervous system, had been put forth before. But here, at last, was proof of structures to go with the theory. When Kupfer was finished, the room erupted in prolonged applause, followed by a barrage of questions.

A decade later these structured synapses formed by immune cells are still generating questions: about how cellular machinery or other forces produce the synaptic architecture, how the architecture, in turn, might regulate cell-to-cell communication, how its malfunction could lead to disease, and even how pathogens might exploit the mechanism to their own advantage.

Discovery of the immune synapse and its ongoing exploration has been made possible by new high-resolution microscopy techniques and computer enhancement of older imaging methods. Now the realization that a thought, the sensation of a touch, or the detection of a virus in the bloodstream all require similar choreography of molecules is providing a compelling new framework for understanding immunity.

SEEKING DIRECTION

Long before the immune synapse was seen, the possibility that immune cells might be able to target their communication was apparent. Scientists knew that immune cells secreted protein molecules called cytokines to talk with one another and with other types of cells. Yet at least some of these molecules did not seem to function like hormones, which diffuse throughout the body broadcasting their message widely. Rather cytokines could barely be detected in the blood and seemed to act only between cells that were touching.

This ability to trade chemical signals with just a particular neighbor is important for immune cells. Unlike neurons, which tend to form stable, long-term junctions with other cells, immune cells make fleeting contacts as they constantly roam the body seeking out signs of disease and exchanging information about present dangers. When an immune cell charged with identifying illness bumps into another cell, it may have only a couple of minutes to decide whether its target is healthy or not. If not, the immune cell, depending on its type, might kill the sick cell directly or raise an alarm, calling other immune soldiers to come do the job. Getting the communication wrong might lead to immune cells mistakenly killing healthy cells, as happens in autoimmune diseases such as multiple sclerosis, or allowing cancer cells to continue growing unchecked. Thus, immunologists have keen interest in figuring out not only which molecules are involved in these dialogues but how they interact to enable such critical decisions.

In the early 1980s scientists in the Laboratory of Immunology at the National Institutes of Health began exploring the idea that a structured interface could allow immune cells to direct their secretion of cytokines to another cell. Because the cellular membranes, made largely from fat and protein molecules, are fluid, proteins could certainly move easily up to the point of contact between two cells and form an organized architecture there, as happens when neurons create a connection to another cell.

The NIH group's hypothesis grew from critical experiments showing that clustering specific proteins together at the surface of immune cells called T cells was sufficient to trigger activation of those cells. In a paper published in 1984, NIH investigator Michael A. Norcross first formally articulated the possibility that the nervous and immune systems have a common mechanism of communication through synapses. Unfortunately, it appeared in a journal that was not widely read, and some of his molecular details were off, so his early synaptic model of immune cell communication was soon forgotten. But curiosity about whether and how immune cells might target their messages remained.

In 1988 the late Charles A. Janeway, Jr., and his colleagues at Yale University performed a beautiful experiment to confirm that immune cells could indeed secrete proteins in a specific direction. They fitted T cells tightly into the pores of a membrane dividing a chamber containing solution. By adding a stimulant to the solution, on only side of the membrane, they activated the T cells, which subsequently started secreting proteins toward the source of the stimulant but not into the stimulant-free solution on the other side of the membrane.

Encouraged by this key observation, in 1994 NIH researchers William E. Paul and Robert A. Seder resurrected the idea that the immune synapse is the communicating junction between immune cells and other cells. They described the synapse as two cell surfaces in close proximity with a structured arrangement of receptor proteins on one cell surface, opposite their binding partners on the contacting cell. Acknowledging that immune cells move about far more than neurons, Paul spoke of the immune synapse as a "make and break" union in contrast with longer-term neuronal connections.

Thus, by the mid-1990s the immune synapse was established as a provocative concept for which a structure still needed to be seen experimentally. Then Avi Kupfer presented his slide show at the Keystone symposium. His images showed interactions between immune cells called antigen-presenting cells (APCs) that specialize in breaking up proteins belonging to an invader, such as a virus, and displaying the protein fragments to T cells, which become activated when they recognize one of the antigens. Hence, Kupfer dubbed the bull's-eye patterns of protein molecules formed at the interface of the two cells supramolecular activation clusters, or SMACs.

Independently Michael L. Dustin, Paul M. Allen and Andrey S. Shaw of the Washington University School of Medicine, in St. Louis, with Mark M.

Davis of Stanford University, had also been imaging T cell activation, but with an interesting twist. Instead of observing two cells interacting together, they replaced the APC with a surrogate membrane composed of lipid molecules from a real cell laid out flat on a glass slide. To this glass-supported lipid membrane, they added the key proteins normally found at the surface of APCs, each tagged with a different colored fluorescent dye. They then watched the organization of these labeled proteins as T cells landed on the membrane.

Dustin's group also saw bull's-eye patterns of proteins emerge as the T cells surveyed the proteins within the supported membrane. Clearly, a structured synapse did not require the effort of two cells; instead it could form as one immune cell contacted and responded to an artificial array of proteins.

This work also revealed that the synapse itself is dynamic: arrangements of proteins change as the cell communication continues. For example, T cell receptors interacting with the antigen were first seen to accumulate in a ring surrounding a central cluster of adhesive proteins, creating an immature T cell synapse. Later, that structure inverted so that in the mature synapse the adhesive molecules formed an outer ring of the bull's-eye, surrounding a central cluster of interacting T cell receptors.

Since Kupfer and Dustin published their initial T cell synapse images, a variety of synapse structure patterns have also been seen between other types of immune cells. Indeed, my own contribution, while working with Jack Strominger of Harvard University in 1999, was to observe a structured synapse formed by a different kind of white blood cell—known as a natural killer (NK) cell—which helped to confirm the generality of their observations. Exploring how such changing arrangements of molecules occur and how they control immune cell communication is the new science opened up by the immune synapse concept.

DECIPHERING THE DANCE

Observations of the structure of immune synapses immediately spurred researchers to investigate what makes the cellular proteins move to the contact point between the cells and organize themselves into specific patterns. One driver of protein movements in all cells is a remarkable network of filaments known as the cytoskeleton, which is made up of long chains of proteins that can extend or shrink in length. Tethered to the cell surface by adapter proteins, the cytoskeleton can push or pull the cell membrane, enabling muscles to contract or sperm to swim.

Experiments showed that when a cell's cytoskeleton was incapacitated by toxins, some proteins were no longer able to move toward the immune synapse, suggesting that movements of cytoskeletal filaments allow cells to control when and where the proteins accumulate at the synapse.

At least two other mechanisms could play a role in organizing proteins at the synapse, but the extent of their influence on immune cell communication is controversial. One set of proposals theorizes that small platforms made up of a few protein molecules each may be clustered in cell membranes and capable of moving around the cell surface together, most likely with help from the cytoskeleton. When these molecular "rafts" are brought together in the synapse with the key receptor proteins that detect disease in an opposing cell, their interaction could be what activates the immune cell. These preexisting platforms are contentious, however, because they are too small to see directly with an optical microscope, so evidence of their existence is somewhat indirect.

Another interesting possibility, with both indirect and direct support, is that the physical size of each type of protein forming the synapse can play an important role in determining where it goes when the cells come into contact. As proteins on one cell bind their counterparts on the opposing cell, the two cell membranes will be drawn together and the remaining gap between them will correspond to the size of the bound proteins. Thus, a central cluster of small proteins could bring the membranes close enough together to squeeze out larger proteins and hence segregate different types of protein to different regions of the synapse.

Arup K. Chakraborty and his colleagues at the University of California, Berkeley, used a mathematical model to test this idea by assessing the consequence of different-size proteins interacting across two opposing cell membranes. Although Chakraborty is not an immunologist, a colleague had shown him images from Dustin's work, and the mathematician says that he became fascinated by the intriguing spatial patterns his immune cells might be forming whenever he had the flu. His group's analysis suggested that in fact the difference in size between proteins could be enough to cause bigger and smaller proteins to cluster in separate regions of the immune synapse.

Of course, immunologists also want to know what, if anything, these protein movements "mean" in the context of immune cell communication. The answer could be "nothing": the earliest conception of the immune synapse being a kind of gasket enabling immune cells to direct their secretion of cytokines to a target cell may be the sole purpose of the structure. Increasingly, however, evidence is suggesting that the synapse may also have other functions that, depending on the cells involved, could include initiating communication, or terminating it, or serving to modulate the volume, so to speak, of signals between two cells.

In 2002 Kupfer (now at the Johns Hopkins School of Medicine) observed, for example, that signaling between T cells and antigen-presenting cells before the SMAC begins to take shape fostered adhesion between the two cells but that a SMAC was necessary for the cells' interaction to produce T cell responses.

Yet Shaw and Allen, along with Dustin, now at New York University, and their co-workers have shown that productive signaling between T cells and APCs starts before the T cell receptors have clustered in their final position at the center of the synapse. In fact, some of the communication is done before the mature structure forms, implying that the mature synapse pattern might signal an end to the conversation.

These investigators and others have also been exploring what role synapse architecture might play in regulating the volume of dialogues between T cells and APCs. By pulling receptors away from their surface membrane during signaling, T cells can prevent themselves from being lethally overstimulated by too much antigen. Experiments have shown that T cells can reduce the number of receptors present in the synapse architecture to dampen signaling, or when only a small amount of antigen is available, T cells may cluster their receptors more closely within the synapse to amplify the signal.

My own research group has been studying similar phenomena in natural killer cells, a type of immune cell that seeks and destroys cells damaged, for example, by a cancerous mutation or infected by a pathogen. These sick cells can lose the expression of some proteins on their surfaces, and NK cells recognize the loss as a sign of disease. We are finding that the amount of these proteins present on the target cell influences the pattern of the immune synapse formed by the NK cell. Different patterns correlate with whether or not the NK cell ultimately decides to kill the target cell, so the patterns may transmit, or at least reflect, information the NK cell uses to determine the extent of the target cell's illness.

Alongside these fascinating new insights into the possible functions of the immune synapse, disturbing news has emerged, too: another very recent observation is that the molecular dance that helps our immune cells communicate can be exploited by some viruses, including HIV. Charles R. M. Bangham of Imperial College London and his collaborators first showed that at the contact point between cells where viral particles are crossing over, proteins aggregate into a structure that resembles the immune synapse. Several researchers have since observed similar "viral synapse" phenomena, and so it seems that viruses, which are known for hijacking cellular machinery to copy their genetic material, may also be able to co-opt cellular mechanisms for communication to propel themselves from one cell to another.

HEALTHY VOYEURISM

The discovery of the immune synapse has triggered a wave of research based on imaging immune cell interactions whose results have yet to be fully understood. But this fertile field is already producing new hypotheses and generating further research to test those. And the very idea of the synapse is already reshaping conceptions of the immune system, revealing it to be a sophisticated information-sharing network more like the nervous system than was previously realized.

Just using the synapse terminology to describe immune cell interactions has also encouraged neuroscientists and immunologists to compare notes, and they are finding that the two types of synapses use many common protein molecules. Agrin, for example, is an important protein involved in clustering other proteins at the synapse between neurons and muscle. Imaging experiments have shown that the same molecule also accumulates at immune synapses and can enhance at least some types of immune responses. Similarly, a receptor called neuropilin-1, known to participate in signaling between neurons, has been discovered at immune synapses. Experiments suggest that neuropilin-1 aids immune cells in their search for disease by helping to establish an immune synapse with other cells, but more research is needed to tease out the receptor's exact role in immunity.

My own team identified yet another intriguing similarity between neurons and immune cells when we observed that long tubes made of cell membrane readily form between immune cells and a variety of other cell types. Our, investigation that led to this discovery was prompted by a report from German and Norwegian researchers of a similar phenomenon observed between neurons. Neither we nor the neuroscientists know the function of these nanotubular highways, but finding out is a new goal for immunology and neuroscience alike.

These membrane nanotubes might, for example, constitute a previously unknown mechanism for immune cell communication by allowing directed secretion of cytokines between cells far apart. Simon C. Watkins and Russell D. Salter of the University of Pittsburgh School of Medicine have found that a population of immune cells could use such nanotubular highways to transmit calcium signals across vast (for cells) distances of hundreds of microns within seconds.

In the future, more studies of interactions among larger groups of immune cells could reveal additional aspects of immune cell communication networks. Imaging immune cell interactions as they traffic inside living organisms, rather than on a slide, is another important frontier for this line of research.

In a recent memoir, Nobel laureate John Sulston described using cutting edge microscopy in the 1970s to understand worm development: "Now to my amazement, I could watch the cells divide. Those Nomarski images of the worm are the most beautiful things imaginable. . . . In one weekend I unraveled most of the postembryonic development of the ventral cord, just by watching.

High-resolution microscopy of immune cell interactions is still a very young field, and more surprises are surely in store. Virtually all the surface proteins involved in immune cells' recognition of disease have been identified and named. But the ability of scientists to now observe as these molecules play out their roles in space and time has revealed the immune synapse mechanism and reconfirmed the value of "just watching" as a scientific method.

FURTHER READING

Three-Dimensional Segregation of Supramolecular Activation Clusters in T Cells. C. R. Monks, B. A. Freiberg, H. Kupfer, N. Sciaky and A. Kupfer in *Nature*, Vol. 395, pages 82–86; September 3, 1998.

Neural and Immunological Synaptic Relations. Michael L. Dustin and David R. Colman in *Science*, Vol. 298, pages 785–789; October 25, 2002.

What Is the Importance of the Immunological Synapse? Daniel M. Davis and Michael L. Dustin in *Trends in Immunology*, Vol. 25, No. 6, pages 323–327; June 2004.

The Language of Life: How Cells Communicate in Health and Disease. Debra Niehoff. Joseph Henry Press, Washington, D.C., 2005. Available online at National Academies Press: www.nap.edu/books/0309089891/html

Peacekeepers of the Immune System

ZOLTAN FEHERVARI AND SHIMON SAKAGUCHI

ORIGINALLY PUBLISHED IN OCTOBER 2006

"Horror autotoxicus."A century ago the visionary bacteriologist Paul Ehrlich aptly coined that term to describe an immune system attack against a person's own tissues. Ehrlich thought such autoimmunity— another term he coined—was biologically possible yet was somehow kept in check, but the medical community misconstrued his two-sided idea, believing instead that autoimmunity had to be inherently impossible. After all, what wrong turn of evolution would permit even the chance of horrendous, built-in self-destruction?

Slowly, though, a number of mysterious ailments came to be recognized as examples of horror autotoxicus—among them multiple sclerosis, insulin-dependent diabetes (the form that commonly strikes in youth) and rheumatoid arthritis. Investigators learned, too, that these diseases usually stem from the renegade actions of white blood cells known as CD4₊ T lymphocytes (so named because they display a molecule called CD4 and mature in the thymus). Normal versions of these cells serve as officers in the immune system's armed forces, responsible for unleashing the system's combat troops against disease-causing microorganisms. But sometimes the cells turn against components of the body.

Ehrlich was correct in another way as well. Recent work has identified cells that apparently exist specifically to block aberrant immune behavior. Called regulatory T cells, they are a subpopulation of CD4₊ T cells, and they are vital for maintaining an immune system in harmony with its host. Increasingly, immunologists are also realizing that these cells do much more than quash autoimmunity; they also influence the immune system's responses to infectious agents, cancer, organ transplants and pregnancy. We and others are working to understand exactly how these remarkable cells carry out their responsibilities and why they sometimes function imperfectly. The findings should reveal ways to regulate the regulators and thus to depress or enhance immune activity as needed and, in so doing, to better address some of today's foremost medical challenges.

IMPERFECT DEFENSES

Like the immunologists of Ehrlich's time, many people today would be dismayed to know that no matter how healthy they may be, their bodies harbor potentially destructive immune system cells quite capable of triggering autoimmune disease. Yet this immunological sword of Damocles can be easily demonstrated. If a mouse, for example, is injected with proteins from its own central nervous system, along with an adjuvant (a generalized immune system stimulus), a destructive immune reaction ensues. Much as in multiple sclerosis, T cells launch an attack on the animal's brain and spinal cord.

By varying the source of the injected self-protein, researchers can provoke other autoimmune diseases in laboratory animals—which indicates that potentially harmful immune system cells can mount self-attacks on a wide variety of tissues. The risk appears to hold true in humans, too, because autoreactive immune system cells can be captured readily from the blood of a healthy person. In a test tube, they react strongly to samples of that person's tissues.

Given such demonstrations of clear and imminent danger, investigators naturally wondered how it is that most animals and humans are untroubled by autoimmune disease. Put another way, they wanted to know how the immune system distinguishes threats such as microbes from a person's own tissues. They found that to achieve self-tolerance—the ability to refrain from attacking one's own organs—the immune system enlists numerous safeguards. The first defense, at least where T cells are concerned, occurs in the thymus, which lies inconspicuously in front of the heart. In the thymus, immature T cells undergo a strict "education" in which they are programmed to not react strongly (and therefore harmfully) to any bodily tissues. Disobedient cells are destroyed. No system is perfect, though, and in fact a small number of autoaggressive T cells slip through. Escaping into the bloodstream and into lymph vessels, they create the immune system's potential for unleashing autoimmune disease.

Blood and lymph vessels are where a second line of defense comes into play. This layer of protection against autoimmunity has several facets. Certain tissues, including those of the brain and spinal cord, are concealed from immune cell patrols simply by having a paucity of blood and lymph vessels that penetrate deep into the tissue. Their isolation, however, is not absolute, and at times, such as when the tissues are injured, self-reactive immune cells can find a way in. Additional modes of protection are more proactive. Immune cells showing an inappropriate interest in the body's own tissues can be targeted for destruction or rendered quiescent by other immune system components.

Among the immune cells that carry out these proactive roles, regulatory T cells may well be the most crucial. The majority, if not all of them, learn their "adult" roles within the thymus, as other T cells do, then go forth and persist throughout the body as a specialized T cell subpopulation.

DISCOVERING THE PEACEKEEPERS

Findings hinting at the existence of regulatory T cells date back surprisingly far. In 1969 Yasuaki Nishizuka and Teruyo Sakakura, working at the Aichi Cancer Center Research Institute in Nagoya, Japan, showed that removing the thymus from newborn female mice had a curious outcome: the animals lost their ovaries. At first it was thought that the thymus must secrete some kind of hormone needed for survival of the developing ovaries. Later, though, it turned out that immune system cells invaded the ovaries. The ovarian destruction was therefore an autoimmune disease, which had presumably been unleashed by the animals' loss of a countervailing regulatory process. If the mice were inoculated with normal T cells, the autoimmune disease was inhibited. T cells, then, could at times police themselves somehow.

In the early 1970s John Penhale of the University of Edinburgh made analogous observations in adult rats, and Richard J Gershon of Yale University became the first to propose the existence of a T cell population capable of damping immune responses, including autoaggressive ones. This hypothetical immune system member was christened the suppressor T cell. At the time, though, no researcher was able to actually find one or pinpoint the molecular action by which one immune system cell could restrain another. Consequently, the concept of the suppressor T cell languished along the fringes of mainstream immunology.

Despite the negative atmosphere, some researchers persisted in trying to identify T cells with an ability to prevent autoimmune disease. The basic hope was to discover a telltale molecular feature at the surface of such cells—a "marker" by which suppressor T cells could be distinguished from other cells. Beginning in the mid-1980s, various candidate markers were explored.

In 1995 one of us (Sakaguchi) finally demonstrated that a molecule called CD25 was a reliable marker. When, in studies of mice, he removed CD4₊ T cells displaying that molecule, organs such as the thyroid, stomach, gonads, pancreas and salivary glands came under an autoimmune attack

MECHANISMS OF TOLERANCE

T-reg cells help to ensure that immune system components—including T cells that fight infections—refrain from attacking normal tissues. The thymus, where all T cell varieties mature, directly eliminates many strongly autoreactive cells (*left*), but its vigilance is imperfect, so T-regs patrol the body in search of renegades (*right*).

T-regs resemble helper T cells, orchestrators of immune responses. Both display a T cell receptor (TCR), which can lock onto a particular antigen: a substance perceived as nonself. Both cell types also exhibit a so-called co-receptor named CD4. But T-regs differ in displaying a molecule called CD25, which is why they are also known as CD4+ CD25+ T cells; CD25 is a component of a receptor for interleukin-2 (IL-2), which promotes T-reg activities. T-regs also contain high amounts of the protein Foxp3, which confers the ability to quiet other T cells. When T-regs encounter autoreactive T cells, they disable them.

characterized by dramatic inflammation: white blood cells swarmed into the organs and damaged them.

In an important confirmatory experiment, T cell populations obtained from normal mice were depleted of their CD4+ CD25+ T cells, which evidently made up only a small proportion (at most, 10 percent) of the overall T cell pool. Then T cells left behind were transferred to mice engineered to lack an immune system of their own. This maneuver caused autoimmune disease. And the more complete the depletion was in the donor animals, the more severe the spectrum of disease became in the recipients—with comprehensive depletion often proving to be fatal. Reintroducing CD4+ CD25+ T cells, even in small numbers, conferred normal immunity and protected the animals from these disorders. Experiments conducted wholly in test tubes also produced valuable confirmatory evidence. Perhaps to absolve "suppressor cells" of any lingering stigma, immunologists started to call them CD25+ regulatory T cells, or simply T-regs.

HOW DO T-REGS WORK?

To this day, the precise ways in which T-regs suppress autoimmune activity have remained mysterious, making their function a continuing subject of intense inquiry. The cells appear capable of suppressing a wide variety of immune system cells, impeding the cells' multiplication and also their other activities, such as secretion of cell-to-cell chemical signals (cytokines). And researchers tend to agree that T-regs are activated by direct cell-to-cell contacts. Beyond that, the picture is rather murky.

Recently, however, our laboratory at Kyoto University and, independently, Alexander Rudensky's group at the University of Washington and Fred Ramsdell's group at CellTech R&D in Bothell, Wash., found a fresh clue as to how T-regs develop and function. The cells contain a large amount of an intracellular molecule called Foxp3. In fact, the enrichment is greater than has been reported for any other T-reg molecular feature.

Foxp3 is a transcription factor: a molecule that regulates the activity of specific genes, thereby controlling a cell's production of the protein that each such gene encodes. Because proteins are the main worker molecules in cells, altered production of one or more of them can affect how a cell functions. In the case of Foxp3, the changes it induces in gene activity apparently turn developing T cells into T-regs. Indeed, artificially introducing Foxp3 into otherwise unremarkable T cells provokes a reprogramming, by which the cells acquire all the suppressive abilities of full-fledged T-regs produced by the thymus. A type of mouse called the Scurfy strain, long known to researchers, has recently been found to have only an inactive, mutant form of the Foxp3 protein, along with a total absence of T-regs. The consequence is an immune system gone haywire, with massive inflammation in numerous organs, leading to the animals' early death.

Of course, investigators study T-regs in animals such as mice so that the knowledge gained may be applied to humans. So what evidence is there that T-regs are indeed important in humans—or that they exist in us at all?

It turns out that the molecular features characteristic of T-regs in rodents are also characteristic of a subset of T cells in humans. In humans, as in rodents, these cells exhibit the CD25 molecule and have a high content of Foxp3. In addition, the cells are immunosuppressive, at least in a test tube.

Perhaps the most compelling indications that they are vital to human health come from a rare genetic abnormality called IPEX (immune dysregulation, polyendocrinopathy, enteropathy, X-linked syndrome). Arising from mutations in a gene on the X chromosome, IPEX affects male children, who unlike females inherit only one X chromosome and hence have no chance of inheriting a second, normal copy of the gene, which would encode a healthy version of the affected protein. In males the mutation results in autoimmune disease affecting multiple organs, including the

HOW DO T-REGS PREVENT AUTOIMMUNITY?

No one fully understands how T-regs block autoimmune attacks. Three reasonable possibilities follow. All three involve interfering with a key step in triggering immune response: signaling between T cells and antigen-presenting cells (APCs). Before helper T cells will call forth other troops and before "cytotoxic" T cells will attack tissue perceived to be infected, APCs must display antigens for the cells' perusal. If the T cell receptor (TCR) of a helper or cytotoxic cell recognizes a displayed antigen and also receives certain other signals from the APC, the T cells will become active against the bearer of that antigen—even if the antigen is from the body itself, instead of from an infectious agent. The TCRs of T-regs also recognize particular antigens, and they specifically suppress T cells that focus on those same antigens.

T-REG OUTCOMPETES OTHER T CELLS

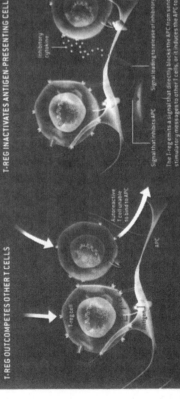

By binding to an APC, the T-reg prevents other T cells from latching on

T-REG INACTIVATES ANTIGEN-PRESENTING CELL

The T-reg emits a signal that directly blocks the APC from sending stimulatory messages to other T cells, or it induces the APC to actively suppress the other cells, for example, by releasing signaling molecules (cytokines) having inhibitory effects.

T-REG QUIETS OTHER T CELLS DIRECTLY

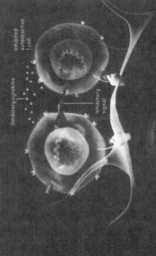

The T-reg uses the APC essentially as a platform for stabilizing contact with another T cell bound to the APC. Then the T-reg sends an inhibiting signal directly into the T cell or emits inhibiting molecules that act at close range.

thyroid and (as happens in insulin-dependent diabetes) the pancreas, and also in chronic intestinal inflammation (inflammatory bowel disease) and uncontrolled allergy (food allergy and severe skin inflammation), all of which can be understood as varied manifestations of the hyperactivity of an immune system unrestrained by T-regs. Death comes in infancy or soon after, with contributing causes ranging from autoimmune diabetes to severe diarrhea. The specific genetic flaw underlying IPEX has recently proved to be mutation in none other than *Foxp3*. IPEX is therefore the human counterpart of the illness in Scurfy mice.

BEYOND SELF-TOLERANCE

The evidence, then, indicates that T-regs do prevent autoimmune disease in humans. But the cells also appear to serve health in other ways, including participating (in some surprising ways) in responses to microbes.

Throughout the 1990s Fiona Powrie and her colleagues at the DNAX Research Institute in Palo Alto, Calif., experimented with transferring T cell populations depleted of T-regs into mice engineered to lack an immune system of their own. In one set of studies, the transfer induced a severe, often fatal form of inflammatory bowel disease. But the aberrant immune activity was not directed primarily at bowel tissue itself.

The bowels of rodents, like those of humans, are home to a vast bacterial population, typically more than a trillion for every gram of intestinal tissue. Although these bacteria are foreign, they are usually far from harmful; indeed, they promote the digestion of food and even displace dangerous bacteria, such as salmonella, that would otherwise try to colonize the intestines. Normally the immune system tolerates the presence of the helpful population. But in Powrie's mice, it attacked. And in doing so, the transplanted immune cells caused collateral damage to the recipient's gut. Yet transfer of T-regs caused no problems. In fact, if the T-regs were transferred along with the other T cells, they prevented the bowel disease that would otherwise have ensued. Overall, the immune system appeared to be on a hair trigger, prepared to assault gut bacteria and held in check only by T-regs.

A similar hair trigger may affect the immune system's responses to harmful foreigners. On the one hand, T-regs might rein in an overemphatic response. On the other hand, the reining in might keep an invader from being totally destroyed, enabling it to persist and potentially flare up again. For example, some findings suggest that failure to clear the stomach of a bacterium called *Helicobacter pylori*, now known to cause stomach ulcer, stems from blunting by T-regs of the immune system's weaponry.

Work by David Sacks and his colleagues at the National Institutes of Health has revealed further complexity. It implies that leaving a few survivors among invading organisms may not be entirely a bad thing. The researchers infected mice with a fairly innocuous parasite. Even when the immune system was fully intact, it allowed a small number of parasites to remain, after which reinfection triggered a prompt, efficient response. If the immune system was depleted of its T-regs, however, the parasite was completely purged, but reinfection was dealt with inefficiently, as if the mice had never before encountered the invader. Hence, T-regs appear to contribute to maintaining immunological memory, a process that is crucial for immunity to repeated infection and that also underlies the success of vaccination.

Research hints, too, at a role for T-regs in protecting pregnancies. Every pregnancy unavoidably poses quite a challenge to the mother's immune defenses. Because the fetus inherits half its genes from the father, it is genetically half-distinct from its mother and thus is in essence an organ transplant. Within the trophoblast, the placental tissue that attaches the fetus to the uterine wall, a number of mechanisms give the fetus some safety from what would amount to transplant rejection. The trophoblast not only presents a physical barrier to would-be attackers in the mother's blood but also produces immunosuppressive molecules.

The mother's immune system seems to undergo changes as well. Reports of women in whom an autoimmune disease such as multiple sclerosis abates during pregnancy provide anecdotal evidence that T-regs become more active. Some recent experiments offer more direct support. At the University of Cambridge, Alexander Betz and his colleagues have shown that during pregnancy in mice, maternal T-regs expand in number. Conversely, an experimentally engineered absence of T-regs leads to fetal rejection marked by a massive infiltration of immune cells across the maternal-fetal boundary. It is tempting to speculate that in some women, insufficient T-reg activity may underlie recurrences of spontaneous abortion.

RECRUITING THE REGULATORS

In T-regs, nature clearly has crafted a potent means of controlling immune responses. Tapping into this control would make T-regs a potentially powerful therapeutic ally against a wide range of medical disorders.

It is still too early to expect to see applications in doctors' offices, but the available data suggest that delivering T-regs themselves, or perhaps medicines that increase or decrease their activity, could provide novel treatments for a variety of conditions. Indeed, some human trials are under way.

The most obvious application would involve enhancing T-reg activity to fight autoimmune diseases, and drug therapy is being explored in patients with multiple sclerosis and psoriasis, among other conditions. Pumping up T-reg activity might also be useful for treating allergies. The ease with which T-regs can keep immune responses at bay suggests that T-reg-based therapies could hold particular promise for preventing rejection of transplanted organs. The ideal would be for transplant recipients to tolerate grafts as well as they do their own tissues. Also ideal would be a tolerance that endures as a permanent state of affairs, without need for immunosuppressive drugs, which can have many side effects.

The opposite type of T-reg-based therapy would be a selective depletion of T-regs to counter unwanted immunosuppression and, consequently, to strengthen beneficial immune responses. In practice, a partial depletion might be preferred to a complete one, because it should pose less risk of inducing autoimmune disease. Best of all would be removal solely of those T-regs that were specifically blocking a useful immune response. The depletion strategy might be especially advantageous against infectious diseases that the immune system, left to itself, tends to combat inadequately—perhaps tuberculosis or even AIDS.

In addition, T-reg reduction might be advantageous for fighting cancer. Much evidence suggests that circulating immune cells keep a lookout for molecular aberrations that occur as a cell becomes cancerous. To the extent that T-regs impede this surveillance, they might inadvertently help a malignancy take root and grow. In fact, some cancers appear to encourage such help: they secrete molecular signals capable of attracting T-regs and of converting non-T-regs into T-regs. Some findings suggest, for example, that cancer patients have abnormally high numbers of active T-regs both in their blood and in the tumors themselves. Much of today's research into therapeutic manipulations of T-regs focuses on cancer.

TECHNICAL CHALLENGES

So far investigators are finding it challenging to develop medicines able to deplete or expand T-reg populations within a patient's body. To be most useful, these drugs would usually need to act on the subsets of

T-regs that have roles in a particular disorder, yet scientists often do not know precisely which T-regs to target.

Devising therapies based on administering T-regs themselves is difficult as well. One of the main obstacles is the need to obtain enough of the cells. Although researchers have found that T-regs can operate at low abundance relative to the cells they are suppressing, control of a human autoimmune disease would probably require tens of millions of T-regs. Acquiring such numbers of these relatively rare cells from a person's circulation might be impossible. Accordingly, some technique to expand their numbers outside the body would seem to be imperative.

Luckily, it also seems that this numbers game can be won. Worldwide, several research groups have reported that cells with immunosuppressive actions can be generated in relatively large numbers by treating ordinary T cells with a well-defined "cocktail" of biochemical signals. Whether the engendered cells, termed Tr1 cells, are identical to T-regs remains unclear, but it is beyond dispute that the cells are profoundly immunosuppressive.

Now that Foxp3 is known to be a key molecule controlling the development and function of T-regs, investigators may also be able to tailor-make large numbers of regulatory cells by using fairly standard laboratory techniques to transfer the *Foxp3* gene into more prevalent, and thus more easily obtainable, types of T cells. We and others are pursuing this approach intently and are also trying to identify the molecular events that switch on Foxp3 production during T-reg development. This knowledge might enable pharmaceutical researchers to fashion drugs specifically for that purpose, so that processing of cells outside the body and then infusing them would not be necessary.

For organ transplant patients, another way to obtain useful T-regs is under consideration. The procedure would involve removing T-regs from a prospective transplant recipient and culturing them with cells from the organ donor in a way that causes the T-regs most capable of suppressing rejection to multiply. In rodents, T-regs generated in this manner have worked well. One of us (Sakaguchi) has shown, for example, that injection of a single dose of such T-regs at the time of skin grafting results in the graft's permanent acceptance, even though transplanted skin typically is rejected strongly. Meanwhile the treatment left the rest of the immune system intact and ready to fend off microbial invaders. The abundant research into T-regs suggests that such an approach can become a reality for humans and could be used to protect new transplant recipients until medications able to produce the same benefit more simply are developed.

One day T-reg-based therapy may help preserve transplanted organs while limiting the amount of time a patient has to take immunosuppressive drugs, which can have undesirable side effects. A protocol might look like the following:

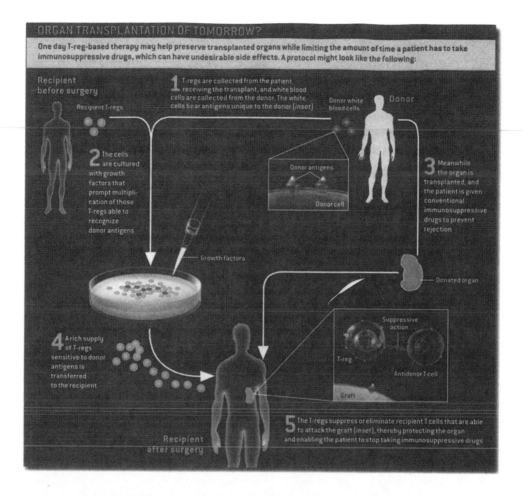

Recipient before surgery

Recipient T-regs

1 T-regs are collected from the patient receiving the transplant, and white blood cells are collected from the donor. The white cells bear antigens unique to the donor (*inset*)

Donor white blood cells

Donor

2 The cells are cultured with growth factors that prompt multiplication of those T-regs able to recognize donor antigens

Donor antigens

Donor cell

3 Meanwhile the organ is transplanted, and the patient is given conventional immunosuppressive drugs to prevent rejection

Growth factors

Donated organ

4 A rich supply of T-regs sensitive to donor antigens is transferred to the recipient

Suppressive action

T-reg

Antidonor T cell

Graft

Recipient after surgery

5 The T-regs suppress or eliminate recipient T cells that are able to attack the graft (*inset*), thereby protecting the organ and enabling the patient to stop taking immunosuppressive drugs

Some T-Reg-Based Therapies under Study

The therapies listed below are among those in or likely to enter human trials. Most of the drugs under study aim to deplete or inhibit T-reg cells, so as to increase antitumor immune responses normally tempered by the cells. Delivery of such agents into the body would need to be managed carefully, though, to ensure that reducing T-reg activity does not lead to autoimmunity.

EFFECT ON T-REGS	EXAMPLES OF DISORDERS BEING TARGETED	TREATMENT APPROACHES
Depletion or inhibition (to enhance immunity)	Cancers of the skin (melanoma), ovary, kidney	A toxin fused to a substance, such as interleukin-2, able to deliver the toxin to T-regs Monoclonal antibodies (which bind to specific molecules) that have shown an ability to induce T-reg death or to block the cells' migration into tumors
Multiplication in patient (to dampen autoimmunity)	Multiple sclerosis, psoriasis, Crohn's disease, insulin-dependent diabetes	Vaccine composed of T cell receptor constituents thought to stimulate T-reg proliferation A monoclonal antibody that appears to stimulate T-regs by binding to a molecule called CD3
Multiplication in the lab, for delivery to patient	Graft-versus-host disease (immune cells in donated bone marrow attack recipient tissue)	Culture donor T-regs with selected antibodies and growth factors, then deliver resulting T-reg population before or at the time of bone marrow transplant (for prevention) or if graft-versus-host disease arises

Over the past decade, researchers' understanding of the immune system and how it governs its own actions has changed profoundly. In particular, it is now recognized that although the system permits potentially autodestructive T cells to circulate, it also deploys T cells capable of controlling them. Knowledge of how they develop and how they perform their remarkable immunosuppressive activities will be key in recruiting them for use against a host of debilitating and even fatal disorders. In permitting destruction of nonself while preventing destruction of self, T-regs may prove to be the ultimate immunological peacekeepers.

FURTHER READING

NATURALLY ARISING CD4+ REGULATORY T CELLS FOR IMMUNOLOGIC SELF-TOLERANCE AND NEGATIVE CONTROL OF IMMUNE RESPONSES. Shimon Sakaguchi in *Annual Review of Immunology,* Vol. 22, pages 531–562; 2004.

REGULATORY T-CELL THERAPY: IS IT READY FOR THE CLINIC? J. A. Bluestone in *Nature Reviews Immunology,* Vol. 5, No. 4, pages 343–349; April 2005.

REGULATORY T CELLS, TUMOUR IMMUNITY AND IMMUNOTHERAPY. Weiping Zou in *Nature Reviews Immunology,* Vol. 6, No. 4, pages 295–307; April 2006.

T LYMPHOCYTES: REGULATORY. Zoltan Fehervari and Shimon Sakaguchi in *Encyclopedia of Life Sciences.* Wiley InterScience, 2006. Available at www.els.net

New Predictors of Disease

ABNER LOUIS NOTKINS

ORIGINALLY PUBLISHED IN MARCH 2007

A middle-aged woman—call her Anne—was taken aback when one day her right hand refused to hold a pen. A few weeks later her right foot began to drag reluctantly behind her left. After her symptoms worsened over months, she consulted a neurologist. Anne, it turned out, was suffering from multiple sclerosis, a potentially disabling type of autoimmune disease. The immune system normally jumps into action in response to bacteria and viruses, deploying antibodies, other molecules and various white blood cells to recognize and destroy trespassers. But in autoimmune disorders, components of the body's immune system target one or more of the person's own tissues. In Anne's case, her defensive system had begun to turn against her nerves, eroding her ability to move.

Every story of autoimmune disease is sad—but collectively the impact of these illnesses is staggering. More than 40 autoimmune conditions have been identified, including such common examples as type 1 (insulin-dependent) diabetes, rheumatoid arthritis and celiac disease. Together they constitute the third leading cause of sickness and death after heart disease and cancer. And they afflict between 5 and 8 percent of the U.S. population, racking up an annual medical bill in the tens of billions of dollars.

Recent findings offer a way to brighten this gloomy picture. In the past 10 years a growing number of studies have revealed that the body makes certain antibodies directed against itself—otherwise known as autoantibodies—years, and sometimes a decade, before autoimmunity causes clinical disease, damaging tissues so much that people begin showing symptoms. This profound insight is changing the way that doctors and researchers think about autoimmune conditions and how long they take to arise. It suggests that physicians might one day screen a healthy person's blood for certain autoantibodies and foretell whether a specific disease is likely to develop years down the line. Armed with such predictions, patients could start fighting the ailment with drugs or other available interventions, thereby preventing or delaying symptoms.

Those interventions may not be easy to find; most likely, preventive therapy would have to be tailored specifically for each condition. In certain disorders, such as myasthenia gravis, autoantibodies participate in the disease process, and so blocking the activity of the particular autoantibodies at fault could be therapeutic. Autoantibodies that presage certain other conditions, though, probably are more siren than fire, announcing brewing disease actually caused by other components of the immune system, such as cells known as T lymphocytes and macrophages. In those cases, preventive treatments would have to target the offending cells.

The revolution in predictive medicine and preventive care will take time and effort to effect. Many autoantibodies have been uncovered, but only a few large-scale trials have been conducted to evaluate how accurately they can predict disease. If inexpensive, quick tests for predictive autoantibodies can be developed, though, they could become as standard a part of routine checkups as cholesterol monitoring.

EARLY INSIGHT FROM DIABETES

People familiar with advances in genetics might wonder why researchers would want to develop tests for predictive autoantibodies when doctors might soon be able to scan a person's genes for those that put the individual at risk of various disorders. The answer is that most chronic diseases arise from a complex interplay between environmental influences and multiple genes (each of which makes but a small contribution to a disease). So detection of susceptibility genes would not necessarily reveal with any certainty whether or when an individual will come down with a particular autoimmune condition. In contrast, detection of specific autoantibodies would signal that a disease-causing process was already under way. Eventually, genetic screening for those with an inherited predisposition to a disease may help reveal those who need early autoantibody screening.

Studies of patients with type 1 diabetes provided the first clues that autoantibodies could be valuable for predicting later illness. In this condition, which typically arises in children or teenagers, the immune system ambushes the beta cells in the pancreas. These cells are the manufacturers of insulin, a hormone that enables cells to take up vital glucose from the blood for energy. When the body lacks insulin, cells starve and blood glucose levels soar, potentially leading to blindness, kidney failure, and a host of other complications.

Forty years ago type 1 diabetes was not yet recognized as an autoimmune disease, and no one knew what caused the beta cells to die. But in the 1970s Willy Gepts of Vrije University of Brussels in Belgium examined the pancreases of children who had died of the disease and found that the islets of Langerhans, where the beta cells reside, had been infiltrated by lymphocytes—a sign of probable autoimmune activity. Soon thereafter Franco Bottazzo of Middle-sex Hospital Medical School in London established that blood from patients with type 1 diabetes reacted to islets but that blood of nondiabetics did not, which suggested that autoantibodies targeted to the diabetics' own beta cells were circulating in the patients' blood. This finding set off a hunt for the autoantigens—the specific molecular targets of the autoantibodies—in the beta cells, because researchers hoped that discovery of the autoantigens would clarify how diabetes arises.

Intensive research over the past 20 years has uncovered three major pancreatic autoantigens produced in people with newly diagnosed type 1 diabetes: insulin itself, an enzyme called glutamic acid decarboxylase (GAD) and a protein known as islet antigen-2 (I A-2), which was discovered by my group at the National Institutes of Health and is a component of the tiny sacs that ferry insulin around in beta cells. Experts still do not know whether the autoantibodies that bind these proteins play a part in killing beta cells. But they do know, based on highly sensitive detection tests, that one or more are present at diagnosis in some 70 to 90 percent of patients with type 1 diabetes. Today research laboratories use these tests to diagnose type 1 diabetes and distinguish it from type 2 diabetes, which usually arises in overweight adults and does not stem from autoimmunity. (Surprisingly, such tests have uncovered autoantibodies in about 5 percent of patients otherwise diagnosed with type 2 diabetes, which suggests that those individuals have been misclassified or have a combination of type 1 and type 2 diabetes.)

Interest in the three autoantibodies escalated with the discovery that they appear long before the onset of diabetic symptoms. In studies conducted by various laboratories, investigators took blood samples from thousands of healthy schoolchildren and then monitored the youngsters' health for up to 10 years. When a child came down with type 1 diabetes, the researchers pulled the individual's blood sample out of storage to see whether it contained autoantibodies. The vast majority of children destined to become diabetic had one or more of the three signature diabetes-related autoantibodies in their blood as long as 10 years before any recognizable symptoms arose.

Before this work, some experts thought that type 1 diabetes developed suddenly, perhaps within a matter of weeks. The new data demonstrated,

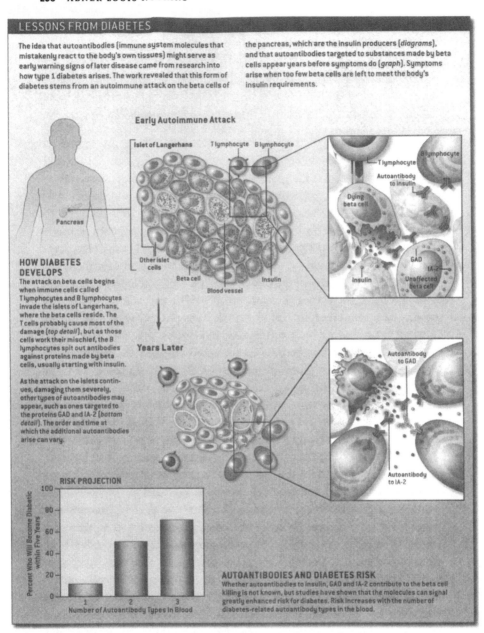

LESSONS FROM DIABETES

The idea that autoantibodies (immune system molecules that mistakenly react to the body's own tissues) might serve as early warning signs of later disease came from research into how type 1 diabetes arises. The work revealed that this form of diabetes stems from an autoimmune attack on the beta cells of the pancreas, which are the insulin producers (*diagrams*), and that autoantibodies targeted to substances made by beta cells appear years before symptoms do (*graph*). Symptoms arise when too few beta cells are left to meet the body's insulin requirements.

Early Autoimmune Attack

Islet of Langerhans · T lymphocyte · B lymphocyte · B lymphocyte · T lymphocyte · Autoantibody to insulin · Dying beta cell · Pancreas · Other islet cells · Beta cell · Blood vessel · Insulin · GAD · IA-2 · Insulin · Unaffected beta cell

HOW DIABETES DEVELOPS

The attack on beta cells begins when immune cells called T lymphocytes and B lymphocytes invade the islets of Langerhans, where the beta cells reside. The T cells probably cause most of the damage (*top detail*), but as those cells work their mischief, the B lymphocytes spit out antibodies against proteins made by beta cells, usually starting with insulin.

As the attack on the islets continues, damaging them severely, other types of autoantibodies may appear, such as ones targeted to the proteins GAD and IA-2 (*bottom detail*). The order and time at which the additional autoantibodies arise can vary.

Years Later

Autoantibody to GAD · Autoantibody to IA-2

RISK PROJECTION

Percent Who Will Become Diabetic within Five Years

100 — 80 — 60 — 40 — 20 — 0

Number of Autoantibody Types in Blood: 1 · 2 · 3

AUTOANTIBODIES AND DIABETES RISK

Whether autoantibodies to insulin, GAD and IA-2 contribute to the beta cell killing is not known, but studies have shown that the molecules can signal greatly enhanced risk for diabetes. Risk increases with the number of diabetes-related autoantibody types in the blood.

instead, that in most cases the immune system silently assails the pancreas for years until so many beta cells die that the organ can no longer make enough insulin for the body's needs. That is the point when the classic early symptoms of diabetes arise, such as excessive hunger, thirst and urination.

More important, these studies also raised the prospect that doctors might forecast whether a child is at risk for type 1 diabetes by testing blood for the presence of these autoantibodies. Clinical researchers found that an individual with one autoantibody has a 10 percent risk of showing symptoms within five years. With two autoantibodies, the chance of disease jumps to 50 percent; with three autoantibodies, the threat rockets to between 60 and 80 percent.

The ability to predict whether a person is likely to fall ill with type 1 diabetes has had major repercussions for medical researchers trying to better understand and prevent the disease. Before the discovery of predictive autoantibodies, for example, it was almost impossible to conduct clinical trials of new preventive therapies, because the disorder is relatively rare, affecting about one individual in 400. Such odds meant that more than 40,000 subjects would have to be entered into a trial in order to assess the effects of an intervention on the 100 who would eventually be affected.

Now scientists can select for study those people whose blood shows two or more of the diabetes-related autoantibodies, because at least half the subjects, if untreated, will most likely come down with the disease within five years. Slashing the number of subjects who must be enrolled in a prevention trial has made such experiments feasible for the first time. In one investigation, doctors identified several thousand individuals at high risk of diabetes and tested whether injections of insulin could avert the disease. Sadly, this treatment proved unsuccessful; efforts to find useful interventions continue.

The discovery that autoantibodies frequently herald the onset of type 1 diabetes prompted scientists to examine whether the same might be true in other autoimmune diseases. One that has been the focus of especially intense research is rheumatoid arthritis, a debilitating condition that is highly prevalent, afflicting about 1 percent of the world's population. In those affected, the immune system attacks and destroys the lining of the joints, causing swelling, chronic pain and eventual loss of movement.

PREDICTING OTHER DISEASES

Immunologists have recently unearthed an autoantibody that is present in 30 to 70 percent of patients diagnosed with rheumatoid arthritis. The antibody latches onto citrulline (a modified version of the amino acid arginine), which is present in certain proteins. Studies have now revealed that the autoantibody appears in the bloodstream before the first symptoms turn up, in some cases more than 10 years before. Further, the likelihood that the illness will develop is as much as 15 times greater in people carrying that antibody than in those who lack it.

The knowledge that the anticitrulline autoantibody might serve as a predictive marker is particularly exciting because, in contrast to the situation in type 1 diabetes, doctors already have medicines that might be delivered to prevent or slow the onset of arthritis. Rheumatologists know that quickly and aggressively treating newly diagnosed patients with certain drugs, such as ones that combat inflammation, can retard or sometimes stop the devastating loss of joint flexibility. It is not unreasonable to think therefore that earlier intervention might be even more protective. The hope now is that doctors will be able to screen the general population, or those with a family history of the condition, and then start treating those who make anticitrulline antibodies before autoimmunity irrevocably harms their tissues. First, however, further clinical trials must be carried out to confirm that these autoantibodies accurately predict the onset of joint symptoms. In addition, a reasonably priced test suitable for screening will have to be introduced, along with protocols for deciding exactly who should be tested, when and how often.

For certain other autoimmune disorders, the detection of predictive autoantibodies could potentially enable people to shut down autoimmune activity by avoiding certain triggers in their environment. A case in point is celiac disease. In people with this condition, the gluten protein found in wheat, rye and barley incites the immune system to attack the lining of the small intestine, which then fails to absorb food properly; diarrhea, weight loss and malnourishment then ensue. Patients must eliminate gluten from their diet, bypassing most bread, pasta and cereal for the rest of their lives.

Investigations into the underpinnings of celiac disease have revealed that many patients make an autoantibody that reacts with tissue transglutaminase, an enzyme that modifies many newly made proteins. This autoantibody emerges up to seven years before symptoms do, suggesting that high-risk individuals might forestall the disease entirely by eliminating gluten from their diet. This idea has not yet been tested, however.

MORE USES FOR AUTOANTIBODY TESTS

Immunologists are exploring whether autoantibodies can serve as early warnings in other ways as well. For instance, some autoantibodies might help doctors to gauge the rate at which an already diagnosed autoimmune condition is likely to progress or how severe it will become.

Patients with multiple sclerosis often start off with relatively mild symptoms that then disappear for a while. Some people continue in remission for a long time or have manageable recurrences. But others grapple with more frequent or severe symptoms, and a few enjoy no remissions at all. Doctors struggle to discern which individuals with early symptoms will go on to suffer from the harshest effects, so that they can counsel the patients accordingly. In 2003 a study of more than 100 individuals with newly identified multiple sclerosis revealed that those who made autoantibodies directed against two proteins that insulate nerve cells were almost four times more likely to suffer a relapse after the initial symptoms abated than were those without the autoantibodies. In addition, the antibody-positive patients relapsed more quickly than the others. These results suggest that testing for these autoantibodies could offer a quick way to predict whether, and how rapidly, multiple sclerosis will advance, although further study is needed before such testing can be put into practice and used to guide therapy.

In the past few years, researchers have made the intriguing finding that autoantibodies can also appear in people with certain disorders not typically thought of as autoimmune conditions, such as some cancers. These autoantibodies probably do not control tumor growth, but laboratories around the world are examining whether they can be useful for the early detection of cancer. In other conditions, such as atherosclerosis, investigators are looking into the possibility that autoantibodies might

Disorders under Study

Investigators have found autoantibodies that might serve as predictors of risk or of progression for a number of autoimmune conditions beyond diabetes, including those listed below.

DISORDER	STATUS OF RESEARCH
Addison's disease (a disorder of the adrenal glands; results in low blood pressure, weakness and weight loss)	Autoantibodies to adrenal tissue and the enzyme 21-hydroxylase are highly predictive in children
Antiphospholipid syndrome (marked by recurrent clots in blood vessels and pregnancy loss)	Autoantibodies to various molecules appear to signal risk for complications of the disorder
Celiac disease (a digestive disorder triggered by gluten in foods)	Predictive autoantibodies have been identified that target an enzyme called tissue transglutaminase
Multiple sclerosis (neurological condition causing loss of movement)	Autoantibodies to proteins in the myelin sheath that insulates nerve cells appear to predict risk of relapse
Rheumatoid arthritis (chronic joint inflammation)	Autoantibodies to citrulline, a component of many modified proteins, have been found to appear as many as 10 years before symptoms occur
Systemic lupus erythematosus (can affect many organs, including the joints, kidneys and skin)	Several related autoantibodies have been found; one or more of these appear in up to 80 percent of patients before symptoms arise

show which patients are more prone to a blockage in the arteries to the brain and therefore to stroke.

SCIENTIFIC CHALLENGES

So far much of the work I have discussed has been confined to a small number of academic laboratories and to a few of the major autoimmune diseases. Investigators and companies, however, are now beginning to recognize the potential value of these proteins for improving patient care. They are trying to extend the findings and unearth predictive auto-antibodies linked to other autoimmune disorders.

This task is challenging, however, in part because researchers will have to follow large populations for years to prove that particular autoantibodies can signal future disease. That is, many thousands of healthy people must be recruited to give blood samples and then tracked carefully for 10 years or more to see if they fall sick. Aside from posing logistical difficulties, these prospective studies can cost tens of millions of dollars.

An alternative to conducting prospective studies from scratch might be to rap into existing health databases and carry out retrospective studies. For example, blood samples and medical information have already been collected over many years from members of the U.S. military and from subjects in the Women's Health Initiative, a vast, ongoing study of more than 100,000 women. Experts in autoimmunity could team up with investigators in these and other projects, identify individuals who have been diagnosed with an autoimmune disease and then examine their stored blood for the presence of predictive autoantibodies. This approach would be relatively inexpensive and could yield rapid results—and a few researchers have already embarked on such collaborations.

A second avenue of attack would involve identifying heretofore unrecognized autoantigens. One could search human genetic databases for the sequences that encode proteins and use this information to manufacture these proteins in the lab. Scientists could pinpoint those that are auto-antigens by mixing each of the manufactured proteins with blood from patients who have an autoimmune disease and allowing complexes of proteins and antibodies to form. Analyses of such complexes could identify both the autoantigens in the collection and the autoantibodies that recognize them. That done, the predictive value of the autoantibodies could be determined in a prospective or retrospective study.

This full-genome approach to isolating autoantigens is difficult. Nevertheless, a handful of research groups are now screening smaller batches of proteins in this way. In my laboratory, for example, we are hunting for new autoantigens involved in type 1 diabetes by manufacturing dozens of selected pancreatic proteins known to be involved in the secretion of insulin and testing whether autoantibodies blood from diabetics bind these proteins.

PRACTICAL CHALLENGES

Medicine as we know it is evolving from diagnosing and treating diseases after they develop to predicting and preventing them. Ten or 20 years from now autoantibody screening for at least some diseases will almost certainly become a familiar part of the standard medical examination.

In the future, patients visiting their doctors for a physical might have their blood tested for multiple predictive autoantibodies in a single test. In one plausible scenario, the doctor would send a blood sample to a lab for an autoantibody analysis, along with standard tests for cholesterol, blood glucose and other health indicators. There a machine would pass the blood over a tiny chip displaying an array of known autoantigens. Autoantibodies in the blood that bound to one or more of these antigens would trigger pulses of light that would be picked up by a detector. Within hours, the doctor would receive a readout translating this information into a health forecast. The presence of predictive antibodies would not mean that a patient will definitely get sick, but would give a percentage risk of diabetes and numerous other conditions developing over some number of months or years.

These tests might even be combined with other biological assays to give more accurate health predictions. In the case of type 1 diabetes, possession of certain forms of genes that regulate the immune system, called HLA genes, are also known to correlate with disease risk. A prognostic assay might combine tests for those HLA variants with tests for predictive autoantibodies.

The vision of prediction is an enticing one, but even after the challenges of identifying predictive autoantibodies have been overcome, other issues will need to be resolved in preparation for their use in the clinic. One critical question relates to cost. At present, lab screening for predictive autoantibodies is cumbersome and labor-intensive. Widespread population screening for multiple autoantibodies will become practical only when rapid, inexpensive automated methods for detecting them are designed. To date, only a few small biotechnology companies are trying to devise such methods.

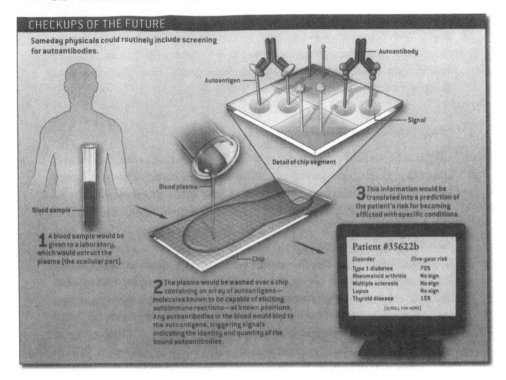

CHECKUPS OF THE FUTURE

Someday physicals could routinely include screening for autoantibodies.

Autoantibody

Autoantigen

Signal

Detail of chip segment

Blood plasma

Blood sample

1 A blood sample would be given to a laboratory, which would extract the plasma (the acellular part).

Chip

2 The plasma would be washed over a chip containing an array of autoantigens— molecules known to be capable of eliciting autoimmune reactions—at known positions. Any autoantibodies in the blood would bind to the autoantigens, triggering signals indicating the identity and quantity of the bound autoantibodies.

3 This information would be translated into a prediction of the patient's risk for becoming afflicted with specific conditions.

Patient #35622b

Disorder	Five-year risk
Type 1 diabetes	70%
Rheumatoid arthritis	No sign
Multiple sclerosis	No sign
Lupus	No sign
Thyroid disease	15%

[SCROLL FOR MORE]

Another issue to be decided is which people should be screened and how often. It is unreasonable to test children for diseases that occur only in adults, and the converse would also hold true. Similarly, the frequency of screening would have to depend in part on whether autoantibodies tend to arise many years or just a few months before the onset of clinical symptoms.

By far the most important factor controlling whether predictive autoantibody screening will become widespread is the availability of therapy. Some would argue that embarking on predictive testing for diseases makes little sense if patients can be offered no preventive or ameliorating treatment. A large and intensive research effort is under way to develop new therapies for autoimmune diseases, but because the conditions are so complicated and varied, progress may not come quickly.

Of course, the ability to forecast someone's life and death raises thorny ethical issues. Some people may choose not to know that they are likely to come down with a given disease, and doctors must be careful to respect that decision. Patients may also be concerned that insurers or employers could obtain medical information and use it to discriminate against them, even while they are healthy. As is true of genetic testing, such issues call for in-depth discussion.

Forecasts of the future have always intrigued and frightened people. Handled properly, though, such knowledge could benefit the millions of patients and doctors destined to battle autoimmune diseases. By making early intervention possible, predictive autoantibodies have the potential to alleviate much misery and to help provide extra years of healthy life.

FURTHER READING

IMMUNOLOGIC AND GENETIC FACTORS IN TYPE 1 DIABETES. Abner Louis Notkins in *Journal of Biological Chemistry*, Vol. 277, No. 46, pages 43545–43548; November 15, 2002.

ANTIMYELIN ANTIBODIES AS A PREDICTOR OF CLINICALLY DEFINITE MULTIPLE SCLEROSIS AFTER A FIRST DEMYELINATING EVENT. Thomas Berger et al. in *New England Journal of Medicine*, Vol. 349, No. 2, pages 139–145; July 10, 2003.

DEVELOPMENT OF AUTOANTIBODIES BEFORE THE CLINICAL ONSET OF SYSTEMIC LUPUS ERYTHEMATOSUS. Melissa R. Arbuckle et al. in *New England Journal of Medicine*, Vol. 349, No. 16, pages 1526–1533; October 16, 2003.

AUTOANTIBODIES AS DIAGNOSTIC AND PREDICTIVE MARKERS OF AUTOIMMUNE DISEASES. Edited by Abner Louis Notkins, Ake Lernmark and David Leslie in *Autoimmunity*, Vol. 37, No. 4, pages 251–368; June 2004.

DIAGNOSTIC AND PREDICTIVE VALUE OF ANTI-CYCLIC CITRULLINATED PROTEIN ANTIBODIES IN RHEUMATOID ARTHRITIS: A SYSTEMATIC LITERATURE REVIEW. J. Avouac, L. Gossec and M. Dougados in *Annals of Rheumatic Diseases*, Vol. 65, No. 7, pages 845–851; July 2006.

Taming Lupus

MONCEF ZOUALI

ORIGINALLY PUBLISHED IN MARCH 2005

A 24-year-old woman undergoes medical evaluation for kidney failure and epilepsy-like convulsions that fail to respond to antiepileptic drugs. Her most visible sign of illness, though, is a red rash extending over the bridge of her nose and onto her cheeks, in a shape resembling a butterfly.

A 63-year-old woman insists on hospitalization to determine why she is fatigued, her joints hurt, and breathing sometimes causes sharp pain. Ever since her teen years she has avoided the sun, which raises painful blistering rashes wherever her skin is unprotected.

A 20-year-old woman is surprised to learn from a routine health exam that her urine has an abnormally high protein level—a sign of disturbed kidney function. A renal biopsy reveals inflammation.

Although the symptoms vary, the underlying disease in all three patients is the same—systemic lupus erythematosus, which afflicts an estimated 1.4 million Americans, including one out of every 250 African-American women aged 18 to 65. It may disrupt almost any part of the body: skin, joints, kidneys, heart, lungs, blood vessels or brain. At times, it becomes life-threatening.

Scientists have long known that, fundamentally, lupus arises from an immunological malfunction involving antibody molecules. The healthy body produces antibodies in response to invaders, such as bacteria. These antibodies latch onto specific molecules that are sensed as foreign (antigens) on an invader and then damage the interloper directly or mark it for destruction by other parts of the immune system. In patients with lupus, however, the body produces antibodies that perceive its own molecules as foreign and then launch an attack targeted to those "self-antigens" on the body's own tissues.

Self-attack—otherwise known as autoimmunity—is thought to underpin many diseases, including type 1 diabetes, rheumatoid arthritis, multiple sclerosis and, possibly, psoriasis. Lupus, however, is at an extreme. The immune system reacts powerfully to a surprising variety of the patient's molecules, ranging from targets exposed at the surface of cells to some inside of cells to even some within a further sequestering chamber, the cell nucleus. In fact, lupus is notorious for the presence of antibodies that take aim at the patient's DNA. In the test tube, these anti-DNA "autoantibodies" can directly digest genetic material.

Until recently, researchers had little understanding of the causes of this multipronged assault. But clues from varied lines of research are beginning to clarify the underlying molecular events. The work is also

THE DIAGNOSTIC CHALLENGE

Physicians who suspect that a patient, whether female or male, has lupus continue to be hampered by the lack of a conclusive test. Because immunological self-attack may underlie many illnesses, even a classic sign of lupus—the presence of antinuclear autoantibodies—does not unmistakably diagnose this disorder.

In the absence of a sure test, doctors might gather information from a variety of sources, including not only laboratory tests but also the patient's description of current symptoms and medical history. To assist, the American College of Rheumatology has issued a list of 11 criteria that could indicate lupus. Seven concern symptoms, such as arthritis, sensitivity to sunlight or a butterfly facial rash. (The butterfly pattern is still unexplained.) The other four describe laboratory findings that include the presence of antinuclear autoantibodies or depressed concentrations of lymphocytes.

Researchers will consider a subject to have lupus if the person meets four of the criteria, but physicians might base a diagnosis on fewer cues, especially if a patient has strong indicators of the disorder, such as clinical evidence of abnormalities in several different organ systems combined with the presence of antinuclear autoantibodies. For more on common manifestations of lupus, visit the Lupus Foundation of America: www.lupus.org/ or the Lupus site: www.uklupus.co.uk/

Current Criteria

Malar rash (a rash, often butterfly-shaped, over the cheeks)

Discoid rash (a type involving red raised patches)

Photosensitivity (reaction to sunlight in which a skin rash arises or worsens)

Nose or mouth ulcers, typically painless

Nonerosive arthritis (which does not involve damage to the bones around the joints) in two or more joints

Inflammation of the lining in the lung or heart (also known as pleuritis or pericarditis)

Kidney disorder marked by high levels in the urine of protein or of abnormal substances derived from red or white blood cells or kidney tubule cells

Neurological disorder marked by seizures or psychosis not explained by drugs or metabolic disturbances (such as an electrolyte imbalance)

Blood disorder characterized by abnormally low concentrations of red or white blood cells or platelets (specifically, hemolytic anemia, leukopenia, lymphopenia or thrombocytopenia) and not caused by medications

Positive test for antinuclear antibodies (ANA) not explained by drugs known to trigger their appearance

Positive test for antibodies against double-stranded DNA or certain phospholipids or a false positive result on a syphilis test probing the most basic, yet still enigmatic, facets of immune system function: the distinction of self from nonself; the maintenance of self-tolerance (nonaggression against native tissues); and the control over the intensity of every immune response. The discoveries suggest tantalizing new means of treating or even preventing not only lupus but also other autoimmune illnesses.

SOME GIVENS

One thing about lupus has long been clear: the auto antibodies that are its hallmark contribute to tissue damage in more than one way. In the blood, an autoantibody that recognizes a particular self-antigen can bind to that antigen, forming a so-called immune complex, which can then deposit itself in any of various tissues. Autoantibodies can also recognize self-antigens already in tissues and generate immune complexes on-site. Regardless of how the complexes accumulate, they spell trouble.

For one, they tend to recruit immune system entities known as complement molecules, which can directly harm tissue. The complexes, either by themselves or with the help of the complement molecules, also elicit an inflammatory response. This response involves an invasion by white blood cells that attempt to wall off and destroy any disease-causing agents. Inflammation is a protective mechanism, but if it arises in the absence of a true danger or goes on for too long, the inflammatory cells and their secretions can harm the tissues they are meant to protect. Inflammation can additionally involve the abnormal proliferation of cells native to an affected tissue, and this cellular excess can disrupt the normal functioning of the tissue. In the kidney, for instance, immune complexes can accumulate in glomeruli, the organ's blood-filtering knots of capillary loops. Excessive deposition then initiates glomerulonephritis, a local inflammatory reaction that can lead to kidney damage.

Beyond inciting inflammation, certain lupus autoantibodies do harm directly. In laboratory experiments, they have been shown to bind to and then penetrate cells. There they become potent inhibitors of cellular function.

The real mystery about lupus is what precedes such events. Genetic predisposition seems to be part of the answer, at least in some people. About 10 percent of patients have close blood relatives with the disease, a pattern that usually implies a genetic contribution. Moreover, investigators have found greater lupus concordance—either shared lupus or a shared absence of it—in sets of identical twins (who are genetically indistinguishable) than in sets of fraternal twins (whose genes generally are no more alike than it, those of other pairs of siblings).

GENETIC HINTS

Spurred by such findings, geneticists are hunting for the genes at fault, including those that confer enhanced susceptibility to the vast majority of patients who have no obvious family history of the disease. Knowledge of the genes, the proteins they encode and the normal roles of these proteins should one day help clarify how lupus develops and could point to ways to better control it.

In mice prone to lupus, the work has identified more than 30 fairly broad chromosomal regions associated to some extent either with lupus or with resistance to it. Some are tied to specific elements of the disease. One region, for example, apparently harbors genes that participate in producing autoantibodies that recognize components of the cell nucleus (although the region itself does not encode antibodies); another influences the severity of the kidney inflammation triggered by lupus-related immune complexes.

In human lupus, the genetic story may be even more mind-boggling. An informative approach scans DNA from families with multiple lupus patients to identify genetic features shared by the patients but not by the other family members. Such work has revealed a connection between lupus and 48 chromosomal regions. Six of those regions (on five different chromosomes) appear to influence susceptibility most. Now investigators have to identify the lupus-related genes within those locales.

Already it seems fair to conclude that multiple human genes can confer lupus susceptibility, although each gene makes only a hard-to-detect contribution on its own. And different combinations of genes might lay the groundwork for lupus in different people. But clearly, single genes are rarely, if ever, the primary driver; if they were, many more children born to a parent with lupus would be stricken. Lupus arises in just about 5 percent of such children, and it seldom strikes in multiple generations of a family.

MANY TRIGGERS

If genes alone rarely account for the disease, environmental contributors must play a role. Notorious among these is ultraviolet light. Some 40 to 60 percent of patients are photosensitive: exposure to sunlight, say for 10 minutes at midday in the summer, may suddenly cause a rash. Prolonged exposure may also cause flares, or increased symptoms. Precisely how it does so is still unclear. In one scenario, ultraviolet irradiation induces changes in the DNA of skin cells, rendering the DNA molecules alien (from the viewpoint of the body's immune defenses) and thus potentially antigenic. At the same time, the irradiation makes the cells prone to breakage, at which point they will release the antigens, which can then provoke an autoimmune response.

Environmental triggers of lupus also include certain medications, among them hydralazine (for controlling blood pressure) and procainamide (for irregular heartbeat). But symptoms usually fade when the drugs are discontinued. In other cases, an infection, mild or serious, may act as a lupus trigger or aggravator. One suspect is Epstein-Barr virus, perhaps best known for causing infectious mononucleosis, or "kissing disease." Even certain vaccines may provoke a lupus flare. Yet despite decades of research, no firm proof of a bacterium, virus or parasite that transmits lupus has been put forth. Other possible factors include diets

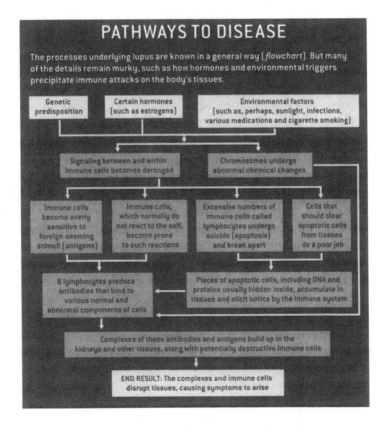

PATHWAYS TO DISEASE

The processes underlying lupus are known in a general way (*flowchart*). But many of the details remain murky, such as how hormones and environmental triggers precipitate immune attacks on the body's tissues.

Genetic predisposition	Certain hormones (such as estrogens)	Environmental factors (such as, perhaps, sunlight, infections, various medications and cigarette smoking)

Signaling between and within immune cells becomes deranged

Chromosomes undergo abnormal chemical changes

Immune cells become overly sensitive to foreign-seeming stimuli (antigens)

Immune cells, which normally do not react to the self, become prone to such reactions

Excessive numbers of immune cells called lymphocytes undergo suicide (apoptosis) and break apart

Cells that should clear apoptotic cells from tissues do a poor job

B lymphocytes produce antibodies that bind to various normal and abnormal components of cells

Pieces of apoptotic cells, including DNA and proteins usually hidden inside, accumulate in tissues and elicit notice by the immune system

Complexes of these antibodies and antigens build up in the kidneys and other tissues, along with potentially destructive immune cells

END RESULT: The complexes and immune cells disrupt tissues, causing symptoms to arise high in saturated fat, pollutants, cigarette smoking, and perhaps extreme physical or psychological stress.

PERILS OF CELL SUICIDE

Another line of research has revealed cellular and molecular abnormalities that could well elicit or sustain autoimmune activity. Whether these abnormalities are usually caused more by genetic inheritance or by environmental factors remains unknown. People may be affected by various combinations of influences.

One impressive abnormality involves a process known as apoptosis, of cell suicide. For the body to function properly, it has to continually eliminate cells that have reached the end of their useful life or turned dangerous. It achieves this pruning by inducing the cells to make proteins that essentially destroy the cell from within—such as by hacking to pieces cellular proteins and the chromosomes in the nucleus. But the rate of apoptosis in certain cells—notably, the B and T lymphocytes of the immune system—is excessive in those who have lupus.

When cells die by apoptosis, the body usually disposes of the remains efficiently. But in those with lupus, the disposal system seems to be defective. This double whammy of increased apoptosis and decreased clearance can promote autoimmunity in a fairly straightforward way: if the material inside the apoptotic cells is abnormal, its ejection from the cells in quantity could well elicit the production of antibodies that mistakenly perceive the aberrant material as a sign of invasion by a disease-causing agent. And such antibody production is especially likely if the ejected material, rather than being removed, accumulates enough to call attention to itself.

Unfortunately, the material that spills from apoptotic cells of those with lupus, especially the chromosomal fragments, is often abnormal. In healthy cells, certain short sequences of DNA carry methyl groups that serve as tags controlling gene activity. The DNA in circulating immune complexes from lupus patients is undermethylated. Scientists have several reasons to suspect that this methylation pattern might contribute to autoimmunity. In the test tube, abnormally methylated DNA can stimulate a number of cell types involved in immunity, including B lymphocytes, which, when mature, become antibodyspewing factories. (Perhaps the body misinterprets these improperly methylated stretches as sign that a disease causing agent is present and must be eliminated.) Also, certain drugs known to cause lupus symptoms lead to undermethylation of DNA in T cells, which leads to T cell autoreactivity in mice.

B CELLS GONE WRONG

B lymphocytes normally respond only to foreign substances, or antigens, such as bacteria. But in people with lupus the cells react to the body's own molecules, generating antibodies that bind to those "self-antigens" and then accumulate in tissues. There the antibody-antigen complexes lead to tissue damage. Several therapies under study for lupus aim to delete B cells or to block one or another of the molecular interactions that lead to antibody production and tissue injury. Red arrows in the diagram point to molecules targeted by the drugs listed in the table on page 254.

The stage for antibody production is set when antigen-presenting cells take up and degrade potential antigens (a). They fit selected fragments into MHC molecules and display the resulting complexes on the cell surface for perusal by helper T cells (b).

If a T cell binds to such a complex and also links to a B7 molecule on the antigen-presenting cell, the signals generated by the binding will activate the T cell, causing it to proliferate (c) and undergo changes that enable it to stimulate B cells (d). In particular, the T cells begin to secrete stimulatory cytokines, or signaling molecules, and to display a molecule called CD154 that can lock onto CD40 on B cells.

B cell activation also depends on several other signaling events (e), including attachment of the T cell receptor to the same antigen-MHC complex it saw on the antigen-presenting cell, stimulation of the B cell by a molecule called BAFF, and binding of antigen by B cell receptors. A receptor known as CD20 may also participate in B cell activation, although its exact function is still unclear.

Having received the needed stimulation, a B cell matures into an antibody-secreting plasma cell (f), and the antibodies go off to attack or mark for destruction any cells or tissues possessing the antigen recognized by the antibodies (g). For instance, antibodies attract complement molecules and inflammatory cells, both of which can be destructive (inset).

All in all, the findings suggest that apoptotic cells are a potential reservoir of autoantigens that are quite capable of provoking an autoantibody response. In further support of this idea, intravenous administration of large quantities of irradiated apoptotic cells is able to induce autoantibody synthesis in normal mice.

Hence, part of the underlying process leading to the formation of destructive immune complexes may involve the body's production of foreign-seeming antigens, which cause the body to behave as if tissues bearing those antigens were alien and threatening. But other work indicates that, in addition, the B lymphocytes of lupus patients are inherently deranged; they are predisposed to generate autoantibodies even when the self-molecules they encounter are perfectly normal. In other words, the mechanisms that should ensure self-tolerance go awry.

DERANGED CELLS

The problem mostly seems to stem from signaling imbalances within B cells. In the healthy body, a B cell matures into an antibody-secreting machine—known as a plasma cell—only after antibodylike projections on the B cell's surface (B cell receptors) bind to a foreign antigen. If a B cell instead attaches to a self-component, this binding normally induces the cell to kill itself, to retreat into a nonresponsive (anergic) state or to "edit" its receptors until they can no longer recognize the self-antigen.

Whether the cell responds appropriately depends in large measure on the proper activity of the internal signaling pathways that react to external inputs. Mouse studies show that even subtle signaling imbalances can predispose animals to produce antibodies against the self. And various lines of evidence indicate that certain signaling molecules (going by such names as Lyn, CD5 and SHP-1) on and in B cells of patients with lupus are present in abnormal amounts.

Other work suggests that it is not only the B cells that are deranged. For a B cell to become an antibody maker, it must do more than bind to an antigen. It must also receive certain stimulatory signals from immune system cells known as helper T lymphocytes. Helper cells of lupus patients are afflicted by signaling abnormalities reminiscent of those in the B cells. The T cell aberrations, though, may lead to autoantibody production indirectly—by causing the T cells to inappropriately stimulate self-reactive B cells.

All theorizing about the causes of lupus must account not only for the vast assortment of autoantibodies produced by patients but also for another striking aspect of the disease: the disorder is 10 times as common in women than in men. It also tends to develop earlier in women (during childbearing years). This female susceptibility—a pattern also seen in some other autoimmune diseases—may stem in part from greater immune reactivity in women. They tend to produce more antibodies and lymphocytes than males and, probably as a result, to be more resistant to infections. Among mice, moreover, females reject foreign grafts more rapidly than the males do. Perhaps not surprisingly, sex hormones seem to play a role in this increased reactivity, which could explain why, in laboratory animals, estrogens exacerbate lupus and androgens ameliorate it.

Estrogens could pump up immune reactivity in a few ways. They augment the secretion of prolactin and growth hormone, substances that contribute to the proliferation of lymphocytes, which bear receptor molecules sensitive to estrogens. Acting through such receptors, estrogens may modulate the body's immune responses and may even regulate lymphocyte development, perhaps in ways that impair tolerance of the self.

TOWARD NEW THERAPIES

Those of us who study the causes of lupus are still pondering how the genetic, environmental and immunological features that have been uncovered so far collaborate to cause the disease. Which events come first, which are most important, and how much do the underlying processes differ from one person to another? Nevertheless, the available clues suggest at least a partial scenario for how the disease could typically develop.

The basic idea is that genetic susceptibilities and environmental influences may share responsibility for an impairment of immune system function—more specifically an impairment of the signaling within lymphocytes and possibly within other cells of the immune system, such as those charged with removing dead cells and debris. Faulty signaling, in turn, results in impaired self-tolerance, accelerated lymphocyte death, and defective disposal of apoptotic cells and the self-antigens they release. Abundantly available to the body's unbalanced immune surveillance, the antigens then misdirect the immune system, inducing it to attack the self.

Drugs do exist for lupus, but so far they focus on dampening the overall immune system. In other words, they are nonspecific: instead of targeting immunological events underlying lupus in particular, they dull the body's broad defenses against infectious diseases. Corticosteroids, for instance, reduce inflammation at the cost of heightening susceptibility to infections.

The challenge is to design new drugs that prevent autoimmune self-attacks without seriously hobbling the body's ability to defend itself against infection. To grasp the logic of the approaches being attempted, it helps to know a bit more about how helper T cells usually abet the transformation of B cells into vigorous antibody makers [see *illustration on page 250*].

First, the helper cells themselves must be activated, which occurs through interactions with so-called professional antigen presenting cells (such as macrophages and dendritic cells). These antigen presenters ingest bacteria, dead cells and cellular debris, chop them up, join the fragments to larger molecules (called MHC class II molecules) and display the resulting MHC-antigen complexes on the cell surface. If the receptor on a helper T cell recognizes a complex and binds to it, the binding conveys an antigen-specific signal into the T cell. If, at the same time, a certain T cell projection near the receptor links to a particular partner (known as a B7 molecule) on the antigen-presenting cell, this binding will convey an antigen-independent, or co-stimulatory, signal into the T cell. Having received both messages, the T cell will switch on; that is, it will produce or display molecules needed to activate B cells and will seek out those cells.

Like the professional antigen-presenting cells, B cells display fragments of ingested material—notably fragments of an antigen they have snared—on MHC class II molecules. If an activated helper T cell binds through its receptor to such a complex on a B cell, and if the T and B cells additionally signal each other through co-stimulatory surface molecules, the B cell will display receptors for small proteins called cytokines. These cytokines, which are secreted by activated helper T cells, induce the B cell to proliferate and mature into a plasma cell, which dispatches antibodies that specifically target the same antigen recognized by the coupled B and T cells.

Of course, any well-bred immune response shuts itself off when the danger has passed. Hence, after an antigen-presenting cell activates a helper T cell, the T cell also begins to display a "shutoff" switch known as CTLA-4. This molecule binds to B7 molecules on antigen-presenting cells so avidly that it links to most or even all of them, thereby putting a break on any evolving helper T and, consequently, B cell responses.

One experimental approach to treating lupus, essentially mimics this shutoff step, dispatching CTLA-4 to cap over B7 molecules. In mice prone to lupus, this method prevents kidney disease from progressing and prolongs life. This substance is beginning to be tested in lupus patients; in those with psoriasis, initial clinical trials have shown that the treatment is safe.

A second approach would directly impede the signaling between helper T cells and B cells. The T cell molecule that has to "clasp hands" with a B cell molecule to send the needed co-stimulatory signal into B cells is called CD154. The helper cells of lupus patients show increased production of CD154, and in mice prone to the disease, antibodies engineered to bind to CD154 can block B cell activation, preserve kidney function and prolong life. So far early human tests of different versions of anti-CD154 antibodies have produced a mixture of good news and bad. One version significantly reduced autoantibodies in the blood, protein in the urine and certain symptoms, but it also elicited an unacceptable degree of blood-clot formation. A different version did not increase thrombosis but worked poorly. Hence, no one yet knows whether this approach to therapy will pan out.

A third strategy would interfere with B cell activity in a different way. Certain factors secreted by immune system cells, such as the cytokine BAFF, promote cell survival after they bind to B cells. These molecules have been implicated in various autoimmune diseases, including lupus and its flares: mice genetically engineered to overproduce BAFF or one of its three receptors on B cells develop signs of autoimmune disease, and BAFF appears to be overabundant both in lupus-prone mice and in human patients. In theory, then, preventing BAFF from binding to its receptors should minimize antibody synthesis. Studies of animals and humans support this notion. In mice, a circulating decoy receptor, designed to mop up BAFF before it can find its true receptors, alleviates lupus and lengthens survival. Findings for a second decoy receptor are also encouraging. Human trials are in progress.

Targeting other cytokines might help as well. Elevated levels of interleukin-10 and depressed amounts of transforming growth factor

Some Treatment Strategies under Study

	TYPE OF AGENT	STATUS
1	Blocker of B7's interaction with CD28, to impede activation of helper T cells	Immune Tolerance Network, a research consortium, and the National Institutes of Health are undertaking a small human trial of a blocker called RG2077
2	Blocker of BAFF's interaction with its receptor, to keep BAFF (also called BLyS) from promoting B cell survival and antibody production	Human Genome Sciences (Rockville, Md.) is evaluating one such drug, LymphoStat-B, in a multicenter trial; ZymoGenetics (Seattle) and Serono S.A. (Geneva, Switzerland) are conducting an early human trial of an agent named TACI-Ig
3	Blocker of B cell receptors and of antibodies that recognize the body's own DNA, to inhibit the production and activity of antibodies that target such DNA	La Jolla Pharmaceuticals (San Diego) is conducting a multicenter trial of abetimus sodium (Riquent) against lupus-related kidney disease
4	Antibody to CD20, to deplete B cells	Genentech (South San Francisco, Calif.) and Biogen Idec (Cambridge, Mass.) are conducting a multicenter lupus trial of rituximab (Rituxan), a drug already approved for B cell cancer
5	Complement inhibitor, to prevent complement-mediated tissue damage	Alexion Pharmaceuticals (Cheshire, Conn.) found evidence of disease amelioration in mice given an inhibitor of complement C5

beta are among the most prominent cytokine abnormalities reported in lupus, and lupus-prone mice appear to benefit from treatments that block the former or boost the latter. Taking a different tack, investigators studying various autoimmune conditions are working on therapies aimed specifically at reducing B cell numbers. An agent called rituximab, which removes B cells from circulation before they are able to secrete antibodies, has shown promise in early trials in patients with systemic lupus.

Some other therapies under investigation include molecules designed to block production of anti-DNA autoantibodies or to induce those antibodies to bind to decoy compounds that would trap them and provoke their degradation. An example of such a decoy is a complex consisting of four short DNA strands coupled to an inert backbone. Although the last idea is intriguing, I have to admit that the effects of introducing such decoys are apt to be complex.

Certain cytokines might be useful as therapies, but these and other protein drugs could be hampered by the body's readiness to degrade circulating proteins. To circumvent such problems, researchers are considering gene therapies, which would give cells the ability to produce useful proteins themselves. DNA encoding transforming growth factor beta has already been shown to treat lupus in mice, but too few tests have been done yet in humans to predict how useful the technique will be in people. Also, scientists are still struggling to perfect gene therapy techniques in general.

As treatment-oriented investigators pursue new leads for helping patients, others continue to probe the central enigmas of lupus. What causes the aberrant signaling in immune cells? And precisely how does such deranged signaling lead to autoimmunity? The answers may well be critical to finally disarming the body's mistaken attacks on itself.

FURTHER READING

DUBOIS' LUPUS ERYTHEMATOSUS. Sixth edition. Daniel J. Wallace and Bevra H. Hahn. Lippincott Williams & Wilkins, 2001.

IMMUNOBIOLOGY: THE IMMUNE SYSTEM IN HEALTH AND DISEASE. Sixth edition. Charles A. Janeway, Paul Travers, Mark Walport and Mark J. Shlomchik. Garland Science, 2004.

B LYMPHOCYTE SIGNALING PATHWAYS IN SYSTEMIC AUTOIMMUNITY: IMPLICATIONS FOR PATHOGENESIS AND TREATMENT. Moncef Zouali and Gabriella Sarmay in Arthritis & Rheumatism, Vol. 50, No. 9, pages 2730–2741; September 2004.

MOLECULAR AUTOIMMUNITY. Edited by Moncef Zouali. Springer Science and Business Media, 2005.

www.lupusresearch.org

GLOBAL MANAGEMENT AND TREATMENT ISSUES

Preparing for a Pandemic

W. WAYT GIBBS AND CHRISTINE SOARES

ORIGINALLY PUBLISHED IN NOVEMBER 2005

When the levees collapsed in New Orleans, the faith of Americans in their government's ability to protect them against natural disasters crumbled as well. Michael Chertoff, the secretary of homeland security who led the federal response, called Hurricane Katrina and the flood it spawned an "ultracatastrophe" that "exceeded the foresight of the planners."

But in truth the failure was not a lack of foresight. Federal, state and local authorities had a plan for how governments would respond if a hurricane were to hit New Orleans with 120-mile-per-hour winds, raise a storm surge that overwhelmed levees and water pumps, and strand thousands inside the flooded city. Last year they even practiced it. Yet when Katrina struck, the execution of that plan was abysmal.

The lethargic, poorly coordinated and undersized response raises concerns about how nations would cope with a much larger and more lethal kind of natural disaster that scientists warn will occur, possibly soon: a pandemic of influenza. The threat of a flu pandemic is more ominous, and its parallels to Katrina more apt, than it might first seem. The routine seasonal upsurges of flu and of hurricanes engender a familiarity that easily leads to complacency and inadequate preparations for the "big one" that experts admonish is sure to come.

The most fundamental thing to understand about serious pandemic influenza is that, except at a molecular level, the disease bears little resemblance to the flu that we all get at some time. An influenza pandemic, by definition, occurs only when the influenza virus mutates into something dangerously unfamiliar to our immune systems and yet is able to jump from person to person through a sneeze, cough or touch.

Flu pandemics emerge unpredictably every generation or so, with the last three striking in 1918, 1957 and 1968. They get their start when one of the many influenza strains that constantly circulate in wild and domestic birds evolves into a form that infects us as well. That virus then adapts further or exchanges genes with a flu strain native to humans to produce a novel germ that is highly contagious among people.

Some pandemics are mild. But some are fierce. If the virus replicates much faster than the immune system learns to defend against it, it will cause severe and sometimes fatal illness, resulting in a pestilence that could easily claim more lives in a single year than AIDS has in 25. Epidemiologists have warned that the next pandemic could sicken one in every three people on the planet, hospitalize many of those and kill tens to hundreds of millions. The disease would spare no nation, race or income group. There would be no certain way to avoid infection.

Scientists cannot predict which influenza strain will cause a pandemic or when the next one will break out. They can warn only that another is bound to come and that the conditions now seem ripe, with a fierce strain of avian flu killing people in Asia and infecting birds in a rapid westward lunge toward Europe. That strain, influenza A (H5N1) does not yet pass readily from one person to another. But the virus is evolving, and some of the affected avian species have now begun their winter migrations.

As a sense of urgency grows, governments and health experts are working to bolster four substantial lines of defense against a pandemic: surveillance, vaccines, containment measures and medical treatments. The U.S. plans to release by October a pandemic preparedness plan that surveys the strength of each of these barricades. Some failures are inevitable, but the more robust those preparations are, the less humanity will suffer. The experience of Katrina forces a question: Will authorities be able to keep to their plans even when a large fraction of their own workforce is downed by the flu?

SURVEILLANCE: WHAT IS INFLUENZA UP TO NOW?

Our first defense against a new flu is the ability to see it coming. Three international agencies are coordinating the global effort to tracks H5N1 and other strains of influenza. The World Health Organization (WHO), with 110 influenza centers in 83 countries, monitors human cases. The World Organization for Animal Health (OIE, formerly the Office International des Épizooties) and the Food and Agriculture Organization (FAO) collect reports on outbreaks in birds and other animals. But even the managers of these surveillance nets acknowledge that they are still too porous and too slow.

Speed is of the essence when dealing with a fast-acting airborne virus such as influenza. Authorities probably have no realistic chance of halting a nascent pandemic unless they can contain it within 30 days [see "Rapid Response," on page 268]. The clock begins ticking the moment that the first victim of a pandemic-capable strain becomes contagious.

The only way to catch that emergence in time is to monitor constantly the spread of each outbreak and the evolution of the virus's abilities. The WHO assesses both those factors to determine where the world is in the pandemic cycle, which a new guide issued in April divides into six phases.

The self-limiting outbreaks of human H5N1 influenza seen so far bumped the alert level up to phase three, two steps removed from outright pandemic (phase six). Virologists try to obtain samples from every new H5N1 patient to scout for signs that the avian virus is adapting to infect humans more efficiently. It evolves in two ways: gradually through random mutation, and more rapidly as different strains of influenza swap genes inside a single animal or person.

The U.S. has a sophisticated flu surveillance system that funnels information on hospital visits for influenzalike illness, deaths from respiratory illness and influenza strains seen in public health laboratories to the Centers for Disease Control and Prevention in Atlanta. "But the system is not fast enough to take the isolation or quarantine action needed to manage avian flu," said Julie L. Gerberding, the CDC director, at a February conference. "So we have been broadening our networks of clinicians and veterinarians."

In several dozen cases where travelers to the U.S. from H5N1-affected Asian countries developed severe flulike symptoms, samples were rushed to the CDC, says Alexander Klimov of the CDC's influenza branch. "Within 40 hours of hospitalization we can say whether the patient has H5N1. Within another six hours we can analyze the genetic sequence of the hemagglutinin gene" to estimate the infectiousness of the strain. (The virus uses hemagglutinin to pry its way into cells.) A two-day test then reveals resistance to antiviral drugs, he says.

The next pandemic could break out anywhere, including in the U.S. But experts think it is most likely to appear first in Asia, as do most influenza strains that cause routine annual epidemics. Aquatic birds such as ducks and geese are the natural hosts for influenza, and in Asia many villagers reside cheek by bill with such animals. Surveillance in the region is still spotty, however, despite a slow trickle of assistance from the WHO, the CDC and other organizations.

A recent H5N1 outbreak in Indonesia illustrates both the problems and the progress. In a relatively wealthy suburb of Jakarta, the eight-year-old

EVOLUTION OF AN EPIDEMIC

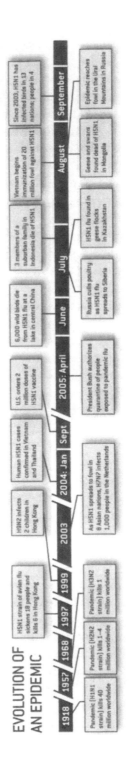

1918 — Pandemic (H1N1 strain) kills 40 million worldwide

1957 — Pandemic (H2N2 strain) kills 1–4 million worldwide

1968 — Pandemic (H3N2 strain) kills 1 million worldwide

1997 — H5N1 strain of avian flu sickens 18 people and kills 6 in Hong Kong

1999 — H9N2 infects 2 children in Hong Kong

2003 — As H5N1 spreads to fowl in 8 Asian nations, H7N7 infects 1,000 people in the Netherlands

2004: Jan — Human H5N1 cases confirmed in Vietnam and Thailand

Sept — U.S. orders 2 million doses of H5N1 vaccine

2005: April — President Bush authorizes quarantine of people exposed to pandemic flu

June — 6,000 wild birds die from H5N1 flu at a lake in central China; Russia culls poultry as H5N1 flu spreads to Siberia

July — 3 members of a suburban family in Indonesia die of H5N1; H5N1 flu found in geese flocks in Kazakhstan

August — Vietnam begins immunization of 20 million fowl against H5N1; Geese and swans found dead of H5N1 in Mongolia

September — Since 2003, H5N1 has infected birds in 13 nations; people in 4; Epidemic reaches fowl in the Ural Mountains in Russia

daughter of a government auditor fell ill in late June. A doctor gave her antibiotics, but her fever worsened, and she was hospitalized on June 28. A week later her father and one-year-old sister were also admitted to the hospital with fever and cough. The infant died on July 9, the father on July 12.

The next day an astute doctor alerted health authorities and sent blood and tissue samples to a U.S. Navy medical research unit in Jakarta. On July 14 the girl died; an internal report shows that on this same day Indonesian technicians in the naval laboratory determined that two of the three family members had H5N1 influenza. The government did not acknowledge this fact until July 22, however, after a WHO lab in Hong Kong definitively isolated the virus.

The health department then readied hospital wards for more flu patients, and I Nyoman Kandun, head of disease control for Indonesia, asked WHO staff to help investigate the outbreak. Had this been the onset of a pandemic, the 30-day containment window would by that time have closed. Kandun called off the investigation two weeks later. "We could not find a clue as to where these people got the infection," he says.

Local custom prohibited autopsies on the three victims. Klaus Stöhr of the WHO Global Influenza Program has complained that the near absence of autopsies on human H5N1 cases leaves many questions unanswered. Which organs does H5N1 infect? Which does it damage most? How strongly does the immune system respond?

Virologists worry as well that they lave too little information about the role of migratory birds in transmitting the disease across borders. In July domestic fowl infected with H5N1 began turning up in Siberia, then Kazakhstan, then Russia. How the birds caught the disease remains a mystery.

Frustrated with the many unanswered questions, Stöhr and other flu scientists have urged the creation of a global task force to supervise pandemic preparations. The OIE in August appealed for more money to support surveillance programs it is setting up with the FAO and the WHO.

"We clearly need to improve our ability to detect the virus," says Bruce G. Gellin, who coordinates U.S. pandemic planning as head of the National Vaccine Program Office at the U.S. Department of Health and Human Services (HHS). "We need to invest in these countries to help them, because doing so helps everybody."

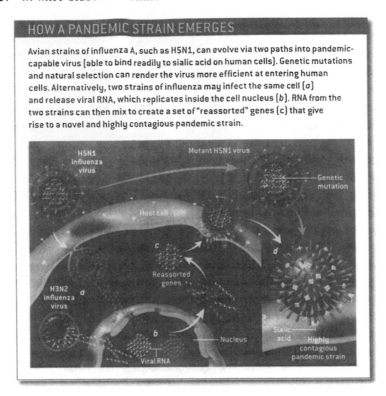

HOW A PANDEMIC STRAIN EMERGES

Avian strains of influenza A, such as H5N1, can evolve via two paths into pandemic-capable virus (able to bind readily to sialic acid on human cells). Genetic mutations and natural selection can render the virus more efficient at entering human cells. Alternatively, two strains of influenza may infect the same cell (*a*) and release viral RNA, which replicates inside the cell nucleus (*b*). RNA from the two strains can then mix to create a set of "reassorted" genes (*c*) that give rise to a novel and highly contagious pandemic strain.

VACCINES: WHO WILL GET THEM—AND HOW QUICKLY?

Pandemics of smallpox and polio once ravaged humanity, but widespread immunization drove those diseases to the brink of extinction. Unfortunately, that strategy will not work against influenza—at least not without a major advance in vaccine technology.

Indeed, if an influenza pandemic arrives soon, vaccines against the emergent strain will be agonizingly slow to arrive and frustratingly short in supply. Biology, economics and complacency all contribute to the problem.

Many influenza strains circulate at once, and each is constantly evolving. "The better the match between the vaccine and the disease virus, the better the immune system can defend against the virus," Gellin explains. So every year manufacturers fashion a new vaccine against the three most threatening strains. Biologists first isolate the virus and then modify it using a process called reverse genetics to make a seed virus. In vaccine factories, robots inject the seed virus into fertilized eggs laid by hens bred under hygienic conditions. The pathogen replicates wildly inside the eggs.

Vaccine for flu shots is made by chemically dissecting the virus and extracting the key proteins, called antigens, that stimulate the human immune system to make the appropriate antibodies. A different kind of vaccine, one inhaled rather than injected, incorporates live virus that has been damaged enough that it can infect but not sicken. The process requires six months to transform viral isolates into initial vials of vaccine.

Because people will have had no prior exposure to a pandemic strain of influenza, everyone will need two doses: a primer and then a booster about four weeks later. So even those first in line for vaccines are unlikely to develop immunity until at least seven or eight months following the start of a pandemic.

And there will undoubtedly be a line. Total worldwide production of flu vaccine amounts to roughly 300 million doses a year. Most of that is made in Europe; only two plants operate in the U.S. Last winter, when contamination shut down a Chiron facility in Britain, Sanofi Pasteur and MedImmune pulled out all stops on their American lines—and produced 61 million doses. The CDC recommends annual flu immunization for high-risk groups that in the U.S. include some 185 million people.

Sanofi now runs its plant at full bore 365 days a year. In July it broke ground for a new facility in Pennsylvania that will double its output—in 2009. Even in the face of an emergency, "it would be very hard to compress that timeline," says James T. Matthews, who sits on Sanofi's pandemic-planning working group. He says it would not be feasible to convert factories for other kinds of vaccines over to make flu shots.

Pascale Wortley of the CDC's National Immunization Program raises another concern. Pandemics typically overlap with the normal flu season, she notes, and flu vaccine plants can make only one strain at a time. Sanofi spokesman Len Lavenda agrees that "we could face a Sophie's choice: whether to stop producing the annual vaccine in order to start producing the pandemic vaccine."

MedImmune aims to scale up production of its inhalable vaccine from about two million doses a year to 40 million doses by 2007. But Gellin cautions that it might be too risky to distribute live vaccine derived from a pandemic strain. There is a small chance, he says, that the virus in the vaccine could exchange genes with a "normal" flu virus in a person and generate an even more dangerous strain of influenza.

Because delays and shortages in producing vaccine against a pandemic are unavoidable, one of the most important functions of national pandemic plans is to push political leaders to decide in advance which groups will be the first to receive vaccine and how the government will enforce its rationing. The U.S. national vaccine advisory committee recommended in July that the first shots to roll off the lines should go to key government leaders, medical caregivers, workers in flu vaccine and drug factories, pregnant women, and those infants, elderly and ill people who are already in the high-priority group for annual flu shots. That top tier includes about 46 million Americans.

Among CDC planners, Wortley says, "there is a strong feeling that we ought to say beforehand that the government will purchase some amount of vaccine to guarantee equitable distribution." Australia, Britain, France and other European governments are working out advance contracts with vaccine producers to do just that. The U.S., so far, has not.

In principle, governments could work around these supply difficulties by stockpiling vaccine. They would have to continually update their stocks as new strains of influenza threatened to go global; even doing so, the reserves would probably always be a step or two behind the disease. Nevertheless, Wortley says, "it makes sense to have H5N1 vaccine on hand, because even if it is not an exact match, it probably would afford some amount of protection" if the H5N1 strain evolved to cause a pandemic.

To that end, the U.S. National Institute of Allergy and Infectious Diseases (NIAID) last year distributed an H5N1 seed virus created from a victim in Vietnam by scientists at St. Jude Children's Research Hospital in Memphis. The HHS then placed an order with Sanofi for two million doses of vaccine against that strain. Human trials began in March, and "the preliminary results from the clinical trial indicate that the vaccine would be protective," says NIAID director Anthony S. Fauci. "HHS Secretary Michael Leavitt is trying to negotiate to get up to 20 million doses," he adds. (Leavitt announced in September that HHS had increased its H5N1 vaccine order by $100 million.) According to Gellin, current vaccine producers could contribute at most 15 million to 20 million doses a year to the U.S. stockpile.

Those numbers are probably over-optimistic, however. The trial tested four different concentrations of antigen. A typical annual flu shot has 45 micrograms of protein and covers three strains of influenza. Officials had expected that 30 micrograms of H5N1 antigen—two shots, with 15 micrograms in each—would be enough to induce immunity. But the preliminary trial results suggest that 180 micrograms of antigen are needed to immunize one person.

NEW VACCINE TECHNOLOGIES

Researchers in industry and academia are testing new immunization methods that would stretch the limited supply of vaccines to cover more people. They are also developing technologies that could allow vaccine production to increase rapidly in an emergency.

Technology	Benefits	Readiness	Companies
Intradermal injectors	Delivering flu vaccine into the skin rather than muscle might cut the required dose per shot by a factor of five	Clinical trials show promise, but few nurses and doctors are trained in the procedure	Iomai, GlaxoSmithKline
Adjuvants	Chemical additives called adjuvants can increase the immune response, so that less protein is needed per shot	One such vaccine is licensed in Europe. Others are in active development	Iomai, Chiron, GlaxoSmithKline
Cell-cultured vaccines	Growing influenza virus for vaccine in cell-filled bioreactors, rather than in eggs, would enable faster increases in production if a flu pandemic broke out	Chiron is conducting a large-scale trial in Europe. Sanofi Pasteur and Crucell are developing a process for the U.S.	Chiron, Baxter, Sanofi Pasteur, Crucell, Protein Sciences
DNA vaccines	Gold particles coated with viral DNA could be injected into the skin with a jet of air. Production of DNA vaccines against a new strain could begin in weeks, rather than months. Stockpiles would last years without refrigeration	No DNA vaccine has yet been proved effective in humans. PowderMed expects results from a small-scale trial of an H5N1 DNA vaccine in late 2006	PowderMed, Vical
All-strain vaccines	A vaccine that raises immunity against a viral protein that rarely mutates might thwart every strain of influenza. Stockpiles could then reliably defend against a pandemic	Acambis began developing a vaccine against the M2e antigen this past summer	Acambis

An order for 20 million conventional doses may thus actually yield only enough H5N1 vaccine for about 3.3 million people. The true number could be even lower, because H5 strains grow poorly in eggs, so each batch yields less of the active antigen than usual. This grim picture may brighten, however, when NIAID analyzes the final results from the trial. It may also be possible to extend vaccine supplies with the use of adjuvants (substances added to vaccines to increase the immune response they induce) or new immunization approaches, such as injecting the vaccine into the skin rather than into muscle.

Caching large amounts of prepandemic vaccine, though not impossible, is clearly a challenge. Vaccines expire after a few years. At current production rates, a stockpile would never grow to the 228 million doses needed to cover the three highest priority groups, let alone to the roughly 600 million doses that would be needed to vaccinate everyone in the U.S. Other nations face similar limitations.

The primary reason that capacity is so tight, Matthews explains, is that vaccine makers aim only to meet the demand for annual immunizations when making business decisions. "We really don't see the pandemic itself as a market opportunity," he says.

To raise manufacturers' interest, "we need to offer a number of incentives, ranging from liability insurance to better profit margins to guaranteed purchases," Fauci acknowledges. Long-term solutions, Gellin predicts, may come from new technologies that allow vaccines to be made more efficiently, to be scaled up more rapidly, to be effective at much lower doses and perhaps to work equally well on all strains of influenza.

RAPID RESPONSE: COULD A PANDEMIC BE STOPPED?

As recently as 1999, WHO had a simple definition for when a flu pandemic began: with confirmation that a new virus was spreading between people in at least one country. Thereafter, stopping the flu's lightningfast expansion was unthinkable—or so it then seemed. But because of recent advances in the state of disease surveillance and antiviral drugs, the latest version of WHO's guidelines recognizes a period on the cusp of the pandemic when a flu virus ready to burst on the world might instead be intercepted and restrained, if not stamped out.

Computer models and common sense indicate that a containment effort would have to be exceptionally swift and efficient. Flu moves with extraordinary speed because it has such a short incubation period—just two days after infection by the virus, a person may start showing symptoms and shedding virus particles that can infect others. Some people may become infectious a day before their symptoms appear. In contrast, people infected by the SARS coronavirus that emerged from China in 2003 took as long as 10 days to become infectious, giving health workers ample time to trace and isolate their contacts before they, too, could spread the disease.

Contact tracing and isolation alone could never contain flu, public health experts say. But computer-simulation results published in August showed when up to 30 million doses of antiviral drugs and a low-efficacy vaccine were added to the interventions a chance emerged to thwart a potential pandemic.

Conditions would have to be nearly ideal. Modeling a population of 85 million based on the demographics and geography of Thailand, Neil M. Ferguson of Imperial College London found that health workers would have at most 30 days from the start of person-to-person viral transmission to deploy antivirals as both treatment and preventives wherever outbreaks were detected.

But even after seeing the model results earlier this year, WHO officials expressed doubt that surveillance in parts of Asia is reliable enough to catch a budding epidemic in time. In practice, confirmation of some human H5N1 cases has taken more than 20 days, WHO flu chief Stöhr warned a gathering of experts in Washington, D.C., this past April. That leaves just a narrow window in which to deliver the drugs to remote areas and dispense them to as many as one million people.

Partial immunity in the population could buy more time, however, according to Ira M. Longini, Jr., of Emory University. He, too, modeled intervention with antivirals in a smaller community based on Thai demographic data, with outcomes similar to Ferguson's. But Longini added scenarios in which people had been vaccinated in advance. He assumed that an existing vaccine, such as the H5N1 prototype version some countries have already developed, would not perfectly match a new variant of the virus, so his model's vaccinees were only 30 percent less likely to be infected. Still, their reduced susceptibility made containing even a highly infectious flu strain possible in simulations. NIAID director Fauci has said that the U.S. and other nations with H5N1 vaccine are still considering whether to direct it toward prevention in the region where a human-adapted version of that virus is most likely to emerge—even if that means less would remain for their own citizens. "If we're smart, we would," Longini says.

Based on patterns of past pandemics, experts expect that once a new strain breaks loose, it will circle the globe in two or three waves, each potentially lasting several months but peaking in individual communities about five weeks after its arrival. The waves could be separated by as long as a season: if the first hit in springtime, the second might not begin until late summer or early fall. Because meaningful amounts of vaccine tailored to the pandemic strain will not emerge from factories for some six months, government planners are especially concerned with bracing for the first wave.

Once a pandemic goes global, responses will vary locally as individual countries with differing resources make choices based on political priorities as much as on science. Prophylactic use of antivirals is an option for a handful of countries able to afford drug stockpiles, though not a very practical one. No nation has enough of the drugs at present to protect a significant fraction of its population for months. Moreover, such prolonged use has never been tested and could cause unforeseen problems.

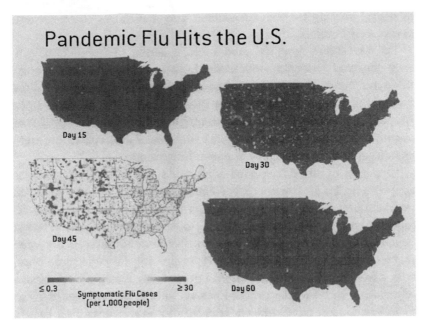

Pandemic Flu Hits the U.S. A simulation created by researchers from Los Alamos National Laboratory and Emory University shows the first wave of a pandemic spreading rapidly with no vaccine or antiviral drugs employed to slow it down. Colors represent the number of symptomatic flu cases per 1,000 people (see scale). Starting with 40 infected people on the first day, nationwide cases peak around day 60, and the wave subsides after four months with 33 percent of the population having become sick. The scientists are also modeling potential interventions with drugs and vaccines to learn if travel restrictions, quarantines and other disruptive disease-control strategies could be avoided.

For these reasons, the U.K. declared this past July that it would use its pandemic stockpile primarily for treating patients rather than for protecting the uninfected. The U.S., Canada and several other countries are still working out their priorities for who will receive antivirals and when.

For most countries there will be no choice: what the WHO calls nonpharmaceutical interventions will have to be their primary defense. Although the effectiveness of such measures has not been extensively researched, the WHO gathered flu specialists in Geneva in March 2004 to try to determine which actions medical evidence does support. Screening incoming travelers for flu symptoms, for instance, "lacks proven health benefit," the group concluded, although they acknowledged that countries might do it anyway to promote public confidence. Similarly, they were skeptical that public fever screening, fever hotlines or fever clinics would do much to slow the spread of the disease.

The experts recommended surgical masks for flu patients and health workers exposed to those patients. For the healthy, hand washing offers more protection than wearing masks in public, because people can be exposed to the virus at home, at work and by touching contaminated surfaces—including the surface of a mask.

Traditional "social distancing" measures, such as banning public gatherings or shutting down mass transit, will have to be guided by what epidemiologists find once the pandemic is under way. If children are especially susceptible to the virus, for example—as was the case in 1957 and 1968—or if they are found to be an important source of community spread, then governments may consider closing schools.

TREATMENT: WHAT CAN BE DONE FOR THE SICK?

If two billion become sick, will 10 million die? Or 100 million? Public health specialists around the world are struggling to quantify the human toll of a future flu pandemic. Casualty estimates vary so widely because until it strikes, no one can be certain whether the next pandemic strain will be mild, like the 1968 virus that some flu researchers call a "wimp"; moderately severe, like the 1957 pandemic strain; or a stone-cold killer, like the "Great Influenza"of 1918.

For now, planners are going by rules of thumb: because no one would have immunity to a new strain, they expect 50 percent of the population to be infected by the virus. Depending on its virulence, between one third and two thirds of those people will become sick, yielding a clinical attack rate of 15 to 35 percent of the whole population. Many governments are therefore trying to prepare for a middle-ground estimate that 25 percent of their entire nation will fall ill.

No government is ready now. In the U.S., where states have primary responsibility for their residents' health, the Trust for America's Health (TFAH) estimates that a "severe" pandemic virus sickening 25 percent of the population could translate into 4.7 million Americans needing hospitalization. The TFAH notes that the country currently has fewer than one million staffed hospital beds.

For frontline health workers, a pandemic's severity will boil down to the sheer number of patients and the types of illness they are suffering. These, in turn, could depend on both inherent properties of the virus and susceptibility of various subpopulations to it, according to Maryland's pandemic planner, Jean Taylor. A so-called mild pandemic, for example, might resemble seasonal flu but with far larger numbers infected.

Ordinarily, those hardest hit by annual flu are people who have complications of chronic diseases, as well as the very young, the very old and others with weak immune systems. The greatest cause of seasonal flu-related deaths is pneumonia brought on by bacteria that invade after flu has depleted the body's defenses, not by the flu virus itself. Modeling a pandemic with similar qualities, Dutch national health agency researchers found that hospitalizations might be reduced by 31 percent merely by vaccinating the usual risk groups against bacterial pneumonia in advance.

In contrast, the 1918 pandemic strain was most lethal to otherwise healthy young adults in their 20s and 30s, in part because their immune systems were so hardy. Researchers studying that virus have discovered that it suppresses early immune responses, such as the body's release of interferon, which normally primes cells to resist attack. At the same time, the virus provokes an extreme immune overreaction known as a cytokine storm, in which signaling molecules called cytokines summon a ferocious assault on the lungs by immune cells.

Doctors facing the same phenomenon in SARS patients tried to quell the storm by administering interferon and cytokine-suppressing corticosteroids. If the devastating cascade could not be stopped in time, one Hong Kong physician reported, the patients' lungs became increasingly inflamed and so choked with dead tissue that pressurized ventilation was needed to get enough oxygen to the bloodstream.

Nothing about the H5N1 virus in its current form offers reason to hope that it would produce a wimpy pandemic, according to Frederick G. Hayden, a University of Virginia virologist who is advising WHO on treating avian flu victims. "Unless this virus changes dramatically in pathogenicity," he asserts, "we will be confronted with a very lethal strain." Many H5N1 casualties have suffered acute pneumonia deep in the lower lungs caused by the virus itself, Hayden says, and in some cases blood tests indicated unusual cytokine activity. But the virus is not always consistent. In some patients, it also seems to multiply in the gut, producing severe diarrhea. And it is believed to have infected the brains of two Vietnamese children who died of encephalitis without any respiratory symptoms.

Antiviral drugs that fight the virus directly are the optimal treatment, but many H5N1 patients have arrived on doctors' doorsteps too late for the drugs to do much good. The version of the strain that has infected most human victims is also resistant to an older class of antivirals called amantadines, possibly as a result of those drugs having been given to poultry in parts of Asia. Laboratory experiments indicate that H5N1 is still susceptible to a newer class of antivirals called neuraminidase inhibitors (NI) that includes two products, oseltamivir and zanamivir, currently on the market under the brand names Tamiflu and Relenza. The former comes in pill form; the latter is a powder delivered by inhaler. To be effective against seasonal flu strains, either drug must be taken within 48 hours of symptoms appearing.

The only formal test of the drugs against H5N1 infection, however, has been in mice. Robert G. Webster of St. Jude Children's Research Hospital reported in July that a mouse equivalent of the normal human dose of two Tamiflu pills a day eventually subdued the virus, but the mice required treatment for eight days rather than the usual five. The WHO is organizing studies of future H5N1 victims to determine the correct amount for people.

Even at the standard dosage, however, treating 25 percent of the U.S. population would require considerably more Tamiflu, or its equivalent, than the 22 million treatment courses the U.S. Department of Health and Human Services planned to stockpile as of September. An advisory committee has suggested a minimum U.S. stockpile of 40 million treatment courses (400 million pills). Ninety million courses would be enough for a third of the population, and 130 million would allow the drugs to also be used to protect health workers and other essential personnel, the committee concluded.

Hayden hopes that before a pandemic strikes, a third NI called peramivir may be approved for intravenous use in hospitalized flu patients. Long-acting NIs might one day be ideal for stockpiling because a single dose would suffice for treatment or offer a week's worth of prevention.

These additional drugs, like a variety of newer approaches to fighting

NEW FLU DRUGS

Today's flu antivirals disable specific proteins on the virus's surface—either M2 (drugs known as amantadines) or neuraminidase (zanamivir and oseltamivir). Some new drugs in development are improved neuraminidase inhibitors. Other novel approaches include blocking the virus's entry into host cells or hobbling its ability to function once inside.

Approach	Drugs	Benefits	Readiness
Inhibition of neuraminidase protein, which the virus uses to detach from one cell and infect another	Peramivir (BioCryst Pharmaceuticals); CS-8958 (Biota/Sankyo)	Neuraminidase inhibitors have fewer side effects and are less likely to provoke viral resistance than the older amantadines. CS-8958 is a long-acting formulation that clings inside lungs for up to a week	Peramivir reached lungs inefficiently in clinical trials of a pill form; trials of intravenous delivery may occur in 2006; initial safety trials are complete on CS-8958
Inhibition of viral attachment to cells	Fludase (NexBio)	Because it blocks the sialic acid receptor that flu viruses use to enter host cells, Fludase should be equally effective on all flu strains	Clinical trials are planned for 2006
Stimulation of RNA interference mechanism	GO0101 (Galenea); unnamed (Alnylam Pharmaceuticals)	Uses DNA to activate a built-in defense mechanism in cells, marking viral instructions for destruction. GO0149B demonstrated effective against avian H5 and H7 flu viruses in mice	Clinical trials are expected within 18 months
Antisense DNA to block viral genes	Neugene (AVI BioPharma)	Synthetic strands of DNA bind to viral RNA that instructs the host cell to make more virus copies. The strategy should be effective against most strains	Animal testing is scheduled for 2006

flu, all have to pass clinical testing before they can be counted on in a pandemic. Researchers would also like to study other treatments that directly modulate immune system responses in flu patients. Health workers will need every weapon they can get if the enemy they face is as deadly as H5N1.

Fatality rates in diagnosed H5N1 victims are running about 50 percent. Even if that fell to 5 percent as the virus traded virulence for transmissibility among people, Hayden warns, "it would still represent a death rate double [that of] 1918, and that's despite modern technologies like antibiotics and ventilators." Expressing the worry of most flu experts at this pivotal moment for public health, he cautions that "we're well behind the curve in terms of having plans in place and having the interventions available."

Never before has the world been able to see a flu pandemic on the horizon or had so many possible tools to minimize its impact once it arrives. Some mysteries do remain as scientists watch the evolution of a potentially pandemic virus for the first time, but the past makes one thing certain: even if the dreaded H5N1 never morphs into a form that can spread easily between people, some other flu virus surely will. The stronger our defenses, the better we will weather the storm when it strikes. "We have only one enemy," CDC director Gerberding has said repeatedly, "and that is complacency."

FURTHER READING

The Great Influenza. Revised edition. John M. Barry. Penguin Books, 2005.

John R. LaMontagne Memorial Symposium on Pandemic Influenza Research: Meeting Proceedings. Institute of Medicine. National Academies Press, 2005.

WHO Global Influenza Preparedness Plan. WHO Department of Communicable Disease Surveillance and Response Global Influenza Program, 2005. www.who.int/csr/resources/publications/influenza/WHo_CDS_CSR_GIP_2005_5/en/index.html

Pandemic influenza Web site of the U.S. Department of Health and Human Services, National Vaccine Program Office: www.hhs.gov/nvpo/pandemics/index.html

If Smallpox Strikes Portland...

CHRIS L. BARRETT, STEPHEN G. EUBANK AND JAMES P. SMITH

ORIGINALLY PUBLISHED IN MARCH 2005

Suppose terrorists were to release plague in Chicago, and health officials, faced with limited resources and personnel, had to quickly choose the most effective response. Would mass administration of antibiotics be the best way to halt an outbreak? Or mass quarantines? What if a chance to nip a global influenza pandemic in the bud meant sending national stockpiles of antiviral drugs to Asia where a deadly new flu strain was said to be emerging? If the strategy succeeded, a worldwide crisis would be averted; if it failed, the donor countries would be left with less protection.

Public health officials have to make choices that could mean life or death for thousands, even millions, of people, as well as massive economic and social disruption. And history offers them only a rough guide. Methods that eradicated smallpox in African villages in the 1970s, for example, might not be the most effective tactics against smallpox released in a U.S. city in the 21st century. To identify the best responses under a variety of conditions in advance of disasters, health officials need a laboratory where "what if" scenarios can be tested as realistically as possible. That is why our group at Los Alamos National Laboratory (LANL) set out to build EpiSims, the largest individual-based epidemiology simulation model ever created.

Modeling the interactions of each individual in a population allows us to go beyond estimating the number of people likely to be infected; it lets us simulate the paths a disease would take through the population and thus where the outbreak could be intercepted most effectively. The networks that support everyday life and provide employment, transportation infrastructure, necessities and luxuries are the same ones that infectious diseases exploit to spread among human hosts. By modeling this social network in fine detail, we can understand its structure and how to alter it to disrupt the spread of disease while inflicting the least damage to the social fabric.

VIRTUAL EPIDEMIOLOGY

Long before the germ theory of disease, London physician John Snow argued that cholera, which had killed tens of thousands of people in England during the preceding 20 years, spread via the water supply. In the summer of 1854 he tested that theory during an outbreak in the Soho district. On a map, he marked the location of the homes of each of the 500 victims who had died in the preceding 10 days and noted where each victim had gotten water. He discovered that every one of them drank water from the Broad Street pump, so Snow convinced officials to remove the pump handle. His action limited the death toll to 616.

Tracing the activities and contacts of individual disease victims, as Snow did, remains an important tool for modern epidemiologists. And it is nothing new for health authorities to rely on models when developing policies to protect the public. Yet most mathematical models for understanding and predicting the course of disease outbreaks describe only the interactions of large numbers of people in aggregate. One reason is that modelers have often lacked detailed knowledge of how specific contagious diseases spread. Another is that they have not had realistic models of the social interactions in which people have contact with one another. And a third is that they have not had the computational and methodological means to build models of diseases interacting with dynamic human populations.

As a result, epidemiology models typically rely on estimates of a particular disease's "reproductive number"—the number of people likely to be infected by one contagious person or contaminated location. Often this reproductive number is a best guess based on historical situations, even though the culture, physical conditions and health status of people in those events may differ greatly from the present situation.

In real epidemics, these details matter. The rate at which susceptible people become infected depends on their individual state of health, the duration and nature of their interactions with contagious people, and specific properties of the disease pathogen itself. Truer models of outbreaks must capture the probability of disease transmission from one person to another, which means simulating not only the properties of the disease and the health of each individual but also detailed interactions between every pair of individuals in the group.

Attempts to introduce such epidemiological models have, until recently, considered only very small groups of 100 to 1,000 people. Their size has been limited because they are based on actual populations, such as the residents, visitors and staff of a nursing home, so they require detailed data about individuals and their contacts over days or weeks. Computing such a large number of interactions also presents substantial technical difficulties.

Our group was able to construct this kind of individual-based epidemic model on a scale of millions of people by using high-performance super-computing clusters and by building on an existing model called TRAN-SIMS developed over more than a decade at Los Alamos for urban planning [see "Unjamming Traffic with Computers," by Kenneth R. Howard; *Scientific American*, October 1997]. The TRANSIMS project started as a means of better understanding the potential effects of creating or rerouting roads and other transportation infrastructure. By giving us a way to simulate the movements of a large population through a realistic urban environment, TRANSIMS provided the foundation we needed to model the interactions of millions of individuals for EpiSims.

Although EpiSims can now be adapted to different cities, the original TRANSIMS model was based on Portland, Ore. The TRANSIMS virtual version of Portland incorporates detailed digital maps of the city, includ-

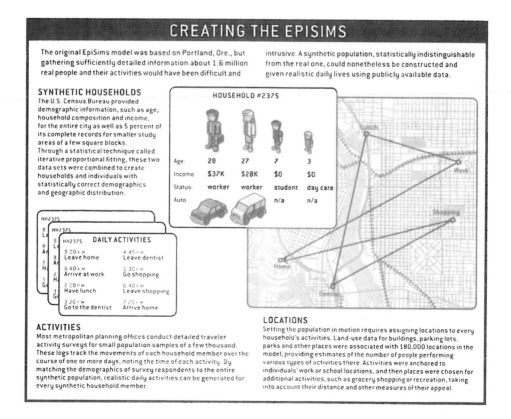

CREATING THE EPISIMS

The original EpiSims model was based on Portland, Ore., but gathering sufficiently detailed information about 1.6 million real people and their activities would have been difficult and intrusive. A synthetic population, statistically indistinguishable from the real one, could nonetheless be constructed and given realistic daily lives using publicly available data.

SYNTHETIC HOUSEHOLDS
The U.S. Census Bureau provided demographic information, such as age, household composition and income, for the entire city as well as 5 percent of its complete records for smaller study areas of a few square blocks. Through a statistical technique called iterative proportional fitting, these two data sets were combined to create households and individuals with statistically correct demographics and geographic distribution.

HOUSEHOLD #2375

Age:	28	27	7	3
Income:	$37K	$28K	$0	$0
Status:	worker	worker	student	day care
Auto:			n/a	n/a

HH2375 **DAILY ACTIVITIES**

8:00 A.M. Leave home	4:45 P.M. Leave dentist
8:40 A.M. Arrive at work	5:30 P.M. Go shopping
2:00 P.M. Have lunch	6:40 P.M. Leave shopping
3:20 P.M. Go to the dentist	7:20 P.M. Arrive home

ACTIVITIES
Most metropolitan planning offices conduct detailed traveler activity surveys for small population samples of a few thousand. These logs track the movements of each household member over the course of one or more days, noting the time of each activity. By matching the demographics of survey respondents to the entire synthetic population, realistic daily activities can be generated for every synthetic household member.

LOCATIONS
Setting the population in motion requires assigning locations to every household's activities. Land-use data for buildings, parking lots, parks and other places were associated with 180,000 locations in the model, providing estimates of the number of people performing various types of activities there. Activities were anchored to individuals' work or school locations, and then places were chosen for additional activities, such as grocery shopping or recreation, taking into account their distance and other measures of their appeal.

ing representations of its rail lines, roads, signs, traffic signals and other transportation infrastructure, and produces information about traffic patterns and travel times. Publicly available data were used to generate 180,000 specific locations, a synthetic population of 1.6 million residents, and realistic daily activities for those people [see box on page 277].

Integrating all this information into a computer model provides the best estimate of physical contact patterns for large human populations ever created. With EpiSims, we can release a virtual pathogen into these populations, watch it spread and test the effects of different interventions. But even without simulating a disease outbreak, the model provides intriguing insights into human social networks, with potentially important implications for epidemic response.

SOCIAL NETWORKS

To understand what a social network really is and how it can be used for epidemiology, imagine the daily activities and contacts of a single hypothetical adult, Ann. She has short brushes with family members during breakfast and then with other commuters or carpoolers on her way to work. Depending on her job, she might meet dozens of people at work, with each encounter having a different duration, proximity and purpose. During lunch or a shopping trip after work, Ann might have additional short contacts with strangers in public places before returning home.

We can visually represent Ann's contacts as a network with Ann in the center and a line connecting Ann to each of them [see box on page 279]. All Ann's contacts engage in various activities and meet other people as well. We can represent these "contacts of contacts" by drawing lines from each—for example, Ann's colleague named Bob—to all his contacts. Unless they are also contacts of Ann, Bob's contacts are two "hops" away from Ann. The number of hops on the shortest path between people is sometimes called the graph distance or degree of separation between those people.

The popular idea that everyone on the earth is connected to everyone else by at most six degrees of separation means that if we continued building our social network until it included everyone on the planet, no two people would be more than six hops from one another. The idea is not strictly true, but it makes for a good story and has even led to the well-known game involving the social network of actors who have appeared in films with Kevin Bacon. In academic circles, another such social network traces mathematicians' co-authorship connections, with one's

BUILDING SOCIAL NETWORKS

TYPICAL HOUSEHOLD'S CONTACTS

Constructing a social network for a household of two adults and two children starts by identifying their contacts with other people throughout a typical day.

This diagram shows where the household members go and what they do all day but reveals little about how their individual contacts might be interconnected or connected to others.

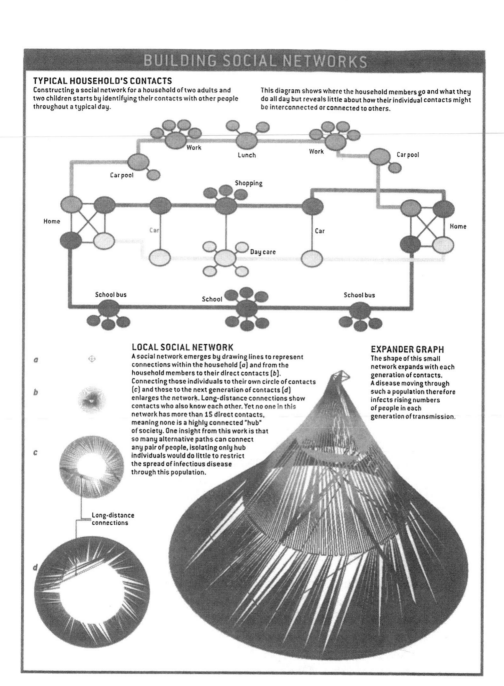

Work

Lunch

Work

Car pool

Car pool

Shopping

Car pool

Home

Car

Car

Home

Day care

School bus

School

School bus

LOCAL SOCIAL NETWORK

a

b

c

d

A social network emerges by drawing lines to represent connections within the household (a) and from the household members to their direct contacts (b). Connecting those individuals to their own circle of contacts (c) and those to the next generation of contacts (d) enlarges the network. Long-distance connections show contacts who also know each other. Yet no one in this network has more than 15 direct contacts, meaning none is a highly connected "hub" of society. One insight from this work is that so many alternative paths can connect any pair of people, isolating only hub individuals would do little to restrict the spread of infectious disease through this population.

Long-distance connections

EXPANDER GRAPH

The shape of this small network expands with each generation of contacts. A disease moving through such a population therefore infects rising numbers of people in each generation of transmission.

"Erdös number" defined by graph distance from the late, brilliant and prolific Paul Erdös.

Other types of networks, including the Internet, the links among scientific article citations and even the interactions among proteins within living cells, have been found to display this same tendency toward having "hubs": certain locations, people or even molecules with an unusually high number of connections to the rest of the network. The shortest path between any two nodes in the network is typically through one of these hubs, much as in a commercial airline's route system. Technically, such networks are called "scale-free" when the number of hubs with exactly k connections, $N(k)$, is proportional to a power of k [see "Scale-Free Networks," by Albert-László Barabasi and Eric Bonabeau; Scientific American, May 2003].

Because a scale-free network can be severely damaged if one or more of its hubs are disabled, some researchers have extrapolated this observation to disease transmission. If infected "hub" individuals, such as the most gregarious people in a population, could somehow be identified and treated or removed from the network, the reasoning goes, then an epidemic could be halted without having to isolate or treat everyone in the population. But our analyses of the social networks used by EpiSims suggest that society is not so easily disabled as physical infrastructure.

The network of physical locations in our virtual Portland, defined by people traveling between them, does indeed exhibit the typical scale-free structure, with certain locations acting as important hubs. As a result, these locations, such as schools and shopping malls, would be good spots for disease surveillance or for placing sensors to detect the presence of biological agents.

The urban social networks in the city also have human hubs with higher than average contacts, many because they work in the physical hub locations, such as teachers or sales clerks. Yet we have also found an unexpectedly high number of "short paths" in the social networks that do not go through hubs, so a policy of targeting only hub individuals would probably do little to slow the spread of a disease through the city.

In fact, another unexpected property we have found in realistic social networks is that everyone but the most devoted recluse is effectively a small hub. That is to say, when we look at the contacts of any small group, such as four students, we find that they are always connected by one hop to a much larger group. Depicting this social network structure results in what is known as an expander graph [see box on page 279], which has a cone shape that widens with each hop. Its most important implication for

SIMULATED SMALLPOX ATTACKS

EpiSims animations depict simulated outbreaks and the effects of official interventions. In the still frames below, vertical lines indicate the number of infected people present at a location, and color shows the percentage of them who are contagious. In both scenarios shown, smallpox is released at a university in central Portland, but the attack is not detected until victims start experiencing symptoms 10 days later. The left-hand images show no public health response as a baseline. In the right-hand images, infected and exposed individuals are targeted for vaccination and quarantine. Results from a series of such simulations (*bottom*) show that people withdrawing to their homes early in an outbreak makes the biggest difference in death toll. The speed of official response, regardless of the strategy chosen, proved to be the second most important factor.

DAY 1: UNDETECTED SMALLPOX RELEASE

NO RESPONSE

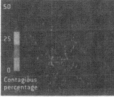

INFECTED: 1,281
DEAD: 0

TARGETED VACCINATION AND QUARANTINE STARTING DAY 14

INFECTED: 1,281
QUARANTINED: 0
VACCINATED: 0
DEAD: 0

DAY 35: SMALLPOX EPIDEMIC

INFECTED: 23,919
DEAD: 551

INFECTED: 2,564
QUARANTINED: 29,910
VACCINATED: 30,560
DEAD: 312

DAY 70: EPIDEMIC UNCONTAINED OR CONTAINED

INFECTED: 380,582
DEAD: 12,499

INFECTED: 2,564
QUARANTINED: 36,725
VACCINATED: 37,207
DEAD: 435

WITHDRAWAL TO HOME

CUMULATIVE DEATHS PER INITIALLY INFECTED PERSON BY DAY 100

Early | Late | Never

10,000
1,000
100
10
1
0.3

OFFICIAL RESPONSE: Nothing, Targeted, Limited, Mass

INTERVENTION KEY
○ No vaccine
● 10-day delay
● 7-day delay
● 4-day delay

RESPONSE EFFECTIVENESS

Simulations allowed people to withdraw to their homes because they felt ill or were following officials' instructions. Withdrawal could be "early," before anyone became contagious, or "never," meaning people continued moving about unless they died. "Late" withdrawal, 24 hours after becoming contagious, was less effective than early withdrawal, which prevented an epidemic without other intervention. Official responses included doing nothing, or targeted vaccination and quarantine with unlimited personnel, or targeted vaccination limited by only half the necessary personnel being available, or mass vaccination of the entire population. The interventions began four, seven or 10 days after the first victims became symptomatic.

epidemiology is that diseases can disseminate exponentially fast because the number of people exposed in each new generation of transmission is always larger than the number in the current generation.

Theoretically, this should mean that whatever health officials do to intervene in a disease outbreak, speed will be one of the most important factors determining their success. Simulating disease outbreaks with Epi-Sims allows us to see whether that theory holds true.

SMALLPOX ATTACK

After we began developing EpiSims in 2000, smallpox was among the first diseases we chose to model because government officials charged with bioterrorism planning and response were faced with several questions and sometimes conflicting recommendations. In the event that smallpox was released into a U.S. population, would mass vaccination be necessary to prevent an epidemic? Or would targeting only exposed individuals and their contacts for vaccination be enough? How effective is mass quarantine? How feasible are any of these options with the existing numbers of health workers, police and other responders?

To answer such questions, we constructed a model of smallpox that we could release into our synthetic population. Smallpox transmission was particularly difficult to model because the virus has not infected humans since its eradication in the 1970s. Most experts agree, though, that the virus normally requires significant physical contact with an infectious person or contaminated object. The disease has an average incubation period of approximately 10 days before flulike symptoms begin appearing, followed by skin rash. Victims are contagious once symptoms have appeared and possibly for a short time before they develop fever. Untreated, some 30 percent of those infected would die, but the rest would recover and be immune to reinfection.

Vaccination before exposure or within four days of infection can stop smallpox from developing. We assumed in all our simulations that health workers and people charged with tracking down the contacts of infected people had already been vaccinated and thus were immune. Unlike many epidemiological models, our realistic simulation also ensures that the chronology of contacts will be considered. If Ann contracted the disease, she could not infect her co-worker Bob a week earlier. Or, if Ann does infect Bob after she herself becomes infected and if Bob in turn infects his family member Cathy, the infection cannot pass from Ann to Cathy in less than twice the minimum incubation period between disease exposure and becoming contagious.

With our disease model established and everyone in our synthetic population assigned an immune status, we simulated the release of smallpox in several hub locations around the city, including a university campus. Initially, 1,200 people were unwittingly infected, and within hours they had moved throughout the city, going about their normal activities.

We then simulated several types of official responses, including mass vaccination of the city's population or contact tracing of exposed individuals and their contacts who could then be targeted for vaccination and quarantine. Finally, we simulated no response at all for the purpose of comparison.

In each of these circumstances, we also simulated delays of four, seven and 10 days in implementing the response after the first victims became known. In addition, we allowed infected individuals to isolate themselves by withdrawing to their homes.

Each simulation ran for a virtual 100 days [*see box on page 281*], and the precise casualty figures resulting from each scenario were less important than the relative effect different responses had on the death tolls. The results upheld our theoretical prediction based on the expander-graph structure of the social network: time was by far the most important factor in limiting deaths. The speed with which people withdrew to their homes or were isolated by health officials was the strongest determinant of the outbreak's extent. The second most influential factor was the length of the delay in officials' response. The actual response strategy chosen made little difference compared with the time element.

In the case of a smallpox outbreak, these simulations indicate that mass vaccination of the population, which carries its own risks, would be unnecessary. Targeted vaccination would be just as effective so long as it was combined with rapid detection of the outbreak and rapid response. Our results also support the importance of measures such as quarantine and making sure that health officials give enforcement adequate priority during highly infectious disease outbreaks.

Of course, appropriate public health responses will always depend on the disease, the types of interventions available and the setting. For example, we have simulated the intentional release of an inhalable form of plague in the city of Chicago to evaluate the costs and effects of different responses. In those simulations we found that contact tracing, school closures and city closures each incurred economic losses of billions of dollars but did not afford many health benefits over voluntary mass use of rapidly available antibiotics at a much lower economic cost.

Most recently, as part of a research network organized by the National Institute of General Medical Sciences called the Models of Infectious

Disease Agent Study (MIDAS), we have been adapting EpiSims to model a naturally occurring disease that may threaten the entire planet: pandemic influenza.

FLU AND THE FUTURE

Over the past year, a highly virulent strain of influenza has raged through bird populations in Asia and has infected more than 40 human beings in Japan, Thailand and Vietnam, killing more than 30 of those people. The World Health Organization has warned that it is only a matter of time before this lethal flu strain, designated H5N1, more easily infects people and spreads between them. That development could spark a global flu pandemic with a death toll reaching tens of millions [see SA Perspectives, Scientific American, January 2005].

MIDAS collaborators will be studying the possibility that an H5N1 virus capable of spreading in humans might be contained or even eradicated by rapid intervention while it is still confined to a small population. To simulate the appropriate conditions in which the strain would likely emerge among humans, we are constructing a model representing a hypothetical Southeast Asian community of some 500,000 people living on farms and in neighboring small towns. Our model of the influenza virus itself will be based both on historical data about pandemic flu strains and information about the H5N1 virus, whose biology is currently a subject of intense investigation.

We know, for example, that H5N1 is sensitive to antiviral drugs that inhibit of its important enzymes, called neuraminidase. In our simulations, we will be able to use neuraminidase inhibitors as both treatment and prophylaxis.(A vaccine against H5N1 has been developed and recently began clinical trials but because the vaccine is not yet proven or available, we will focus our simulations on seeing whether the antiviral drugs together with traditional public health measures might stop an epidemic.)

Preliminary results announced in late February are reported at www.sciam.com. In April, we will complete similar flu pandemic simulations in the EpiSims Portland model.

Our hope is that the ability to realistically model populations and disease outbreaks can help health officials make difficult decisions based on the best possible answers to "what if" questions.

The creation of models such as TRANSIMS that simulate human movements through urban environments was the computational breakthrough that made EpiSims possible, and epidemiology is only one potential application for this kind of individual-based modeling. We are also in the process of creating and linking simulations of other sociotechnical systems, including environmental and atmospheric pollution, telecommunications, transportation, commodity markets, water supplies and power grids, to provide virtual laboratories for exploring solutions to a wide variety of real-world problems.

FURTHER READING

SCALABLE, EFFICIENT EPIDEMIOLOGICAL SIMULATION. Stephen Eubank in *Proceedings of the 2002 ACM Symposium on Applied Computing*, pages 139–145; 2002.

SIX DEGREES: THE SCIENCE OF A CONNECTED AGE. Duncan J. Watts. W. W. Norton, 2004.

CONTAINING PANDEMIC INFLUENZA WITH ANTIVIRAL AGENTS. Ira M. Longini, Jr., et al. in *American Journal of Epidemiology*, Vol. 159, No. 7, pages 623–633; April 1, 2004.

MODELLING DISEASE OUTBREAKS IN REALISTIC URBAN SOCIAL NETWORKS. Stephen Eubank et al. in *Nature*, Vol. 429, pages 180–184; May 13, 2004.

A sample EpiSims animation and additional data from the Portland smallpox simulations can be viewed at http://episims.lanl.gov

Is Global Warming Harmful to Health?

PAUL R. EPSTEIN

ORIGINALLY PUBLISHED IN AUGUST 2000

Today few scientists doubt the atmosphere is warming. Most also agree that the rate of heating is accelerating and that the consequences of this temperature change could become increasingly disruptive. Even high school students can reel off some projected outcomes: the oceans will warm, and glaciers will melt, causing sea levels to rise and salt water to inundate settlements along many low-lying coasts. Meanwhile the regions suitable for farming will shift. Weather patterns should also become more erratic and storms more severe.

Yet less familiar effects could be equally detrimental. Notably, computer models predict that global warming, and other climate alterations it induces, will expand the incidence and distribution of many serious medical disorders. Disturbingly, these forecasts seem to be coming true.

Heating of the atmosphere can influence health through several routes. Most directly, it can generate more, stronger and hotter heat waves, which will become especially treacherous if the evenings fail to bring cooling relief. Unfortunately, a lack of nighttime cooling seems to be in the cards; the atmosphere is heating unevenly and is showing the biggest rises at night, in winter and at latitudes higher than about 50 degrees. In some places, the number of deaths related to heat waves is projected to double by 2020. Prolonged heat can, moreover, enhance production of smog and the dispersal of allergens. Both effects have been linked to respiratory symptoms.

Global warming can also threaten human well-being profoundly, if somewhat less directly, by revising weather patterns—particularly by pumping up the frequency and intensity of floods and droughts and by causing rapid swings in the weather. As the atmosphere has warmed over the past century, droughts in arid areas have persisted longer, and massive bursts of precipitation have become more common. Aside from causing death by drowning or starvation, these disasters promote by various means the emergence, resurgence and spread of infectious disease.

That prospect is deeply troubling, because infectious illness is a genie that can be very hard to put back into its bottle. It may kill fewer people in one fell swoop than a raging flood or an extended drought, but once it takes root in a community, it often defies eradication and can invade other areas.

The control issue looms largest in the developing world, where resources for prevention and treatment can be scarce. But the technologically advanced nations, too, can fall victim to surprise attacks—as happened last year when the West Nile virus broke out for the first time in North America, killing seven New Yorkers. In these days of international commerce and travel, an infectious disorder that appears in one part of the world can quickly become a problem continents away if the disease-causing agent, or pathogen, finds itself in a hospitable environment.

Floods and droughts associated with global climate change could undermine health in other ways as well. They could damage crops and make them vulnerable to infection and infestations by pests and choking weeds, thereby reducing food supplies and potentially contributing to malnutrition. And they could permanently or semipermanently displace entire populations in developing countries, leading to overcrowding and the diseases connected with it, such as tuberculosis.

Weather becomes more extreme and variable with atmospheric heating in part because the warming accelerates the water cycle: the process in which water vapor, mainly from the oceans, rises into the atmosphere before condensing out as precipitation. A warmed atmosphere heats the oceans (leading to faster evaporation), and it holds more moisture than a cool one. When the extra water condenses, it more frequently drops from the sky as larger downpours. While the oceans are being heated, so is the land, which can become highly parched in dry areas. Parching enlarges the pressure gradients that cause winds to develop, leading to turbulent winds, tornadoes and other powerful storms. In addition, the altered pressure and temperature gradients that accompany global warming can shift the distribution of when and where storms, floods and droughts occur.

I will address the worrisome health effects of global warming and disrupted climate patterns in greater detail, but I should note that the consequences may not all be bad. Very high temperatures in hot regions may reduce snail populations, which have a role in transmitting schistosomiasis, a parasitic disease. High winds may at times disperse pollution. Hotter winters in normally chilly areas may reduce cold-related heart attacks and respiratory ailments. Yet overall, the undesirable effects of more variable weather are likely to include new stresses and nasty surprises that will overshadow any benefits.

MOSQUITOES RULE IN THE HEAT

Diseases relayed by mosquitoes—such as malaria, dengue fever, yellow fever and several kinds of encephalitis—are among those eliciting the greatest concern as the world warms. Mosquitoes acquire disease-causing microorganisms when they take a blood meal from an infected animal or person. Then the pathogen reproduces inside the insects, which may deliver disease-causing doses to the next individuals they bite.

Mosquito-borne disorders are projected to become increasingly prevalent because their insect carriers, or "vectors," are very sensitive to meteorological conditions. Cold can be a friend to humans, because it limits mosquitoes to seasons and regions where temperatures stay above certain minimums. Winter freezing kills many eggs, larvae and adults outright. *Anopheles* mosquitoes, which transmit malaria parasites (such as *Plasmodium falciparum*), cause sustained outbreaks of malaria only where temperatures routinely exceed 60 degrees Fahrenheit. Similarly, *Aedes aegypti* mosquitoes, responsible for yellow fever and dengue fever, convey virus only where temperatures rarely fall below 50 degrees F.

Excessive heat kills insects as effectively as cold does. Nevertheless, within their survivable range of temperatures, mosquitoes proliferate faster and bite more as the air becomes warmer. At the same time, greater heat speeds the rate at which pathogens inside them reproduce and mature. At 68 degrees F, the immature *P. falciparum* parasite takes 26 days to develop fully, but at 77 degrees F, it takes only 13 days. The *Anopheles* mosquitoes that spread this malaria parasite live only several weeks; warmer temperatures raise the odds that the parasites will mature in time for the mosquitoes to transfer the infection. As whole areas heat up, then, mosquitoes could expand into formerly forbidden territories, bringing illness with them. Further, warmer nighttime and winter temperatures may enable them to cause more disease for longer periods in the areas they already inhabit.

The extra heat is not alone in encouraging a rise in mosquito borne infections. Intensifying floods and droughts resulting from global warming can each help trigger outbreaks by creating breeding grounds for insects whose dessicated eggs remain viable and hatch in still water. As floods recede, they leave puddles. In times of drought, streams can become stagnant pools, and people may put out containers to catch water; these pools and pots, too, can become incubators for new mosquitoes. And the insects can gain another boost if climate change or other processes (such as alterations of habitats by humans) reduce the populations of predators that normally keep mosquitoes in check.

MOSQUITOES ON THE MARCH

Malaria and dengue fever are two of the mosquito-borne diseases most likely to spread dramatically as global temperatures head upward. Malaria (marked by chills, fever, aches and anemia) already kills 3,000 people, mostly children, every day. Some models project that by the end of the 21st century, ongoing warming will have enlarged the zone of potential malaria transmission from an area containing 45 percent of the world's population to an area containing about 60 percent. That news is bad indeed, considering that no vaccine is available and that the causative parasites are becoming resistant to standard drugs.

True to the models, malaria is reappearing north and south of the tropics. The U.S. has long been home to *Anopheles* mosquitoes, and malaria circulated here decades ago. By the 1980s mosquito-control programs and other public health measures had restricted the disorder to California. Since 1990, however, when the hottest decade on record began, outbreaks of locally transmitted malaria have occurred during hot spells in Texas, Florida, Georgia, Michigan, New Jersey and New York (as well as in

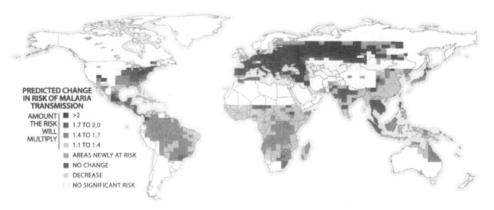

PREDICTED CHANGE
IN RISK OF MALARIA
TRANSMISSION

AMOUNT THE RISK WILL MULTIPLY
- >2
- 1.7 TO 2.0
- 1.4 TO 1.7
- 1.1 TO 1.4
- AREAS NEWLY AT RISK
- NO CHANGE
- DECREASE
- NO SIGNIFICANT RISK

Risk of malaria transmission will have risen in many parts of the world by 2020 (relative to the average risk in the years 1961 to 1990), according to projections assuming a temperature increase of about two degrees Fahrenheit. The analysis was based solely on temperature threshold and did not assess other factors that could influence malaria's spread.

Toronto). These episodes undoubtedly started with a traveler or stowaway mosquito carrying malaria parasites. But the parasites clearly found friendly conditions in the U.S.—enough warmth and humidity, and plenty of mosquitoes able to transport them to victims who had not traveled. Malaria has returned to the Korean peninsula, parts of southern Europe and the former Soviet Union and to the coast of South Africa along the Indian Ocean.

Dengue, or "break bone," fever (a severe flulike viral illness that sometimes causes fatal internal bleeding) is spreading as well. Today it afflicts an estimated 50 million to 100 million in the tropics and subtropics (mainly in urban areas and their surroundings). It has broadened its range in the Americas over the past 10 years and had reached down to Buenos Aires by the end of the 1990s. It has also found its way to northern Australia. Neither a vaccine nor a specific drug treatment is yet available.

Although these expansions of malaria and dengue fever certainly fit the predictions, the cause of that growth cannot be traced conclusively to global warming. Other factors could have been involved as well—for instance, disruption of the environment in ways that favor mosquito proliferation, declines in mosquito-control and other public health programs, and rises in drug and pesticide resistance. The case for a climatic contribution becomes stronger, however, when other projected consequences of global warming appear in concert with disease outbreaks.

Such is the case in highlands around the world. There, as anticipated, warmth is climbing up many mountains, along with plants and butterflies, and summit glaciers are melting. Since 1970 the elevation at which temperatures are always below freezing has ascended almost 500 feet in the tropics. Marching upward, too, are mosquitoes and mosquito-borne diseases.

In the 19th century, European colonists in Africa settled in the cooler mountains to escape the dangerous swamp air ("*malaria*") that fostered disease in the lowlands. Today many of those havens are compromised. Insects and insect-borne infections are being reported at high elevations in South and Central America, Asia, and east and central Africa. Since 1980 *Ae. aegypti* mosquitoes, once limited by temperature thresholds to low altitudes, have been found above one mile in the highlands of northern India and at 1.3 miles in the Colombian Andes. Their presence magnifies the risk that dengue and yellow fever may follow. Dengue fever itself has struck at the mile mark in Taxco, Mexico. Patterns of insect migration change faster in the mountains than they do at sea level. Those alterations can thus serve as indicators of climate change and of diseases likely to expand their range.

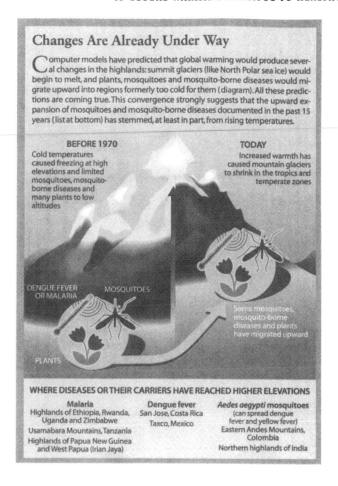

Changes Are Already Under Way

Computer models have predicted that global warming would produce several changes in the highlands: summit glaciers (like North Polar sea ice) would begin to melt, and plants, mosquitoes and mosquito-borne diseases would migrate upward into regions formerly too cold for them (diagram). All these predictions are coming true. This convergence strongly suggests that the upward expansion of mosquitoes and mosquito-borne diseases documented in the past 15 years (list at bottom) has stemmed, at least in part, from rising temperatures.

BEFORE 1970
Cold temperatures caused freezing at high elevations and limited mosquitoes, mosquito-borne diseases and many plants to low altitudes

TODAY
Increased warmth has caused mountain glaciers to shrink in the tropics and temperate zones

DENGUE FEVER OR MALARIA MOSQUITOES

Some mosquitoes, mosquito-borne diseases and plants have migrated upward

PLANTS

WHERE DISEASES OR THEIR CARRIERS HAVE REACHED HIGHER ELEVATIONS

Malaria	Dengue fever	*Aedes aegypti* mosquitoes
Highlands of Ethiopia, Rwanda, Uganda and Zimbabwe	San Jose, Costa Rica	(can spread dengue fever and yellow fever)
Usamabara Mountains, Tanzania	Taxco, Mexico	Eastern Andes Mountains, Colombia
Highlands of Papua New Guinea and West Papua (Irian Jaya)		Northern highlands of India

OPPORTUNISTS LIKE SEQUENTIAL EXTREMES

The increased climate variability accompanying warming will probably be more important than the rising heat itself in fueling unwelcome outbreaks of certain vector-borne illnesses. For instance, warm winters followed by hot, dry summers (a pattern that could become all too familiar as the atmosphere heats up) favor the transmission of St. Louis encephalitis and other infections that cycle among birds, urban mosquitoes and humans.

This sequence seems to have abetted the surprise emergence of the West Nile virus in New York City last year. No one knows how this virus found its way into the U.S. But one reasonable explanation for its persistence and amplification here centers on the weather's effects on *Culex*

El Niño's Message

Scientists often gain insight into the workings of complicated systems by studying subsystems. In that spirit, investigators concerned about global warming's health effects are assessing outcomes of the El Niño/Southern Oscillation (ENSO), a climate process that produces many of the same meteorological changes predicted for a warming world. The findings are not reassuring.

"El Niño" refers to an oceanic phenomenon that materializes every five years or so in the tropical Pacific. The ocean off Peru becomes unusually warm and stays that way for months before returning to normal or going to a cold extreme (La Niña). The name "Southern Oscillation" refers to atmospheric changes that happen in tandem with the Pacific's shifts to warmer or cooler conditions.

During an El Niño, evaporation from the heated eastern Pacific can lead to abnormally heavy rains in parts of South America and Africa; meanwhile other areas of South America and Africa and parts of Southeast Asia and Australia suffer droughts. Atmospheric pressure changes over the tropical Pacific also have ripple effects throughout the globe, generally yielding milder winters in some northern regions of the U.S. and western Canada. During a La Niña, weather patterns in the affected areas may go to opposite extremes.

The incidence of vector-borne and waterborne diseases climbs during El Niño and La Niña years, especially in areas hit by floods or droughts. Long-term studies in Colombia, Venezuela, India and Pakistan reveal, for instance, that malaria surges in the wake of El Niños. And my colleagues and I at Harvard University have shown that regions stricken by flooding or drought during the El Niño of 1997–98 (the strongest of the century) often had to contend as well with a convergence of diseases borne by mosquitoes, rodents and water (see p). Additionally, in many dry areas, fires raged out of control, polluting the air for miles around.

ENSO is not merely a warning of troubles to come; it is likely to be an engine for those troubles. Several climate models predict that as the atmosphere and oceans heat up, El Niños themselves will become more common and severe—which means that the weather disasters they produce and the diseases they promote could become more prevalent as well.

Indeed, the ENSO pattern has already begun to change. Since 1976 the intensity, duration and pace of El Niños have increased. And during the 1990s, every year was marked by an El Niño or La Niña extreme. Those trends bode ill for human health in the 21st century.

Disease Outbreaks Accompanying Extreme Weather during the 1997–98 El Niño

Extreme Weather
Abnormally wet areas
Abnormally dry areas

Disease Outbreaks

Mosquito-borne: Dengue fever, Encephalitis, Malaria, Rift Valley fever
Rodent-borne: Hantavirus pulmonary syndrome
Waterborne: Cholera
Noninfectious: Respiratory illness resulting from fire and smoke

pipiens mosquitoes, which accounted for the bulk of the transmission. These urban dwellers typically lay their eggs in damp basements, gutters, sewers and polluted pools of water.

The interaction between the weather, the mosquitoes and the virus probably went something like this: The mild winter of 1998-99 enabled many of the mosquitoes to survive into the spring, which arrived early. Drought in spring and summer concentrated nourishing organic matter in their breeding areas and simultaneously killed off mosquito predators, such as lacewings and ladybugs, that would otherwise have helped limit mosquito populations. Drought would also have led birds to congregate

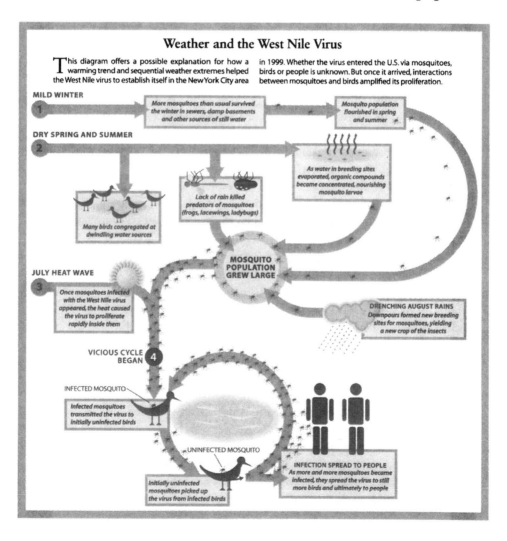

Weather and the West Nile Virus

This diagram offers a possible explanation for how a warming trend and sequential weather extremes helped the West Nile virus to establish itself in the New York City area in 1999. Whether the virus entered the U.S. via mosquitoes, birds or people is unknown. But once it arrived, interactions between mosquitoes and birds amplified its proliferation.

MILD WINTER
1

More mosquitoes than usual survived the winter in sewers, damp basements and other sources of still water

Mosquito population flourished in spring and summer

DRY SPRING AND SUMMER
2

As water in breeding sites evaporated, organic compounds became concentrated, nourishing mosquito larvae

Lack of rain killed predators of mosquitoes (frogs, lacewings, ladybugs)

Many birds congregated at dwindling water sources

MOSQUITO POPULATION GREW LARGE

JULY HEAT WAVE
3

Once mosquitoes infected with the West Nile virus appeared, the heat caused the virus to proliferate rapidly inside them

DRENCHING AUGUST RAINS
Downpours formed new breeding sites for mosquitoes, yielding a new crop of the insects

VICIOUS CYCLE BEGAN
4

INFECTED MOSQUITO

Infected mosquitoes transmitted the virus to initially uninfected birds

UNINFECTED MOSQUITO

Initially uninfected mosquitoes picked up the virus from infected birds

INFECTION SPREAD TO PEOPLE
As more and more mosquitoes became infected, they spread the virus to still more birds and ultimately to people

more, as they shared fewer and smaller watering holes, many of which were frequented, naturally, by mosquitoes.

Once mosquitoes acquired the virus, the heat wave that accompanied the drought would speed up viral maturation inside the insects. Consequently, as infected mosquitoes sought blood meals, they could spread the virus to birds at a rapid clip. As bird after bird became infected, so did more mosquitoes, which ultimately fanned out to infect human beings. Torrential rains toward the end of August provided new puddles for the breeding of *C. pipiens* and other mosquitoes, unleashing an added crop of potential virus carriers.

Like mosquitoes, other disease-conveying vectors tend to be "pests"— opportunists that reproduce quickly and thrive under disturbed conditions unfavorable to species with more specialized needs. In the 1990s climate variability contributed to the appearance in humans of a new rodent-borne ailment: the hantavirus pulmonary syndrome, a highly lethal infection of the lungs. This infection can jump from animals to humans when people inhale viral particles hiding in the secretions and excretions of rodents. The sequential weather extremes that set the stage for the first human eruption, in the U.S. Southwest in 1993, were long-lasting drought interrupted by intense rains.

First, a regional drought helped to reduce the pool of animals that prey on rodents—raptors (owls, eagles, prairie falcons, red-tailed hawks and kestrels), coyotes and snakes. Then, as drought yielded to unusually heavy rains early in 1993, the rodents found a bounty of food, in the form of grasshoppers and piñon nuts. The resulting population explosion enabled a virus that had been either inactive or isolated in a small group to take hold in many rodents. When drought returned in summer, the animals sought food in human dwellings and brought the disease to people. By fall 1993, rodent numbers had fallen, and the outbreak abated.

Subsequent episodes of hantavirus pulmonary syndrome in the U.S. have been limited, in part because early-warning systems now indicate when rodent-control efforts have to be stepped up and because people have learned to be more careful about avoiding the animals' droppings. But the disease has appeared in Latin America, where some ominous evidence suggests that it may be passed from one person to another.

As the natural ending of the first hantavirus episode demonstrates ecosystems can usually survive occasional extremes. They are even strengthened by seasonal changes in weather conditions, because the species that live in changeable climates have to evolve an ability to cope with a broad range of conditions. But long-lasting extremes and very wide fluctuations in weather can overwhelm ecosystem resilience. (Persistent ocean heating, for instance, is menacing coral reef systems, and drought-driven forest fires are threatening forest habitats.) And ecosystem upheaval is one of the most profound ways in which climate change can affect human health. Pest control is one of nature's underappreciated services to people; well-functioning ecosystems that include diverse species help to keep nuisance organisms in check. If increased warming and weather extremes result in more ecosystem disturbance, that disruption may foster the growth of opportunist populations and enhance the spread of disease.

UNHEALTHY WATER

Beyond exacerbating the vector-borne illnesses mentioned above, global warming will probably elevate the incidence of waterborne diseases, including cholera (a cause of severe diarrhea). Warming itself can contribute to the change, as can a heightened frequency and extent of droughts and floods. It may seem strange that droughts would favor waterborne disease, but they can wipe out supplies of safe drinking water and concentrate contaminants that might otherwise remain dilute. Further, the lack of clean water during a drought interferes with good hygiene and safe re-hydration of those who have lost large amounts of water because of diarrhea or fever.

Floods favor waterborne ills in different ways. They wash sewage and other sources of pathogens (such as *Cryptosporidium*) into supplies of drinking water. They also flush fertilizer into water supplies. Fertilizer and sewage can each combine with warmed water to trigger expansive blooms of harmful algae. Some of these blooms are directly toxic to humans who inhale their vapors; others contaminate fish and shellfish, which, when eaten, sicken the consumers. Recent discoveries have revealed that algal blooms can threaten human health in yet another way. As they grow bigger, they support the proliferation of various pathogens, among them *Vibrio cholerae*, the causative agent of cholera.

Drenching rains brought by a warmed Indian Ocean to the Horn of Africa in 1997 and 1998 offer an example of how people will be affected as global warming spawns added flooding. The downpours set off epidemics of cholera as well as two mosquito-borne infections: malaria and Rift Valley fever (a flulike disease that can be lethal to livestock and people alike).

To the west, Hurricane Mitch stalled over Central America in October 1998 for three days. Fueled by a heated Caribbean, the storm unleashed torrents that killed at least 11,000 people. But that was only the beginning of its havoc. In the aftermath, Honduras reported thousands of cases of cholera, malaria and dengue fever. Beginning in February of this year, unprecedented rains and a series of cyclones inundated large parts of southern Africa. Floods in Mozambique and Madagascar killed hundreds, displaced thousands and spread both cholera and malaria. Such events can also greatly retard economic development, and its accompanying public health benefits, in affected areas for years.

SOLUTIONS

The health toll taken by global warming will depend to a large extent on the steps taken to prepare for the dangers. The ideal defensive strategy would have multiple components.

One would include improved surveillance systems that would promptly spot the emergence or resurgence of infectious diseases or the vectors that carry them. Discovery could quickly trigger measures to control vector proliferation without harming the environment, to advise the public about self-protection, to provide vaccines (when available) for at-risk populations and to deliver prompt treatments.

This past spring, efforts to limit the West Nile virus in the northeastern U.S. followed this model. On seeing that the virus had survived the winter, public health officials warned people to clear their yards of receptacles that can hold stagnant water favorable to mosquito breeding. They also introduced fish that eat mosquito larvae into catch basins and put insecticide pellets into sewers.

Sadly, however, comprehensive surveillance plans are not yet realistic in much of the world. And even when vaccines or effective treatments exist, many regions have no means of obtaining and distributing them. Providing these preventive measures and treatments should be a global priority.

A second component would focus on predicting when climatological and other environmental conditions could become conducive to disease outbreaks, so that the risks could be minimized. If climate models indicate that floods are likely in a given region, officials might stock shelters with extra supplies. Or if satellite images and sampling of coastal waters indicate that algal blooms related to cholera outbreaks are beginning, officials could warn people to filter contaminated water and could advise medical facilities to arrange for additional staff, beds and treatment supplies.

Research reported in 1999 illustrates the benefits of satellite monitoring. It showed that satellite images detecting heated water in two specific ocean regions and lush vegetation in the Horn of Africa can predict outbreaks of Rift Valley fever in the Horn five months in advance. If such assessments led to vaccination campaigns in animals, they could potentially forestall epidemics in both livestock and people.

A third component of the strategy would attack global warming itself. Human activities that contribute to the heating or that exacerbate its effects must be limited. Little doubt remains that burning fossil fuels for energy is playing a significant role in global warming, by spewing carbon dioxide and other heat-absorbing, or "greenhouse," gases into the air. Cleaner energy sources must be put to use quickly and broadly, both in the energy-guzzling industrial world and in developing nations, which cannot be expected to cut back on their energy use. (Providing sanitation, housing, food, refrigeration and indoor fires for cooking takes energy, as do the pumping and purification of water and the desalination of seawater for irrigation.) In parallel, forests and wetlands need to be restored, to absorb carbon dioxide and floodwaters and to filter contaminants before they reach water supplies.

The world's leaders, if they are wise, will make it their business to find a way to pay for these solutions. Climate, ecological systems and society can all recoup after stress, but only if they are not exposed to prolonged challenge or to one disruption after another. The Intergovernmental Panel on Climate Change, established by the United Nations, calculates that halting the ongoing rise in atmospheric concentrations of greenhouse gases will require a whopping 60 to 70 percent reduction in emissions.

I worry that effective corrective measures will not be instituted soon enough. Climate does not necessarily change gradually. The multiple factors that are now destabilizing the global climate system could cause it to jump abruptly out of its current state. At any time, the world could suddenly become much hotter or even much colder. Such a sudden, catastrophic change is the ultimate health risk—one that must be avoided at all costs.

FURTHER READING

THE EMERGENCE OF NEW DISEASE. Richard Levins, Tamara Auerbuch, Uwe Brinkmann, Irina Eckardt, Paul R. Epstein, Tim Ford, Najwa Makhoul, Christina dePossas, Charles Puccia, Andrew Spielman and Mary E. Wilson in *American Scientist*, Vol. 82, No. 1, pages 52–60; January/February 1994.

CLIMATE CHANGE AND HUMAN HEALTH. Edited by Anthony J. McMichael, Andrew Haines, Rudolf Slooff and Sari Kovats. World Health Organization, World Meteorological Organization, United Nations Environmental Program, 1996.

THE REGIONAL IMPACTS OF CLIMATE CHANGE: AN ASSESSMENT OF VULNERABILITY, 1997. Edited by R. T. Watson, M. C. Zinyowera and R. H. Moss. Cambridge University Press, 1997. Summary from the Intergovernmental Panel on Climate Change available at www.ipcc.ch/pub/reports.htm

BIOLOGICAL AND PHYSICAL SIGNS OF CLIMATE CHANGE: FOCUS ON MOSQUITO-BORNE DISEASES. Paul R. Epstein, Henry F. Diaz, Scott Elias, Georg Grabherr, Nicholas E. Graham, Willem J. M. Martens, Ellen Mosley-Thompson and Joel Susskind in *Bulletin of the American Meteorological Society*, Vol. 79, pages 409–417; 1998.

Other Web sites of interest: www.heatisonline.org and www.med.harvard.edu/chge

The Challenge of Antibiotic Resistance

STUART B. LEVY

ORIGINALLY PUBLISHED IN MARCH 1998

Last year an event doctors had been fearing finally occurred. In three geographically separate patients, an often deadly bacterium, *Staphylococcus aureus*, responded poorly to a once reliable antidote—the antibiotic vancomycin. Fortunately, in those patients, the staph microbe remained susceptible to other drugs and was eradicated. But the appearance of *S. aureus* not readily cleared by vancomycin foreshadows trouble.

Worldwide, many strains of *S. aureus* are already resistant to all antibiotics except vancomycin. Emergence of forms lacking sensitivity to vancomycin signifies that variants untreatable by every known antibiotic are on their way. *S. aureus*, a major cause of hospital-acquired infections, has thus moved one step closer to becoming an unstoppable killer.

The looming threat of incurable *S. aureus* is just the latest twist in an international public health nightmare: increasing bacterial resistance to many antibiotics that once cured bacterial diseases readily. Ever since antibiotics became widely available in the 1940s, they have been hailed as miracle drugs—magic bullets able to eliminate bacteria without doing much harm to the cells of treated individuals. Yet with each passing decade, bacteria that defy not only single but multiple antibiotics—and therefore are extremely difficult to control—have become increasingly common.

What is more, strains of at least three bacterial species capable of causing life-threatening illnesses (*Enterococcus faecalis, Mycobacterium tuberculosis* and *Pseudomonas aeruginosa*) already evade every antibiotic in the clinician's armamentarium, a stockpile of more than 100 drugs. In part because of the rise in resistance to antibiotics, the death rates for some communicable diseases (such as tuberculosis) have started to rise again, after having declined in the industrial nations.

How did we end up in this worrisome, and worsening, situation? Several interacting processes are at fault. Analyses of them point to a number of actions that could help reverse the trend, if individuals, businesses and governments around the world can find the will to implement them.

One component of the solution is recognizing that bacteria are a natural, and needed, part of life. Bacteria, which are microscopic, single-cell entities, abound on inanimate surfaces and on parts of the body that make contact with the outer world, including the skin, the mucous membranes and the lining of the intestinal tract. Most live blamelessly. In fact, they often protect us from disease, because they compete with, and thus limit the proliferation of, pathogenic bacteria—the minority of species that can multiply aggressively (into the millions) and damage tissues or otherwise cause illness. The benign competitors can be important allies in the fight against antibiotic-resistant pathogens.

People should also realize that although antibiotics are needed to control bacterial infections, they can have broad, undesirable effects on microbial ecology. That is, they can produce long-lasting change in the kinds and proportions of bacteria—and the mix of antibiotic-resistant and antibiotic-susceptible types—not only in the treated individual but also in the environment and society at large. The compounds should thus be used only when they are truly needed, and they should not be administered for viral infections, over which they have no power.

A BAD COMBINATION

Although many factors can influence whether bacteria in a person or in a community will become insensitive to an antibiotic, the two main forces are the prevalence of resistance genes (which give rise to proteins that shield bacteria from an antibiotic's effects) and the extent of antibiotic use. If the collective bacterial flora in a community have no genes conferring resistance to a given antibiotic, the antibiotic will successfully eliminate infection caused by any of the bacterial species in the collection. On the other hand, if the flora possess resistance genes and the community uses the drug persistently, bacteria able to defy eradication by the compound will emerge and multiply.

Antibiotic-resistant pathogens are not more virulent than susceptible ones: the same numbers of resistant and susceptible bacterial cells are required to produce disease. But the resistant forms are harder to destroy. Those that are slightly insensitive to an antibiotic can often be eliminated by using more of the drug; those that are highly resistant require other therapies.

To understand how resistance genes enable bacteria to survive an attack by an antibiotic, it helps to know exactly what antibiotics are and how they harm bacteria. Strictly speaking, the compounds are defined as natural substances (made by living organisms) that inhibit die growth, or proliferation, of bacteria or kill them directly. In practice, though, most commercial antibiotics have been chemically altered in the laboratory to improve their potency or to increase the range of species they affect. Here I will also use the term to encompass completely synthetic medicines, such as quinolones and sulfonamides, which technically fit under the broader rubric of antimicrobials.

Whatever their monikers, antibiotics, by inhibiting bacterial growth, give a host's immune defenses a chance to outflank the bugs that remain. The drugs typically retard bacterial proliferation by entering the microbes and interfering with die production of components needed to form new bacterial cells. For instance, die antibiotic tetracycline binds to ribosomes (internal structures that make new proteins) and, in so doing, impairs protein manufacture; penicillin and vancomycin impede proper synthesis of the bacterial cell wall.

Certain resistance genes ward off destruction by giving rise to enzymes that degrade antibiotics or that chemically modify, and so inactivate, the drugs. Alternatively, some resistance genes cause bacteria to alter or replace molecules that are normally bound by an antibiotic—changes that essentially eliminate the drug's targets in bacterial cells. Bacteria might also eliminate entry ports for the drugs or, more effectively, may manufacture pumps that export antibiotics before the medicines have a chance to find their intracellular targets.

MY RESISTANCE IS YOUR RESISTANCE

Bacteria can acquire resistance genes through a few routes. Many inherit the genes from their forerunners. Other times, genetic mutations, which occur readily in bacteria, will spontaneously produce a new resistance trait or will strengthen an existing one. And frequently, bacteria will gain a defense against an antibiotic by taking up resistance genes from other bacterial cells in the vicinity. Indeed, the exchange of genes is so pervasive that the entire bacterial world can be thought of as one huge multicellular organism in which the cells interchange their genes with ease.

Bacteria have evolved several ways to share their resistance traits with one another [see "Bacterial Gene Swapping in Nature," by Robert V. Miller: Scientific American, January 1998] Resistance genes commonly are carried on plasmids, tiny loops of DNA that can help bacteria survive various hazards in the environment. But the genes may also occur on the bacterial chromosome, the larger DNA molecule that stores the genes needed for reproduction and routine maintenance of a bacterial cell.

Often one bacterium will pass resistance traits to others by giving them a useful plasmid. Resistance genes can also be transferred by viruses that occasionally extract a gene from one bacterial cell and inject it into a different one. In addition, after a bacterium dies and releases its contents into the environment, another will occasionally take up a liberated gene for itself.

In the last two situations, the gene will survive and provide protection from an antibiotic only if integrated stably into a plasmid or chromosome. Such integration occurs frequently, though, because resistance genes are often embedded in small units of DNA, called transposons, that readily hop into other DNA molecules. In a regrettable twist of fate for human beings, many bacteria play host to specialized transposons, termed integrons, that are like flypaper in their propensity for capturing new genes. These integrons can consist of several different resistance genes, which are passed to other bacteria as whole regiments of antibiotic-defying guerrillas.

Many bacteria possessed resistance genes even before commercial antibiotics came into use. Scientists do not know exactly why these genes evolved and were maintained. A logical argument holds that natural antibiotics were initially elaborated as the result of chance genetic mutations. Then the compounds, which turned out to eliminate competitors, enabled the manufacturers to survive and proliferate—if they were also

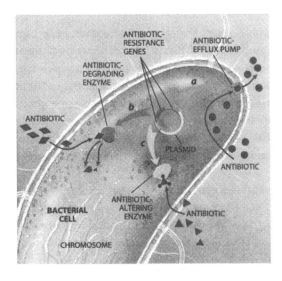

Antibiotic-resistant bacteria owe their drug insensitivity to resistance genes. For example, such genes might code for "efflux" pumps that eject antibiotics from cells (a). Or the genes might give rise to enzymes that degrade the antibiotics (b) or that chemically alter—and inactivate—the drugs (c). Resistance genes can reside on the bacterial chromosome or, more typically, on small rings of DNA called plasmids. Some of the genes are inherited, some emerge through random mutations in bacterial DNA, and some are imported from other bacteria.

lucky enough to possess genes that protected them from their own chemical weapons. Later, these protective genes found their way into other species, some of which were pathogenic.

Regardless of how bacteria acquire resistance genes today, commercial antibiotics can select for—promote the survival and propagation of—antibiotic-resistant strains. In other words, by encouraging the growth of resistant pathogens, an antibiotic can actually contribute to its own undoing.

HOW ANTIBIOTICS PROMOTE RESISTANCE

The selection process is fairly straightforward. When an antibiotic attacks a group of bacteria, cells that are highly susceptible to the medicine will die. But cells that have some resistance from the start, or that acquire it later (through mutation or gene exchange), may survive, especially if too little drug is given to overwhelm the cells that are present. Those cells, facing reduced competition from susceptible bacteria, will then go on to proliferate. When confronted with an antibiotic, the most resistant cells in a group will inevitably outcompete all others.

Promoting resistance in known pathogens is not the only self-defeating activity of antibiotics. When the medicines attack disease-causing bacteria, they also affect benign bacteria—innocent bystanders—in their path. They eliminate drug-susceptible bystanders that could otherwise limit the expansion of pathogens, and they simultaneously encourage the growth of resistant bystanders. Propagation of these resistant, nonpathogenic bacteria increases the reservoir of resistance traits in the bacterial population as a whole and raises the odds that such traits will spread to pathogens. In addition, sometimes the growing populations of bystanders themselves become agents of disease.

Widespread use of cephalosporin antibiotics, for example, has promoted the proliferation of the once benign intestinal bacterium *E. faecalis*, which is naturally resistant to those drugs. In most people, the immune system is able to check the growth of even multidrug-resistant *E. faecalis*, so that it does not produce illness. But in hospitalized patients with compromised immunity, the enterococcus can spread to the heart valves and other organs and establish deadly systemic disease.

Moreover, administration of vancomycin over the years has turned *E. faecalis* into a dangerous reservoir of vancomycin-resistance traits. Recall that some strains of the pathogen *S. aureus* are multidrug-resistant and are responsive only to vancomycin. Because vancomycin-resistant *E. faecalis* has become quite common, public health experts fear that it will soon deliver strong vancomycin resistance to those *S. aureus* strains, making them incurable.

The bystander effect has also enabled multidrug-resistant strains of *Aci-neto-bacter* and *Xanthomonas* to emerge and become agents of potentially fatal blood—borne infections in hospitalized patients. These formerly innocuous microbes were virtually unheard of just five years ago.

As I noted earlier, antibiotics affect the mix of resistant and nonresistant bacteria both in the individual being treated and in the environment. When resistant bacteria arise in treated individuals, these microbes, like other bacteria, spread readily to the surrounds and to new hosts. Investigators have shown that when one member of a household chronically takes an antibiotic to treat acne, the concentration of antibiotic-resistant bacteria on the skin of family members rises. Similarly, heavy use of antibiotics in such settings as hospitals, day care centers and farms (where the drugs are often given to livestock for nonmedicinal purposes) increases the levels of resistant bacteria in people and other organisms who are not being treated—including in individuals who live near those epicenters of high consumption or who pass through the centers.

Given that antibiotics and other antimicrobials, such as fungicides, affect the kinds of bacteria in the environment and people around the individual being treated, I often refer to these substances as societal drugs— the only class of therapeutics that can be so designated. Anticancer drugs, in contrast, affect only the person taking the medicines.

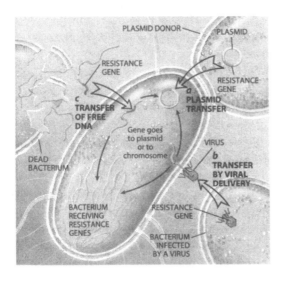

Bacteria pick up resistance genes from other bacterial cells in three main ways. Often they receive whole plasmids bearing one or more such genes from a donor cell (a). Other times, a virus will pick up a resistance gene from one bacterium and inject it into a different bacterial cell (b). Alternatively, bacteria sometimes scavenge gene-bearing snippets of DNA from dead cells in their vicinity (c). Genes obtained through viruses or from dead cells persist in their new owner if they become incorporated stably into the recipient's chromosome or into a plasmid.

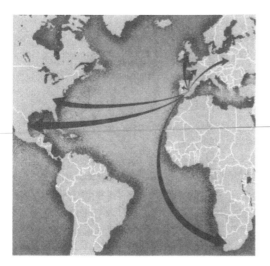

Spread of resistant bacteria, which occurs readily, can extend quite far. In one example, investigators traced a strain of multidrug-resistant Streptococcus pneumoniae from Spain to Portugal, France, Poland, the U.K., South Africa, the U.S. and Mexico.

On a larger scale, antibiotic resistance that emerges in one place can often spread far and wide. The ever increasing volume of international travel has hastened transfer to the U.S. of multidrug-resistant tuberculosis from other countries. And investigators have documented the migration of a strain of multidrug-resistant *Streptococcus pneumoniae* from Spain to the U.K., the U.S., South Africa and elsewhere. This bacterium, also known as the pneumococcus, is a cause of pneumonia and meningitis, among other diseases.

ANTIBIOTIC USE IS OUT OF CONTROL

For those who understand that antibiotic delivery selects for resistance, it is not surprising that the international community currently faces a major public health crisis. Antibiotic use (and misuse) has soared since the first commercial versions were introduced and now includes many nonmedicinal applications. In 1954 two million pounds were produced in the U.S.; today the figure exceeds 50 million pounds.

Human treatment accounts for roughly half the antibiotics consumed every year in the U.S. Perhaps only half that use is appropriate, meant to cure bacterial infections and administered correctly—in ways that do not strongly encourage resistance.

Notably, many physicians acquiesce to misguided patients who demand antibiotics to treat colds and other viral infections that cannot be cured by the drugs. Researchers at the Centers for Disease Control and

Prevention have estimated that some 50 million of the 150 million out-patient prescriptions for antibiotics every year are unneeded. At a seminar I conducted, more than 80 percent of the physicians present admitted to having written antibiotic prescriptions on demand against their better judgment.

In the industrial world, most antibiotics are available only by prescription, but this restriction does not ensure proper use. People often fail to finish the full course of treatment. Patients then stockpile the leftover doses and medicate themselves, or their family and friends, in less than therapeutic amounts. In both circumstances, the improper dosing will fail to eliminate the disease agent completely and will, furthermore, encourage growth of the most resistant strains, which may later produce hard-to-treat disorders.

In the developing world, antibiotic use is even less controlled. Many of the same drugs marketed in the industrial nations are available over the counter. Unfortunately, when resistance becomes a clinical problem, those countries, which often do not have access to expensive drugs, may have no substitutes available.

The same drugs prescribed for human therapy are widely exploited in animal husbandry and agriculture. More than 40 percent of the antibiotics manufactured in the U.S. are given to animals. Some of that amount goes to treating or preventing infection, but the lion's share is mixed into feed to promote growth. In this last application, amounts too small to combat infection are delivered for weeks or months at a time. No one is entirely sure how the drugs support growth. Clearly, though, this long-term exposure to low doses is the perfect formula for selecting increasing numbers of resistant bacteria in the treated animals—which may then pass the microbes to caretakers and, more broadly, to people who prepare and consume undercooked meat.

In agriculture, antibiotics are applied as aerosols to acres of fruit trees, for controlling or preventing bacterial infections. High concentrations may kill all the bacteria on the trees at the time of spraying, but lingering antibiotic residues can encourage the growth of resistant bacteria that later colonize the fruit during processing and shipping. The aerosols also hit more than the targeted trees. They can be carried considerable distances to other trees and food plants, where they are too dilute to eliminate full-blown infections but are still capable of killing off sensitive bacteria and thus giving the edge to resistant versions. Here, again, resistant bacteria can make their way into people through the food chain, finding a home in the intestinal tract after the produce is eaten.

The amount of resistant bacteria people acquire from food apparently is not trivial. Denis E. Corpet of the National Institute for Agricultural Research in Toulouse, France, showed that when human volunteers went on a diet consisting only of bacteria-free foods, the number of resistant bacteria in their feces decreased 1,000-fold. This finding suggests that we deliver a supply of resistant strains to our intestinal tract whenever we eat raw or undercooked items. These bacteria usually are not harmful, but they could be if by chance a disease-causing type contaminated the food.

The extensive worldwide exploitation of antibiotics in medicine, animal care and agriculture constantly selects for strains of bacteria that are resistant to the drugs. Must all antibiotic use be halted to stem the rise of intractable bacteria? Certainly not. But if the drugs are to retain their power over pathogens, they have to be used more responsibly. Society can accept some increase in the fraction of resistant bacteria when a disease needs to be treated; the rise is unacceptable when antibiotic use is not essential.

REVERSING RESISTANCE

A number of corrective measures can be taken right now. As a start, farmers should be helped to find inexpensive alternatives for encouraging animal growth and protecting fruit trees. Improved hygiene, for instance, could, go a long way to enhancing livestock development.

The public can wash raw fruit and vegetables thoroughly to clear off both resistant bacteria and possible antibiotic residues. When they receive prescriptions for antibiotics, they should complete the full course of therapy (to ensure that all the pathogenic bacteria die) and should not "save" any pills for later use. Consumers also should refrain from demanding antibiotics for colds and other viral infections and might consider seeking nonantibiotic therapies for minor conditions, such as certain cases of acne. They can continue to put antibiotic ointments on small cuts, but they should think twice about routinely using hand lotions and a proliferation of other products now imbued with antibacterial agents. New laboratory findings indicate that certain of the bacteria-fighting chemicals being incorporated into consumer products can select for bacteria resistant both to the antibacterial preparations and to antibiotic drugs.

Physicians, for their part, can take some immediate steps to minimize any resistance ensuing from required uses of antibiotics. When possible,

The Antibacterial Fad
A New Threat

Antibiotics are not the only antimicrobial substances being overexploited today. Use of antibacterial agents—compounds that kill or inhibit bacteria but are too toxic to be taken internally—has been skyrocketing as well. These compounds, also known as disinfectants and antiseptics, are applied to inanimate objects or to the skin.

Historically, most antibacterials were used in hospitals, where they were incorporated into soaps and surgical clothes to limit the spread of infections. More recently, however, those substances (including triclocarbon, triclosan and such quaternary ammonium compounds as benzalkonium chloride) have been mixed into soaps, lotions and dishwashing detergents meant for general consumers. They have also been impregnated into such items as toys, high chairs, mattress pads and cutting boards.

There is no evidence that the addition of antibacterials to such household products wards off infection. What is clear, however, is that the proliferation of products containing them raises public health concerns.

Like antibiotics, antibacterials can alter the mix of bacteria: they simultaneously kill susceptible bacteria and promote the growth of resistant strains. These resistant microbes may include bacteria that were present from the start. But they can also include ones that were unable to gain a foothold previously and are now able to thrive thanks to the destruction of competing microbes. I worry particularly about that second group—the interlopers—because once they have a chance to proliferate, some may become new agents of disease.

The potential overuse of antibacterials in the home is troubling on other grounds as well. Bacterial genes that confer resistance to antibacterials are sometimes carried on plasmids (circles of DNA) that also bear antibiotic-resistance genes. Hence, by promoting the growth of bacteria bearing such plasmids, antibacterials may actually foster double resistance—to antibiotics as well as antibacterials.

Routine housecleaning is surely necessary. But standard soaps and detergents (without added antibacterials) decrease the numbers of potentially troublesome bacteria perfectly well. Similarly, quickly evaporating chemicals—such as the old standbys of chlorine bleach, alcohol, ammonia and hydrogen peroxide—can be applied beneficially. They remove potentially disease-causing bacteria from, say, thermometers or utensils used to prepare raw meat for cooking, but they do not leave long-lasting residues that will continue to kill benign bacteria and increase the growth of resistant strains long after target pathogens have been removed.

If we go overboard and try to establish a sterile environment, we will find ourselves cohabiting with bacteria that are highly resistant to antibacterials and, possibly, to antibiotics. Then, when we really need to disinfect our homes and hands—as when a family member comes home from a hospital and is still vulnerable to infection—we will encounter mainly resistant bacteria. It is not inconceivable that with our excessive use of antibacterials and antibiotics, we will make our homes, like our hospitals, havens of ineradicable disease-producing bacteria.   —S.B.L.

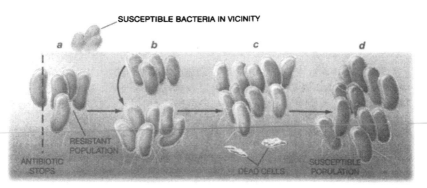

SUSCEPTIBLE BACTERIA IN VICINITY

Resistant population of bacteria will disappear naturally only if susceptible bacteria live in the vicinity. After antibiotic therapy stops (a), resistant bacteria can persist for a while. If susceptible bacteria are nearby, however, they may recolonize the individual (b). In the absence of the drug, the susceptible bugs will have a slight survival advantage because they do not have to expend energy maintaining resistance genes. After a time, then, they may outcompete the resistant microbes (c and d). For this reason, protecting susceptible bacteria needs to be a public health priority.

they should try to identify the causative pathogen before beginning therapy, so they can prescribe an antibiotic targeted specifically to that microbe instead of having to choose a broad-spectrum product. Washing hands after seeing each patient is a major and obvious, but too often overlooked, precaution.

To avoid spreading multidrug-resistant infections between hospitalized patients, hospitals place the affected patients in separate rooms, where they are seen by gloved and gowned health workers and visitors. This practice should continue.

Having new antibiotics could provide more options for treatment. In the 1980s pharmaceutical manufacturers, thinking infectious diseases were essentially conquered, cut back severely on searching for additional antibiotics. At the time, if one drug failed, another in the arsenal would usually work (at least in the industrial nations, where supplies are plentiful). Now that this happy state of affairs is coming to an end, researchers are searching for novel antibiotics again. Regrettably, though, few drugs are likely to pass soon all technical and regulatory hurdles needed to reach the market. Furthermore, those that are close to being ready are structurally similar to existing antibiotics; they could easily encounter bacteria that already have defenses against them.

With such concerns in mind, scientists are also working on strategies that will give new life to existing antibiotics. Many bacteria evade penicillin and its relatives by switching on an enzyme, penicillinase, that degrades those compounds. An antidote already on pharmacy shelves contains an inhibitor of penicillinase; it prevents the breakdown of penicillin and so frees the antibiotic to work normally. In one of the strategies under study, my laboratory at Tufts University is developing a compound to jam a microbial pump that ejects tetracycline from bacteria; with the pump inactivated, tetracycline can penetrate bacterial cells effectively.

CONSIDERING THE ENVIRONMENTAL IMPACT

As exciting as the pharmaceutical research is, overall reversal of the bacterial resistance problem will require public health officials, physicians, farmers and others to think about the effects of antibiotics in new ways. Each time an antibiotic is delivered, the fraction of resistant bacteria in the treated individual and, potentially, in others, increases. These resistant strains endure for some time—often for weeks—after the drug is removed.

The main way resistant strains disappear is by squaring off with susceptible versions that persist in—or enter—a treated person after antibiotic use has stopped. In the absence of antibiotics, susceptible strains have a slight survival advantage, because the resistant bacteria have to divert some of their valuable energy from reproduction to maintaining antibiotic-fighting traits. Ultimately, the susceptible microbes will win out, if they are available in the first place and are not hit by more of the drug before they can prevail.

Correcting a resistance problem, then, requires both improved management of antibiotic use and restoration of the environmental bacteria susceptible to these drugs. If all reservoirs of susceptible bacteria were eliminated, resistant forms would face no competition for survival and would persist indefinitely.

In the ideal world, public health officials would know the extent of antibiotic resistance in both the infectious and benign bacteria in a community. To treat a specific pathogen, physicians would favor an antibiotic most likely to encounter little resistance from any bacteria in the community. And they would deliver enough antibiotic to clear the infection completely but would not prolong therapy so much as to destroy all susceptible bystanders in the body.

Prescribers would also take into account the number of other individuals in the setting who are being treated with the same antibiotic. If many patients in a hospital ward were being given a particular antibiotic, this high density of use would strongly select for bacterial strains unsubmissive to that drug and would eliminate susceptible strains. The ecological effect on the ward would be broader than if the total amount of the antibiotic were divided among just a few people. If physicians considered the effects beyond their individual patients, they might decide to prescribe different antibiotics for different patients, or in different wards, thereby minimizing the selective force for resistance to a single medication.

Put another way, prescribers and public health officials might envision an "antibiotic threshold": a level of antibiotic usage able to correct the infections within a hospital or community but still falling below a threshold level that would strongly encourage propagation of resistant strains or would eliminate large numbers of competing, susceptible microbes. Keeping treatment levels below the threshold would ensure that the original microbial flora in a person or a community could be restored rapidly by susceptible bacteria in the vicinity after treatment ceased.

The problem, of course, is that no one yet knows how to determine where that threshold lies, and most hospitals and communities lack detailed data on the nature of their microbial populations. Yet with some dedicated work, researchers should be able to obtain both kinds of information.

Control of antibiotic resistance on a wider, international scale will require co-operation among countries around the globe and concerted efforts to educate the world's populations about drug resistance and the impact of improper antibiotic use. As a step in this direction, various groups are now attempting to track the emergence of resistant bacterial strains. For example, an international organization, the Alliance for the Prudent Use of Antibiotics (P.O. Box 1372, Boston, MA 02117), has been monitoring the worldwide emergence of such strains since 1981. The group shares information with members in more than 90 countries. It also produces educational brochures for the public and for health professionals.

The time has come for global society to accept bacteria as normal, generally beneficial components of the world and not try to eliminate them—except when they give rise to disease. Reversal of resistance requires a new awareness of the broad consequences of antibiotic use—a perspective that concerns itself not only with curing bacterial disease at the moment but also with preserving microbial communities in the long run, so that bacteria susceptible to antibiotics will always be there to outcompete resistant strains. Similar enlightenment should influence the use of drugs to combat parasites, fungi and viruses. Now that consumption of those medicines has begun to rise dramatically, troubling resistance to these other microorganisms has begun to climb as well.

FURTHER READING

THE ANTIBIOTIC PARADOX: HOW MIRACLE DRUGS ARE DESTROYING THE MIRACLE. S. B. Levy. Plenum Publishers, 1992.

DRUG RESISTANCE: THE NEW APOCALYPSE. Special issue of *Trends in Microbiology,* Vol. 2, No. 10, pages 341–425; October 1,1994.

ANTIBIOTIC RESISTANCE: ORIGINS, EVOLUTION, SELECTION AND SPREAD. Edited by D. J. Chadwick and J. Goode. John Wiley & Sons, 1997.

Behind Enemy Lines

K. C. NICOLAOU AND CHRISTOPHER N. C. BODDY

ORIGINALLY PUBLISHED IN MAY 2001

In the celebrated movie *Crouching Tiger, Hidden Dragon,* two warriors face each other in a closed courtyard whose walls are lined with a fantastic array of martial-arts weaponry, including iron rods, knives, spears and swords.

The older, more experienced warrior grabs one instrument after another from the arsenal and battles energetically and fluidly with them. But one after another, the weapons prove useless. Each, in turn, is broken or thrown aside, the shards of an era that can hold little contest against a young, triumphant, upstart warrior who has learned not only the old ways but some that are new.

One of the foundations of the modern medical system is being similarly overcome. Health care workers are increasingly finding that nearly every weapon in their arsenal of more than 150 antibiotics is becoming useless. Bacteria that have survived attack by antibiotics have learned from the enemy and have grown stronger; some that have not had skirmishes themselves have learned from others that have. The result is a rising number of antibiotic-resistant strains. Infections—including tuberculosis, meningitis and pneumonia—that would once have been easily treated with an antibiotic are no longer so readily thwarted. More and more bacterial infections are proving deadly.

Bacteria are wily warriors, but even so, we have given them—and continue to give them—exactly what they need for their stunning success. By misusing and overusing antibiotics, we have encouraged super-races of bacteria to evolve. We don't finish a course of antibiotics. Or we use them for viral and other inappropriate infections—in fact, researchers estimate that one third to one half of all antibiotic prescriptions are unnecessary. We put 70 percent of the antibiotics we produce in the U.S. each year into our livestock. We add antibiotics to our dishwashing liquid and our hand soap. In all these ways, we encourage the weak to die and the strong to become stronger [see "The Challenge of Antibiotic Resistance," by Stuart B. Levy, pages 299–312].

Yet even absent the massive societal and medical misuse of these medications, the unavoidable destiny of any antibiotic is obsolescence. Bacteria—which grow quickly through many cell divisions a day—will always learn something new, some of the strongest will always survive and thrive. So we have had to become ever more wily ourselves.

In the past 10 years, long-standing complacency about vanquishing infection has been replaced by a dramatic increase in antibacterial research in academic, government and industrial laboratories. Scientists the world over are finding imaginative strategies to attack bacteria. Although they will have a limited life span, new antibiotics are being developed using information gleaned from genome and protein studies. This exciting research and drug development is no panacea, but if combined with the responsible use of antibiotics, it can offer some hope. Indeed, in April 2000 the Food and Drug Administration approved the first new kind of clinical antibiotic in 35 years—linezolid—and several agents are already in the pharmaceutical pipeline.

DISMANTLING THE WALL

Almost all the antibiotics that have been developed so far have come from nature. Scientists have identified them and improved on them, but they certainly did not invent them. Since the beginning of life on this planet, organisms have fought over limited resources. These battles resulted in the evolution of antibiotics. The ability to produce such powerful compounds gives an organism—a fungus or plant or even another species of bacteria—an advantage over bacteria susceptible to the antibiotic. This selective pressure is the force driving the development of antibiotics in nature.

Our window onto this biological arms race first opened with the discovery of penicillin in 1928. Alexander Fleming of St. Mary's Hospital Medical School at London University noticed that the mold *Penicillium notatum* was able to kill nearby *Staphylococcus* bacteria growing in agar in a petri dish. Thus was the field of antibiotics born. By randomly testing compounds, such as other molds, to see if they could kill bacteria or retard their growth, later researchers were able to identify a whole suite of antibiotics.

One of the most successful of these has been vancomycin, first identified by Eli Lilly and Company in 1956. Understanding how it works—a feat that has taken three decades to accomplish—has allowed us insight into the mechanism behind a class of antibiotics called the glycopeptides, one of the seven or so major kinds of antibiotics. This insight is proving important because vancomycin has become the antibiotic of last resort, the only remaining drug effective against the most deadly of all hospital-acquired infections: methicillin-resistant *Staphylococcus aureus*. And yet vancomycin's power—like that of the great, experienced warrior—is itself in jeopardy.

Vancomycin works by targeting the bacterial cell wall, which surrounds the cell and its membrane, imparting structure and support. Because human and other mammalian cells lack such a wall (instead their cells are held up by an internal structure called a cytoskeleton), vancomycin and related drugs are not dangerous to them. This bacterial wall is composed mostly of peptidoglycan, a material that contains both peptides and sugars (hence its name). As the cell assembles this material—a constant process, because old peptidoglycan needs to be replaced as it breaks down—sugar units are linked together by an enzyme called transglycosidase to form a structuralcore. Every other sugar unit along this core has a short peptide chain attached to it. Each peptide chain is composed of five amino acids, the last three being an L-lysine and two D-alanines. An enzyme called transpeptidase then hooks these peptide chains together, removing the final D-alanine and attaching the penultimate D-alanine to an L-lysine from a different sugar chain. As a result, the sugar chains are crocheted together through their peptide chains. All this linking and cross-linking creates a thickly woven material essential for the cell's survival: without it, the cell would burst from its own internal pressure.

Vancomycin meddles in the formation of this essential material. The antibiotic is perfectly suited to bind to the peptide chains before they

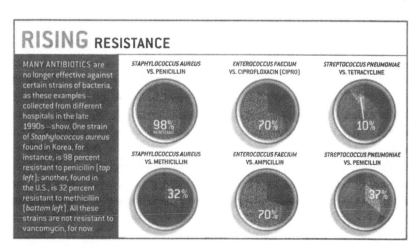

RISING RESISTANCE

MANY ANTIBIOTICS are no longer effective against certain strains of bacteria, as these examples— collected from different hospitals in the late 1990s—show. One strain of *Staphylococcus aureus* found in Korea, for instance, is 98 percent resistant to penicillin [*top left*]; another, found in the U.S., is 32 percent resistant to methicillin [*bottom left*]. All these strains are not resistant to vancomycin, for now.

STAPHYLOCOCCUS AUREUS VS. PENICILLIN	*ENTEROCOCCUS FAECIUM* VS. CIPROFLOXACIN (CIPRO)	*STREPTOCOCCUS PNEUMONIAE* VS. TETRACYCLINE
98%	70%	10%

STAPHYLOCOCCUS AUREUS VS. METHICILLIN	*ENTEROCOCCUS FAECIUM* VS. AMPICILLIN	*STREPTOCOCCUS PNEUMONIAE* VS. PENICILLIN
32%	70%	37%

are linked to one another by transpeptidase. The drug fastens onto the terminal D-alanines, preventing the enzyme from doing its work. Without the thicket of cross-linking connections, peptidoglycan becomes weak, like an ill-woven fabric. The cell wall rends, and cell death rapidly occurs.

RESISTING RESISTANCE

Vancomycin's lovely fit at the end of the peptide chain is the key to its effectiveness as an antibiotic. Unfortunately, its peptide connection is also the key to resistance on the part of bacteria. In 1998 vancomycin-resistant S. aureus emerged in three geographic locations. Physicians and hospital workers are increasingly worried that these strains will become widespread, leaving them with no treatment for lethal staph infections.

Understanding resistance offers the possibility of overcoming it, and so scientists have focused on another bacterium that has been known to be resistant to the powerful drug since the late 1980s: vancomycin-resistant enterococci (VRE). In most enterococci bacteria, vancomycin does what it does best: it binds to the terminal two D-alanines. At a molecular level, this binding entails five hydrogen bonds—think of them as five fingers clasping a ball. But in VRE, the peptide chain is slightly different. Its final D-alanine is altered by a simple substitution: an oxygen replaces a pair of atoms consisting of a nitrogen bonded to a hydrogen. In molecular terms, this one substitution means that vancomycin can bind to the peptide chain with only four hydrogen bonds. The loss of that one bond makes all the difference. With only four fingers grasping the ball, the drug cannot hold on as well, and enzymes pry it off, allowing the peptide chains to link up and the peptidoglycan to become tightly woven once again. One atomic substitution reduces the drug's activity by a factor of 1,000.

Researchers have turned to other members of the glycopeptide class of antibiotics to see if some have a strategy that vancomycin could adopt against VRE. It turns out that some members of the group have long, hydrophobic—that is, oily—chains attached to them that have proved useful. These chains prefer to be surrounded by other hydrophobic molecules, such as those that make up the cell membrane, which is hidden behind the protective peptidoglycan shield. Researchers at Eli Lilly have borrowed this idea and attached hydrophobic chains to vancomycin, creating an analogue called LY333328. The drug connects to the cell membrane in high concentrations, allowing it more purchase and, as a consequence, more power against peptidoglycan. This analogue is effective against VRE and is now in clinical trials.

Other glycopeptide antibiotics use a different strategy: dimerization. This process occurs when two molecules bind to each other to form a single complex. By creating couples, or dimers, of vancomycin, researchers can enhance the drug's strength. One vancomycin binds to peptidoglycan, bringing the other half of the pair—the other molecule of vancomycin—into proximity as well. The drug is more effective because more of it is present. One of the aims of our laboratory is to alter vancomycin so it pairs up more readily, and we have recently developed a number of dimeric vancomycin molecules with exceptional activity against VRE.

Even so, the good news may be short-lived. A second mechanism by which VRE foils vancomycin has recently been discovered. Rather than substituting an atom in the final D-alanine, the bacterium adds an amino acid that is much larger than D-alanine to the very end of the peptide chain. Like a muscular bouncer blocking a doorway, the amino acid prevents vancomycin from reaching its destination.

One method by which the deadly *S. aureus* gains resistance is becoming clear as well. The bacterium thickens the peptidoglycan layer but simultaneously reduces the linking between the peptide fragments. So it makes no difference if vancomycin binds to D-alanine: thickness has replaced interweaving as the source of the peptidoglycan's strength. Vancomycin's meddling has no effect.

THE CUTTING EDGE

As the story of vancomycin shows, tiny molecular alterations can make all the difference, and bacteria find myriad strategies to outwit drugs. Obviously, the need for new, improved or even revived antibiotics is enormous. Historically, the drug discovery process identified candidates using whole-cell screening, in which molecules of interest were applied to living bacterial cells. This approach has been very successful and underlies the discovery of many drugs, including vancomycin. Its advantage lies in its simplicity and in the fact that every possible drug target in the cell is screened. But screening such a large number of targets also has a drawback. Various targets are shared by both bacteria and humans; compounds that act against those are toxic to people. Furthermore, researchers gain no information about the mechanism of action: chemists know that an agent worked, but they have no information about how. Without this critical information it is virtually impossible to bring a new drug all the way to the clinic.

Molecular-level assays provide a powerful alternative. This form of screen identifies only those compounds that have a specified mechanism of action. For instance, one such screen would look specifically for inhibitors of the transpeptidase enzyme. Although these assays are difficult to design, they yield potential drugs with known modes of action. The trouble is that only one enzyme is usually investigated at a time. It would be a vast improvement in the drug discovery process if researchers could review more than one target simultaneously, as they do in the whole-cell process, but also retain the implicit knowledge of the way the drug works. Scientists have accomplished this feat by figuring out how to assemble the many-enzyme pathway of a certain bacterium in a test tube. Using this system, they can identify molecules that either strongly disrupt one of the enzymes or subtly disrupt many of them.

Automation and miniaturization have also significantly improved the rate at which compounds can be screened. Robotics in so-called high-throughput machines allow scientists to review thousands of compounds per week. At the same time, miniaturization has cut the cost of the process by using ever smaller amounts of reagents. In the new ultrahigh-throughput screening systems, hundreds of thousands of compounds can be looked at cost-effectively in a single day. Accordingly, chemists have to work hard to keep up with the demand for molecules. Their work is made possible by new methods in combinatorial chemistry, which allows them to design huge libraries of compounds quickly [see "Combinatorial Chemistry and New Drugs," by Matthew J. Plunkett and Jonathan A. Ellman; *Scientific American*, April 1997]. In the future, some of these new molecules will most likely come from bacteria themselves. By understanding the way these organisms produce antibiotics, scientists can genetically engineer them to produce new related molecules.

THE GENOMIC ADVANTAGE

The methodology of drug design and screening has benefited tremendously from recent developments in genomics. Information about genes and the synthesis of their proteins has allowed geneticists and chemists to go behind enemy lines and use inside information against the organism itself. This microbial counterintelligence is taking place on several fronts, from sabotaging centrally important genes to putting a wrench in the production of a single protein and disrupting a bacterium's ability to infect an organism or to develop resistance.

Studies have revealed that many of the known targets of antibiotics are essential genes, genes that cause cell death if they are not functioning smoothly. New genetic techniques are making the identification of these essential genes much faster. For instance, researchers are systematically analyzing all 6,000 or sogenes of the yeast *Saccharomyces cerevisiae* for essential genes. Every gene can be experimentally disrupted and its effect on yeast determined. This effort will ultimately catalogue all the essential genes and will also provide insight into the action of other genes that could serve as targets for new antibiotics.

The proteins encoded by essential genes are not the only molecular-level targets that can lead to antibiotics. Genes that encode for virulence factors are also important. Virulence factors circumvent the host's immune response, allowing bacteria to colonize. In the past, it has been quite hard to identify these genes because they are "turned on," or transcribed, by events in the host's tissue that are very difficult to reproduce in the laboratory. Now a technique called in vivo expression technology (IVET) can insert a unique sequence of DNA, a form of tag that deactivates a gene, into each bacterial gene. Tagged bacteria are then used to infect an organism. The bacteria are later recovered and the tags identified. The disappearance of any tags means that the genes they were attached to were essential for the bacteria's survival—so essential that the bacteria could not survive in the host without the use of those genes.

Investigators have long hoped that by identifying and inhibiting these virulence factors, they can allow the body's immune system to combat pathogenic bacteria before they gain a foothold. And it seems that the hypothesis is bearing fruit. In a recent study, an experimental molecule that inhibited a virulence factor of the dangerous *S. aureus* permitted infected mice to resist and overcome infection.

In addition to identifying essential genes and virulence factors, researchers are discovering which genes confer antibiotic resistance. Targeting them provides a method to rejuvenate previously ineffective antibiotics. This is an approach used with ß-lactam antibiotics such as penicillin. The most common mechanism of resistance to ß-lactam antibiotics is the bacterial production of an enzyme called ß-lactamase, which breaks one of the antibiotic's chemical bonds, changing its structure and preventing it from inhibiting the enzyme transpeptidase. If ß-lactamase is silenced, the antibiotics remain useful. A ß-lactamase inhibitor called clavulanic acid does just that and is mixed with amoxicillin to create an antibiotic marketed as Augmentin.

In the near future, with improvements in the field of DNA transcriptional profiling, it will become routine to identify resistance determinants, such as ß-lactamase, and virulence factors. Such profiling allows scientists to identify all the genes that are in use under different growth

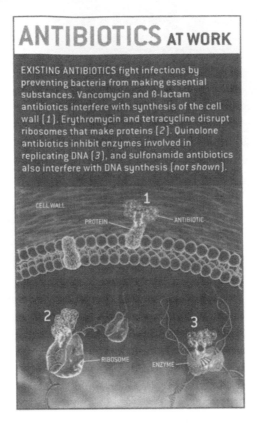

ANTIBIOTICS AT WORK

EXISTING ANTIBIOTICS fight infections by preventing bacteria from making essential substances. Vancomycin and ß-lactam antibiotics interfere with synthesis of the cell wall [1]. Erythromycin and tetracycline disrupt ribosomes that make proteins [2]. Quinolone antibiotics inhibit enzymes involved in replicating DNA [3], and sulfonamide antibiotics also interfere with DNA synthesis [not shown].

CELL WALL

1

PROTEIN   ANTIBIOTIC

2   3

RIBOSOME

ENZYME

conditions in the cell. Virulencegenes can be determined by identifying bacterial genes whose expression increases on infecting a host. Genes that code for antibiotic resistance can be determined by comparing expression levels in bacteria treated with the antibiotic and those not treated. Though still in its infancy, this technique monitored tiny changes in the number of transcription events. With DNA transcriptional profiling, researchers should also be able to determine whether certain drugs have entirely new mechanisms of action or cellular targets that could open up new fields of antibiotic research.

KILLING THE MESSENGER

Another interesting line of genomic research entails interfering with bacterial RNA. Most RNA is ribosomal RNA (rRNA), which forms a major structural component of ribosomes, the cellular factories where proteins are assembled. Ribosomal RNA is vulnerable because it has various places where drugs can attach and because it lacks the ability to repair itself. In 1987 scientists determined t-that antibiotics in the amino glycoside group—which includes streptomycin—bind to rRNA, causing the ribosome to misread the genetic code for protein assembly. Many of these antibiotics, however, are toxic and have only limited usefulness. Recently scientists at the Scripps Research Institute in La Jolla, Calif., have reported a new synthetic aminoglycoside dimer that may have less toxicity.

Investigators can also interfere with messenger RNA (mRNA), which directs the assembly of proteins and travels between the genetic code and the ribosome. Messenger RNA is created by reading one strand of the DNA, using the same nucleic acid, or base pair, interactions that hold the double helix together. The mRNA molecule then carries its message to the ribosome, where a protein is assembled through the process of translation. Because each mRNA codes for a specific protein and is distinct from other mRNAs, researchers have the opportunity to create interactions between small organic molecules—that is, not proteins—and specific mRNAs. Parke-Davis chemists have been able to use such an approach to combat HTV infection. They identified molecules that bind to a part of an mRNA sequence and prevent it from interacting with a required protein activator, thus inhibiting the replication of HIV. This proof-of-principle experiment should help pave the way for further studies of mRNA as a drug development target.

Scientific interest has been intense in another approach, called antisense therapy. By generating sequences of nucleotides that bind perfectly with a specific mRNA sequence, investigators can essentially straitjacket the mRNA. It cannot free itself from the drug, which either destroys it or inhibits it from acting. Although the FDA has recently approved the first antisense drug to treat human cytomegalovirus infections, antisense for bacterial infections has not succeeded yet for several reasons, including toxicity and the challenge of getting enough of the drug to the right spot. Nevertheless, the approach holds promise.

As is clear, all these genomic insights are making it possible to identify and evaluate a range of new biological targets against which chemists can direct their small, bulletlike molecules. A number of antibiotics developed in the past century cannot be used, because they harm us. But by comparing a potential target's genetic sequence with the genes found in humans, researchers can identify genes that are unique to bacteria and can focus on those. Similarly, by comparing a target's genetic sequence to those of other bacteria, they are able to evaluate the selectivity of a drug that would be generated from it. A target sequence that appears in all bacteria would very likely generate an antibiotic active against many different bacteria: a broad-spectrum antibiotic. In contrast, a target sequence that appears in only a few bacterial genomes would generate a narrow-spectrum antibiotic.

If physicians can identify early on which strain is causing an infection, they can hone their prescription to a narrow-spectrum antibiotic. Because this drug would affect only a subset of the bacterial population, selective pressure for the development of resistance would be reduced. Advances in the high-speed replication of DNA and transcriptional profiling may soon make identification of bacterial strains a routine medical procedure.

Although the picture looks brighter than it has for several decades, it is crucial that we recognize that the biological arms race is an ancient one. For every creative counterattack we make, bacteria will respond in kind—changing perhaps one atom in one amino acid. There will always be young warriors to challenge the old ones. The hope is that we exercise restraint and that we use our ever expanding arsenal of weapons responsibly, not relegating them so quickly to obsolescence.

FURTHER READING

The Coming Plague: Newly Emerging Diseases in a World out of Balance. Laurie Garrett. Penguin USA, 1995.

The Chemistry, Biology, and Medicine of the Glycopeptide Antibiotics. K. C. Nicolaou, Christopher N. C. Boddy, Stefan Bräse and Nicolas Winssinger in Angewandte Chemie International Edition, Vol. 38, No. 15, pages 2096–2152; August 2, 1999.

Genome Prospecting. Barbara R. Jasny and Pamela J. Hines in Science, Vol. 286, pages 443–491; October 15, 1999.

An Antibiotic Resistance Fighter

GARY STIX

ORIGINALLY PUBLISHED IN APRIL 2006

Floyd E. Romesberg received his doctorate from Cornell University in 1994, a degree granted for the study of salts called lithium dialkyl-amides. Synthetic chemists routinely use these compounds to remove protons from substances. Romesberg, the son of a chemist, spent his days looking at how these chemicals react and the rate at which the reactions took place. "It wasn't so much that the project was interesting," Romesberg says. "As a matter of fact, the project was pretty boring."

As soon as he finished his degree, he changed course, heading straight for postdoctoral studies in a different field at the University of California, Berkeley. There he extracted a promise from his prospective adviser—noted biochemist Peter G. Schultz—that he would be able to say good-bye to physical chemistry and immerse himself entirely in immunology. Although Romesberg harbors no regrets about his initial decision to study something rudimentary and dull, he remarks that going directly into biology would probably have been a mistake. The complexity of the field ensures that all too many graduate projects end up with only desultory results. "At Cornell, I was fortunately able to work on a small system that was amenable to a complete description. It was something that was actually solvable and allowed me to worry about fundamental questions," he concedes. "I always have this tendency to reduce things to basic molecular-, chemical-level questions, and I think that has served me very well."

Laboring in Schultz's group on antibodies, Romesberg became intrigued by the molecular processes that underlie evolution. Today his laboratory at the Scripps Research Institute in La Jolla, Calif., consisting of 19 researchers (12 graduate students and seven postdocs), has been organized into separate teams that are each tackling different questions related to evolution. One group is using high-powered lasers to look at how antibodies evolve. Another hopes to determine how DNA would function in the presence of an unnatural nucleotide, or letter, added to the genetic code. The research with perhaps the most immediate practical importance attempts to examine how evolution sometimes goes into overdrive. Using this knowledge to block a fundamental process that allows a bacterium to undergo rapid mutation could provide a never-before-tried approach to overcoming bacterial resistance to antibiotics.

SENDING OUT AN SOS

Genetic mutations usually result from errors that occur when a cell reproduces itself. Often mutations hurt cells, and thus they have generally evolved to make as few as possible. Cells come equipped with their own built-in proofreaders and repair equipment to make sure that DNA is copied with as few mistakes as possible. Still, at times a cell may actually embrace the process of genetic mutation—in essence, evolution in fast-forward.

Since the 1970s scientists have known of a process—the SOS response—that occurs in bacteria and actually takes advantage of mutation as a form of self-defense. When bacteria are under extreme stress, they try various means of fixing the damage as an initial step. They then switch on genes whose protein products precipitate a spate of mutations that occur 10,000 times as fast as those arising during normal cell replication. In essence, the cells undergo a quick identity change. The bacterium *Escherichia coli*, for instance, responds to sustained DNA damage induced by ciprofloxacin (commonly shortened to cipro) and other antibiotics by sending out an SOS. The mutations prevent ciprofloxacin from binding to its target, a protein called gyrase, which is required for DNA replication. If the bacterium did not protect gyrase, the antibiotic would link to the protein, impede normal replication and cause breaks in DNA that would lead to the bacterial cell's death.

Having read about SOS, Romesberg hypothesized that switching off the system—shutting down hyperevolution—could prevent the cascade of mutations that allows antibiotic resistance to develop in *E. coli*. In experiments—published last June in the online journal *PLoS Biology*—Romesberg and his co-workers Ryan T. Cirz, Jodie K. Chin and their collaborators at the University of Wisconsin–Madison found that ciprofloxacin induced SOS hypermutation in *E. coli* by triggering the clipping of a protein called LexA that keeps the SOS response repressed. Once cleaved, the repressor protein allowed three enzymes, DNA polymerases, to start producing mutations. Resistance quickly developed.

The researchers then created a strain of *E. coli* in which LexA could not be cut and found that the SOS response did not materialize. Mice infected with the pathogenic *E. coli* and given ciprofloxacin did not develop resistance. The group has achieved similar results for another antibiotic, rifampicin. And it is now testing whether blocking LexA cleavage in *E. coli* might prevent resistance to still other antibiotics and whether it might undermine drug effectiveness in other microbes. Ciprofloxacin, however, is by itself an important drug. Some strains of the bacterium that causes epidemic dysentery—*Shigella dysenteriae*—have become resistant to all antibiotics but ciprofloxacin. Dysentery is capable of causing tens of thousands of deaths in developing countries.

When his laboratory first started getting these results in 2002, Romesberg immediately saw the potential for a pharmaceutical—a small molecule that could be administered orally along with an antibiotic. The drug would function as an off switch for LexA cleavage. He and two partners had little difficulty raising more than $15 million to start a company called Achaogen. ("Achao" means "against chaos"—the "gen" was added by one partner, Ned David, because companies that incorporate this suffix in their name have sometimes flourished.)

Venture capitalists have turned cautious about funding early-stage start-ups. But they were open to a novel approach to antibiotic resistance— most solutions to date have involved new types of antibiotics—and the company is now testing several leads. But, as always, the ingenuity of the microbe could turn a lot of hard work into an exercise in futility. Stuart B. Levy, an expert on antibiotic resistance at Tufts University, commented that Romesberg's work provides new insight but added that its effect may be circumscribed. "We are always looking for novel approaches, especially those that counteract resistance," he says. "This finding suggests one, but it is focused on a limited genetic mechanism of drug resistance—that is, chromosomal mutation." Levy added that other types of resistance can crop up to directly attack the antibiotic. They can be acquired by transferring resistance genes from one species of bacterium to another and also within the same species.

Romesberg and his researchers, however, made their decision to focus on fluoroquinolones because resistance to them develops only through chromosomal mutations (the SOS response) and because they are predicted to become the largest-selling class of antibiotics by 2011. An unexpected—and undesired—aside to the publication of the *PLoS Biology* paper emerged when the intelligent-design community embraced the results as confirmation of its unorthodox worldview. The frenzied mutations in Romesberg's experiments were not random, its members contend, but were set off under deliberate direction of the bacterium:

"Life takes *control* of its fate. Living things are not passive participants of the interplay between stochastic events and environmental pressures," writes the pseudonymous Mike Gene on the Web site idthink.net, while adding, "That evolution may be under some form of intrinsic control is only a piece of the teleological [design in nature] puzzle. But it is a significant piece, in that the ability to adapt, at least to these two antibiotics, is under control."

Romesberg has avoided immersion in this debate but dismisses any basis for these assertions. The SOS response "says nothing about religion," he says. "It does speak to how successful and creative evolution can be. But there's nothing magic about it. It's totally mechanistic."

As with his choice of a seemingly mundane graduate project, Romesberg says he decided to explore antibiotic resistance because the steps to drug development were straightforward and relatively simple: "If we win, then we will be better equipped to handle the next level of complexity." That next level is cancer, a disease for which drug resistance is also a daunting problem. Both his research team and his company plan to craft a treatment that might supplement chemotherapy to prevent the mutations that lead to resistance to cancer drugs. Romesberg's gradualism— evolving from a salt's stripping of protons to the imposing challenge of cancer—may take longer than a headlong push for cures. But this measured path to scientific advance may lead more surely to success.

FURTHER READING

REVENGE OF THE MICROBES: HOW BACTERIAL RESISTANCE IS UNDERMINING THE ANTIBIOTIC MIRACLE. Abigail A. Salyers and Dixie D. Whitt. American Society for Microbiology Press, 2005.

INHIBITION OF MUTATION AND COMBATING THE EVOLUTION OF ANTIBIOTIC RESISTANCE. Ryan T. Cirz et al. in *PLoS Biology,* Vol. 3, No. 6, pages 1024–1033; June 2005.

INDUCTION AND INHIBITION OF CIPROFLOXACIN RESISTANCE-CONFERRING MUTATIONS IN HYPERMUTATOR BACTERIA. Ryan T. Cirz and Floyd E. Romesberg in *Antimicrobial Agents and Chemotherapy,* Vol. 50, No. 1, pages 220–225; January 2006.

ILLUSTRATION CREDITS

Page 3: Johnny Johnson. Page 20: Bryan Christie Design. Page 26: Lucy Reading-Ikkanda. Source: Umesh D. Barashar, CDC (map). Page 29: Andrew Swift. Source: Harihan Jayaram, M. K. Estes and B. V. Venkatarm Prasad, "Emerging Themes in Rotavirus Cell Entry," *Virus Research* 101 (2004). Page 30: Andrew Swift. Source: Philip R. Dormitzer, Harvard Medical School. Page 33: Andrew Swift. Page 39: George Retseck. Page 44: Johnny Johnson. Pages 50–51: Laurie Grace. Pages 52–53: Keith Kasnot. Source: Charles M. Rice, Washington University School of Medicine. Page 54: Laurie Grace. Page 59: Laurie Grace. Sources: UNAIDS (statistics) and Vadim Zalunin, Los Alamos National Laboratory (clade boundaries). Page 64: Terese Winslow. Pages 69, 75, and 80: Tami Tolpa. Page 79: Daniela Naomi Molnar (timeline); Tami Tolpa (illustration). Page 87: Johnny Johnson. Pages 109 and 111: Dimitry Schidlovsky. Page 115: Robert Osti (drawings); Lisa Burnett. Page 145: Jen Christiansen. Page 161: Dimitry Schidlovsky. Pages 172 and 176: Lucy Reading. Page 174: Data as of April 30, 2004. Sources: International Organization for Epizodtic Diseases; World Health Organization; University of California, San Francisco. Pages 196, 199, and 200: Tami Tolpa. Pages 224, 226, and 231 (top): Amadeo Bachar. Page 236: Tami Tolpa (illustrations); Johnny Johnson. Source: Abner Louis Notkins (graph). Page 242: Tami Tolpa. Pages 248, 250, and 254: Jen Christiansen. Page 262: W. Wayt Gibbs. Page 264: Alice Y. Chen. Page 270: Timothy C. Germann, Kai Kadau and Catherine A. Macken, Los Alamos National Laboratory; and Ira M. Longini, Jr., Emory University/ Models of Infectious Disease Agent Study, National Institute of General Medical Sciences. Page 277: Adept Vormgeving (human figures); Lucy Reading-Ikkanda (map). Page 279: Lucy Reading-Ikkanda (top); Stephen G. Eubank (bottom). Page 281: Phillip Romero, Los Alamos National Laboratory (simulation series); Lucy Reading-Ikkanda (graph). Page 289: Bryan Christie. Source: Pim Martens, Maastricht University. Pages 291, 292, and 293: Bryan Christie. Pages 302, 304, and 309: Tomo Narashima. Page 305: Laurie Grace. Source: Christopher G. Dowson, Tracey J. Coffey and Brian G. Spratt, University of Sussex. Page 315: Slim Films. Page 320: Jeff Johnson.